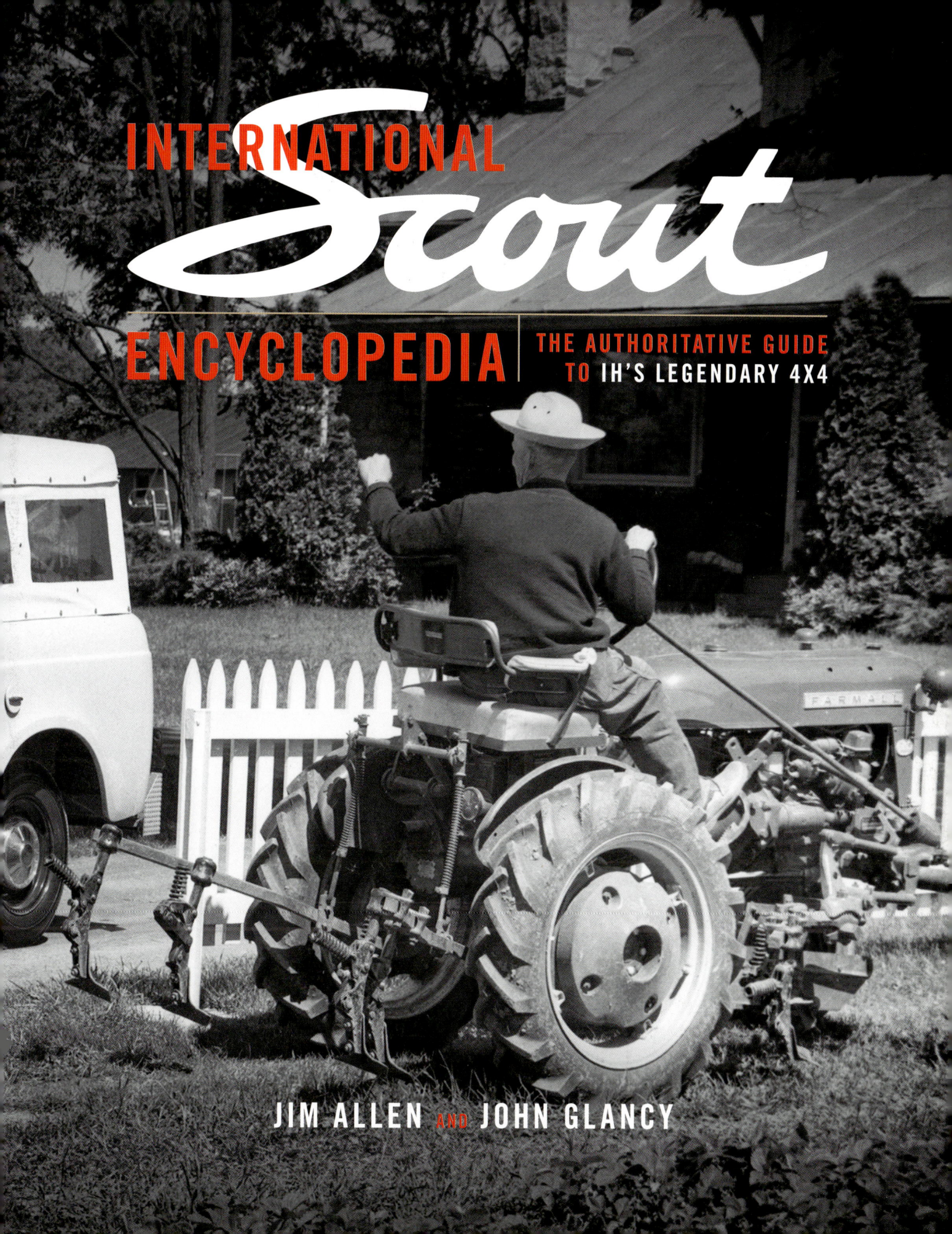

Octane Press, Edition 2.1, September 2023
Edition 2.0, August 2020
Edition 1.0, August 2016
© 2016 by Jim Allen and John Glancy

All rights reserved. With the exception of quoting brief passages for the purposes of review, no part of this publication may be reproduced without prior written permission from the publisher.

ISBN: 978-1-64234-020-4
Library of Congress Control Number: 2020940405

Cover and Interior Design by Tom Heffron
Copyedit by John Koharski
Proofread by Leah Noel

On the front cover: This Scout II IH promotional image was taken in 1978. *WHS 110948*

On the back cover: *(left)* The 1965 Champagne Series was a limited production special, the second of what would be a long string of special edition Scouts. *WHS 54307 (right)* The fiberglass-bodied Supplementary Scout Vehicle (SSV) was slated to be a limited production addition to the '81 and later Scout model line. This is one of the early 95-inch wheelbase versions that was used as demonstrator.

On the front endpapers: In December of 1960, the Scout line was run for the first time and engineers followed along checking the process. Photographers were assigned to document every step. The front endpapers start with welding the stamped pieces of the body structure together in a jig and end with some of the final steps of assembling the body before it was mated to the chassis. Also shown are workers installing the ancillary parts onto the engines, which were fresh from production at the Indy Engine Plant.

On the back endpapers: The back endpapers pick up the assembly line as the chassis is joined with the engine, gearboxes, suspension, and axles. From there, the completed body meets the rolling chassis, the remainder of the parts are installed, and the Scout goes to the roll test facility. The Scout was then run on a rolling road under load to make sure everything worked as designed. That was essentially the last station on the assembly line. If all was copacetic, a final inspection would take place and the Scout would go off for shipment.

On the frontispiece: Dick Hatch's first concept of what would become the SSII. Dubbed "Mountaineer" by Hatch, this rendering as completed in October of 1975. The production SSII followed this rendering very closely.

On the title page: Announced on July 19, 1961, the new soft tops were called Sport-Tops. This is the full-length version, but a pickup-like coupe soft top was also available. They had snapped-in rear and side curtains and were made of vinyl-coated cotton twill. The full top (part number 244123R91) had a $168 list price in 1961. The coupe version (244143R91) listed at $98. Also shown here is the brush guard, which was offered as early as January 1961 (244084R91, $24). *WHS 111948*

On the dedication page: Bob Glancy Sr. doing business in 1973 at the Springfield Ohio, Lagonda Avenue IH Truck Sales and Service Center. The factory-run store was moved to a new location shortly after this picture was taken.

On the contents page: The Valley Forge Military Academy had a fleet of Scout 80s, both Roadsters and Cab-Tops. Shown here in 1963, a group of students get a lesson on military preventative maintenance.

octanepress.com

Printed in China

Dedications

As always, to Linda, for indulging the insanity known as writing a book. — *Jim*

To my late father, Robert T. "Bob" Glancy Sr. — *John*

Contents

	Foreword By Rod Phillips	8
	Introduction	9
One	**Prototypes: Scouting Ahead**	11
	Nuts and Bolts	18
	Last-Minute Changes	22
	The Name of the Game	23
	Ted Ornas: The Scout's Driving Force	26
Two	**Scout 80: Humble Beginnings**	31
	Changes	34
	1962	35
	1963	40
	1964	44
	1965	44
	Scout 80 Data	48
	Comanche: The Engine That Made the Scout	53
Three	**Scout 800, 800A, and 800B: Moving Upmarket**	61
	Meet the Scout 800	65
	1966	66
	1967	68
	1968	70
	Scout 800 LST Anomalies	70
	The Scout 800 Custom	70
	Scout 800s with Fold-Down Windshields	74
	Interim, Interim: The 1969–1970 800A	75
	Interim, Interim, Interim: The 1971 800	79
	Scout 800 Data	84
	Scout 800A and 800B Data	90
	The 800 Engine Evolution	96
Four	**Scout II, Traveler, and Terra: Anything Less Is Just a Car**	109
	Big Shoes to Fill: The Terra and Traveler Story	118
	1971	123
	1972	126
	1973	128
	1974	130
	1975	134
	1976	137
	1977	149
	1978	156
	1979	158
	1980	161
	End of the Road: The Corporate Story	166
	End of the Road: The Scout Story	170
	The Last Day	171
	Scout II, Traveler, and Terra Data	174
	Final Engine Evolution	188
	The Last Scout	200

Five	**Scout Specials Encyclopedia**	205
	Scout Camper	206
	Red Carpet Series	211
	Champagne Series	216
	Sportop	218
	Aristocrat	222
	Scout 800A SR-2	228
	Scout 800B Comanche	231
	Scout 800B Sno-Star	236
	U.S. Ski Team, Spirit of '76, Patriot, and Sno-Star Scouts	241
	SSII	248
	SSII ID	256
	The Midas Touch	257
	Selective Edition	268
	Specials Potpourri	270
	CVI Specials	278
	Special Edition RS	299
	Shawnee	302
Six	**SSV, HVS, and 1981–1985 Scouts: What Might Have Been**	311
	The 1981 Models	312
	SSV: The Scout Corvette?	322
	Future Hopes: The High-Volume Scout (HVS)	335
Seven	**Racing Scouts**	337
	Jimmy Jones	340
	Frank Howarth	348
	Sherman Balch	352
	Jerry Boone	357
	Other Desert Racers	362
Appendix 1	Line Setting Tickets	363
Appendix 2	Serial Number and VIN Locations	371
Appendix 3	Resources	373
Appendix 4	Guide to Graphic Treatments (Appliqués)	374
Appendix 5	Abbreviations and Terms	379
Acknowledgments		380
Index		383

Foreword *By Rod Phillips*

It has been said that timing is everything. Those words could not be truer as they relate to the story of the International Scout.

When the Scout was introduced in 1961, the timing, on the one hand, could not have been more opportune. International Harvester Company (the company has always been "IH" to me and always will be), well known for building durable trucks and farm equipment, was entering a market where Jeep, the only serious competitor, had a several-years head start on everyone else.

On the other hand, the timing could be considered too soon on two counts. First, those who weren't clamoring for performance and horsepower wanted large, comfortable sedans and station wagons. The Scout did not remotely fit either category. It was a Spartan and not particularly highway-worthy vehicle.

The second count to the downside is related to the first. The technology of four-wheel drive was in its infancy. Drivetrains were light-duty and power was inadequate. This had to be compensated for by low gear ratios (high numerically), which necessarily limited the appeal to ranchers, hunters, and the like for off-road use. The term "sport utility vehicle" had not yet been coined—the public did not yet know they wanted one, and the makers did not yet have one to offer that would be widely accepted. But if the Scout wasn't a breakthrough in the world of four-wheel drive, it was a start.

In 1971 the Scout 810 (soon to be the Scout II) entered the scene. The four-wheel-drive market was beginning to come of age and even the unbiased observer surely had to conclude that the Scout was worthy of standing toe to toe with the big boys. Market appeal was broadening as performance, durability, and comfort gradually improved.

In 1979, the Scout enjoyed its best year of sales ever. However, the U.S. economy was in the doldrums. Farm equipment sales were sinking precipitously and the truck business was faring no better. The company's financial condition was weak when it became embroiled in a six-month strike with the United Auto Workers union. As an employee who had to cross the picket lines every workday during that time, I can tell you firsthand that nobody won. Emotions were raw. Pocketbooks were drained. IH emerged a broken entity. Those in charge were forced to make agonizing decisions as to what business units to divest, what to dissolve, and what to keep. Their conclusion was that the only hope for survival was in the production of medium and heavy trucks. There were no resources to devote to the Scout. It became a casualty.

In the early '80s, the economy slowly began to recover from the malaise of the '70s. Also, the motoring public slowly began to realize that they wanted an SUV. It was time for the next generation of Scout. Sadly, IH was in no position to answer the call.

Timing is everything.

Rod Phillips
March 2014
RIP
1942–2015

It's always great when you can get a well-known person to write a foreword, and we were happy Rod Phillips was willing to do so. Sadly, Rod succumbed to cancer before the book was complete. He was a major player in the promotion and preservation of Scout history, and while his contributions weren't always readily visible, they were foundational nonetheless. He is missed. —Jim and John

Introduction

Welcome to the wonderful world of Scouts! When International Harvester discontinued the Scout in October 1980, the loss of detailed records began almost immediately as they were shuffled to backrooms and warehouses. Some ended up in the Dumpster. Some went into storage. Some went home with ex-employees of both IH and the dealer network. As time passed, even more was lost, but as the Scout gained collectible status, the recovery began.

Navistar International, formed from the remaining core of IH when its various divisions were sold off, eventually began donating records to the Wisconsin Historical Society in Madison, where today they are preserved and catalogued for study. When Navistar finds more records in dusty corners, they go to the Wisconsin archives for preservation. Unfortunately, even the vast IH collection is incomplete on the Scout story, leaving large sections of it to be explored in detail.

We have taken great pains to use primary sources. Because there has been relatively little written about Scout, and relatively little hard research done, this book may challenge what you think you know about Scout. The Scout historian's credo is "Never say never, never say always." If you dig around enough, you will find documentary contradictions. Some of that is corporate confusion typical of all large companies (brochures or data books that don't get updated, clerical errors, et cetera). Most, however, is due to missing information, meaning some contradictions cannot be resolved because the needed information is long gone. You can make a guess, either in the "wild-assed" or the "educated" category, but after a while you learn to keep an open mind and not get invested in a guess until it's proven by documentary fact.

Where information was available only via personal recollection, which after thirty-five to sixty years can be a little fuzzy, we tried to back it up with documentary evidence. Where that was not possible, we make it clear the information came from the personal recollections of someone who lived those thrilling days. In a few cases, we connect dots based on a lot of study, but when we do, it's made clear that we have done so.

We aren't sorry for busting myths, however tightly held they may be, but if you have documentary proof that challenges something in this book, we are happy to examine and consider it. We will keep track of any valid corrections and update the book in subsequent editions. In the meantime, we hope you like this book and that it enhances your enjoyment of Scouts and their rich history.

Chapter 1

PROTOTYPES:
SCOUTING AHEAD

International Harvester was an icon in American manufacturing. The innovations it made in nearly 80 years of business under that name, and almost the same amount of time under other names going back to the early nineteenth century, can be credited with helping this nation grow and prosper in significant and specific ways. International Harvester's legacy can still be seen working the nation's farmland and on its highways, even though the company drew its final breath as IH in 1985.

Companies such as Navistar International, Case IH, Cub Cadet, and others grew from the ashes of the old International Harvester and all are still vital concerns. Navistar retained IH's engine and truck manufacturing, as well as what remained of the old IH Company (IHC) corporate roots. The agricultural division was sold to Tenneco, which combined it with JI Case to form Case IH. The Cub Cadet line of lawn tractors became its own company in 1981, while the company's other smaller divisions—and there were many—were sold off to various buyers.

IHC's 1907 entry into motor vehicle production marked the starting point on the road that led to the introduction of the legendary Scout. International would long be a player in the light-truck game, competing against the Big Three and others in that very competitive market. For many of its 73 years building light trucks, IH was a major player, but stumbles in the American economy, market changes, labor troubles, and management errors all contributed to reductions of IH's light-truck market share and profitability, and an eventual abandonment of the light-truck and SUV markets altogether.

◀ This is the earliest known rendering of the Scout idea from the hand of Ted Ornas himself. According to Ornas, this drawing inspired just enough excitement with executives to revive waning enthusiasm for the project. The original drawing was donated to the National Auto & Truck Museum at Auburn, Indiana, and is in a special display Ornas himself made in 2003.
National Auto & Truck Museum

▲ These design sketches from January 17, 1959, are among the earliest known to survive. They were done by Chuck McGrew, a talented designer and one of the people Ornas trusted most to make his visions a reality. All the way through the Scout styling process, McGrew's ideas were everpresent. McGrew worked in the styling department even longer than Ted Ornas himself. *Ted Ornas Collection/Betsy Blume*

The Motor Truck Division of International Harvester had just finished celebrating its Golden Anniversary when the decision was made to build a small 4x4 vehicle. This concept was a bold departure for the IHC of 1958, just as the first motor trucks were a bold departure for the newly formed International Harvester back in 1907.

The official decision to pursue what was then called the "small 4x4 unit" was made by the Motor Truck Product Committee (MTC) on December 9 and 10, 1958. It was put onto paper in MTC report number 571, dated December 22, 1958, giving permission and a budget to start a formal study. Informal pondering had been ongoing several months prior to that. Soon after, the "new one-quarter ton 4x2 and 4x4," as it was called at first for lack of a working name, became a heavily discussed topic in MTC meetings.

That's the dry-as-toast version. Behind the scenes, there was a lot more "color." Unfortunately, not much of it is documented and all the first-hand players are gone, not having written down the full story. Most of what we have comes from the late Ted Ornas, generally given credit for shepherding the new 4x4 into production. In the 1980s, when interest in the history of the Scout finally peaked, Ornas was still around to answer questions and tell some of the backstory.

W. D. Reese ran the IH Truck Engineering Department and worked out of the Fort Wayne, Indiana,

▶ A page of Ted Ornas' sketches and notes from February 1959 about the details of construction with a Royalite body. This would have gone out to the other members of the team submitting ideas. Several designers would be involved, all submitting their takes on the project. Most of those ideas were eventually homogenized into one design.

Ted Ornas Collection/Betsy Blume

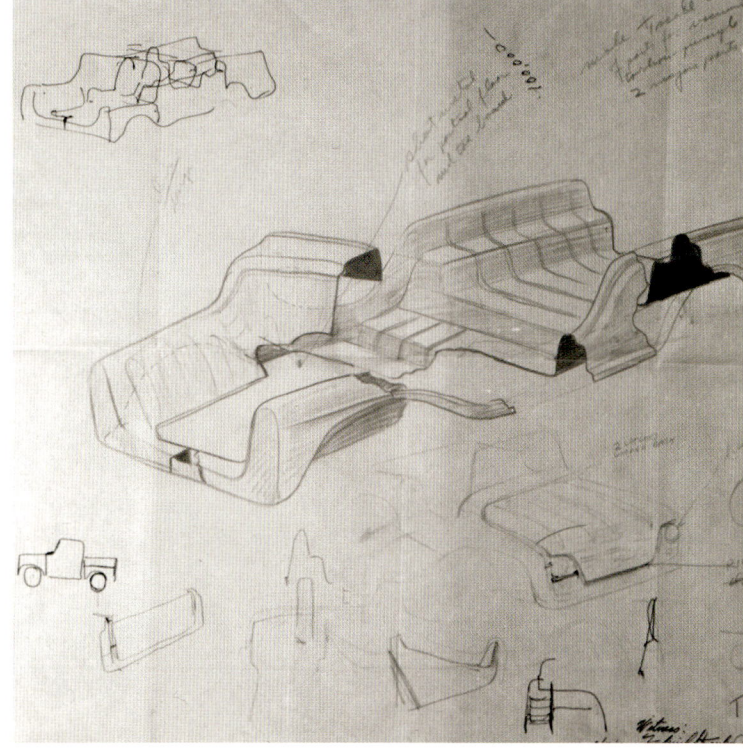

offices. He was frequently in the engineering facility that housed Ornas' Styling Department. At that time, Reese was nearing the end of a very long career in the truck industry and at IH. He had shepherded his department through many of the best eras for IHC trucks and would be one of the better remembered supporters of the Scout project.

Ornas didn't recollect the exact date or wording, but in interviews decades later, he recalled Reese giving him vague instructions early in 1958 to develop ideas for something in the light-duty pickup realm that would "replace the horse." Ornas made it clear he wasn't given many guidelines. Among them, though, was the desire for a utilitarian appearance that would be inexpensive to build. Ornas claimed the inspiration for the idea came from members of upper management who had made a trip out West and noted the common use of small utility vehicles such as Jeeps in rural areas. The IH Sales Department conducted some market analysis and concluded there was a growing demand for small, economical trucks in both two- and four-wheel drive. It needed to be smaller than the smallest IH 4x4 pickup but bigger than the Jeep CJ. They saw the potential for sales of 6,000 to 10,000 units per year.

Ornas was immediately enthusiastic and presented a number of ideas, all of which were rejected. We don't know much about the earliest of them, but according to Ornas, they consisted mostly of vaguely Jeep-like concepts, none of which generated the least bit of enthusiasm. It was clear management was fixated on

▲ Here is another McGrew rendering from February 1959 showing a progression of the fiberglass body. McGrew explored quite a few different ideas.

Ted Ornas Collection/Betsy Blume

PROTOTYPES: SCOUTING AHEAD 13

a low-cost approach and the logical interpretation dictated an angular body with minimal curves or compound bends, very much the embodiment of the original Jeep. Unfortunately, the concepts produced by the stylists using those guidelines brought forth barely concealed yawns. Management interest in the project began to wane.

Good stylists have a sense for what sells, and Ornas knew precisely what was needed. Given the

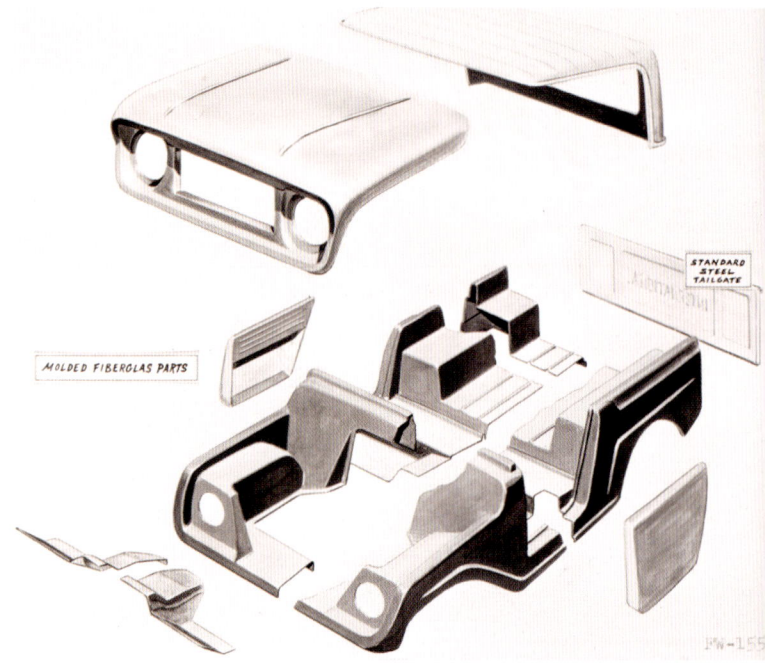

▶ A full-bodied concept was created from the rough ideas and the various component parts of the proposed plastic body are shown here. The exact date of this drawing isn't known, but it's likely from spring 1959.

Ted Ornas Collection/Betsy Blume

▲ On April 4, 1959, a designer named Krippendorf rendered this hardtop version of the fiberglass idea. Krippendorf's work shows up often in the late 1950s, then disappears. The designs have a European flair; perhaps Krippendorf was one of a number of Europeans working in the design department at that time. *Ted Ornas Collection/Betsy Blume*

limitations, others might have just left the project to wither and moved on to the next. Not Ted Ornas. Simple-but-classy is a very high standard to meet, but Ornas was enthusiastic and continued to drive his designers on the project and lobby every executive he could corner. Though Ornas reported a growing frustration, much like a parent trying to prepare a meal for a child who doesn't know what he wants to eat, he continued to push everyone for ideas. He said the breakthrough came in late '58 while sitting at his kitchen table in front of a blank sheet of paper. Ornas sketched an idea that would be met by what he later described as "controlled enthusiasm." Finally, Ornas and his department had a starting point.

Independently of Ornas' team, the MTC had some ideas, too. One of the more cogent was to take the B-Series line (or the new XB-Line, which would become the C-Line in 1961) and shorten it to a 101-inch wheelbase. They discussed using the existing front wrap and cab (narrowed), plus the Bonus Load bed (shortened to five feet) on an adapted chassis using narrowed axles. In many ways, this is exactly what was done. Certainly the final bed design was similar in style to the full-sized C-Line truck's Bonus Load bed, and it helped assure the new design would mirror IH's current styling trends.

Achieving a shapely body with minimal tooling was the difficult part, but once momentum was achieved, the ideas flowed quickly and eventually hit upon plastic. Royalite, an ABS/PVC thermoplastic invented by U.S. Rubber in the 1930s, is one of the best known of the early plastics that have been used for car bodies over the years. Both U.S. Rubber and Goodyear offered a similar plastic and both companies were researched as suppliers. Ornas and crew fleshed out a solid design and in July 1959 solicited cost estimates from Goodyear that shocked everyone away from the idea of using plastic. Unfortunately, the design work

▲ One of the final steps was a small-scale model. Some of the lines of the final Scout can be seen in there, even though this was a fiberglass design. *Ted Ornas Collection/ Betsy Blume*

▲ Once the fiberglass and plastic ideas were abandoned, IH converted to a steel-bodied design. Most of the in-between renderings are missing, but this is a small-scale clay model from August 6, 1959. Most likely taken at the Creative Industries facility in Detroit, this image shows the lines of the final design. *Ted Ornas Collection/Betsy Blume*

PROTOTYPES: SCOUTING AHEAD   15

on the plastic body was well advanced at that point in 1959 and the road to the Scout had again become very rough.

Ralph M. Buzard, the vice president of the Motor Truck Division, came to the rescue by loosening the purse strings just enough so a steel body could be developed. The low-cost directive was still in place, but the company was willing to cut loose with enough tooling money to make the truck at least minimally stylish. From there, the basic design progressed rapidly.

With the switch from plastic to metal construction, Ornas and his team had to start with a new sheet of paper. Fortunately, many of the concepts and project goals had been worked out by this time. Though Ornas is not recorded as having said so, it seems likely his department already had a steel design roughed out. Once a paper design was created, it went off to get a clay model made.

At this point, IH did not have the ability to create clay models in-house. Ornas used his connections to bring Creative Industries (CI) in Detroit on the job. CI was a fully equipped design studio that had begun in 1950 and had done significant work for virtually all the carmakers. They were best known for producing some of the most dazzling concept cars of all time, as well as prototype work on well-known cars such as the Corvette. Starting in July 1959, Ornas regularly found himself commuting to the CI studio to turn the paper designs into clay for final approval by the IH board.

Ornas reported that his dealings with CI weren't always satisfactory. A major sticking point occurred early on when CI wanted to build the model on a concrete floor instead of on a precision surface plate, as is the more common and accurate practice. Perhaps CI did not think of IH as one of the "big boys" or perhaps IH had beaten up CI enough on the costs that the studio wanted to economize. Either way,

▲ By August 27, 1959, McGrew had completed this rendering, which more or less shows a fully featured Scout. *Ted Ornas Collection/Betsy Blume*

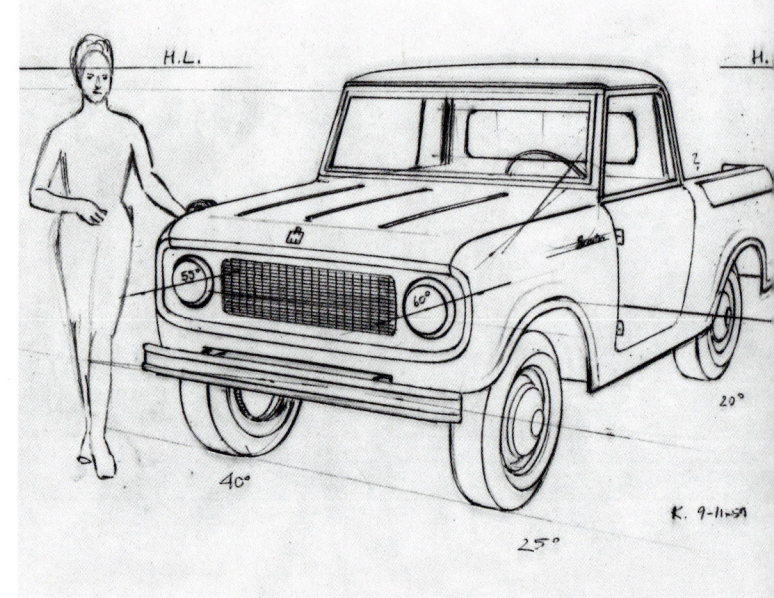

▲ Another drawing by Krippendorf, whose trademark seems to be including a human figure in the rendering. This one is dated September 9, 1959. Note that the badge on the side reads "Scouter." *Ted Ornas Collection/Betsy Blume*

▶ Soon after Ralph Buzard's comment about a cover to protect booze deliveries, the styling department came up with this fiberglass tonneau idea for the Scout bed. It was drawn up on October 22, 1959, by the productive Krippendorf. Again, this rendering is marked "Scouter." *Ted Ornas Collection/Betsy Blume*

▲ A Krippendorf rendering from October 22, 1959, shows a fully formed Scout, still wearing a "Scouter" badge, with a hardtop idea. This was one of many ideas that led to the final version of what was called the Travel-Top. It wasn't until April of 1960 that a final design was approved. *Ted Ornas Collection/Betsy Blume*

▲ This Krippendorf was rendered on November 2, 1959, and shows a very final design. It's possible this very complete rendering was done as part of a presentation to the various committees that approved the final design.
Ted Ornas Collection/Betsy Blume

Ornas didn't stand for it and CI revised their practices forthwith.

By early August, the new design was largely worked out and a full-sized model of a small IH pickup was created. The execs met and viewed the new design. It isn't clear how many changes were ordered from that point on, but available images and dates seem to indicate that not many were made before the final approval in November.

The new vehicle was intended as a light truck with a removable cab top and suitable for any task from a delivery rig in town to an open-topped, off-road utility. One of the early conceptual changes came from Ralph Buzard, who thought a hard bed cover would be necessary and noted that a liquor store making deliveries would want to secure the load from theft. Such a cover was prototyped and discussed but eventually led to a station wagon top that was developed and finalized as the rest of the Scout design was almost complete. Drawings had been made and approved by April 1960 and the new top was ready by the time the Scout was revealed later in the year.

NUTS AND BOLTS

From the engineering point of view, the nitty-gritty part of the new 4x4's development was easy but severely limited by budgets. In June 1959 one of the first complete chassis layouts was outlined:

Gross Vehicle Weight: 2,950 pounds based on a 500-pound payload

Wheelbase: 100 inches

Frame: Both box and channel types were investigated

Front Axle:
4x2 - Conventional construction (forged I-beam)
4x4 - Salisbury 25 (Spicer)

Springs: Conventional for 500-pound load

Brakes: 9- or 11-inch (11-inch if justified by savings in tooling)

Steering Gear: New C-line box

Driveline: Cleveland Steel Products T-55 (same as AM-80 Metro-Mite)

Electrical: 12-volt, 25-amp generator, single headlamps

Clutch: 8.5- or 10-inch (10-inch if justified by costs)

Engine: Four-cylinder of approximately 60 horsepower

Transmission: Stick-control Warner Gear three-speed

Transfer Case: Salisbury Model 18 (Spicer)

Rear Axle: Salisbury Model 27 (Spicer)

Fuel Tank: 10 gallons

Tires: 6.40x15 four-ply (no spare as standard)

Body: One-piece welded major assembly with cargo area 5 feet in length

▲ This full-size clay model made at Creative Industries was the basis of the final design approval after a viewing by execs in November 1959. *Ted Ornas Collection/Betsy Blume*

This list consisted largely of off-the-shelf products from familiar vendors. The big mystery at this point was the engine—a major source of concern. IH made nothing suitable in-house, but this road had been walked on the recently completed AM-80 Metro-Mite van development. As with that product, a large number of domestic and import engines were considered. The natural first choice for the new rig was the engine selected for the AM-80: the British Austin B-Series engine, a small, inexpensive engine used in a large number of cars in England. Displacing a modest 1.5L (91 ci) and making 52 horsepower in IH trim, it had been powering the Nash Metropolitan since 1955.

▶ The first hand-built prototype steel Scout body is moved off its construction jig to the paint shop. Lynn Smith, one of the journeyman sheet-metal fabricators in the Sheet Metal Development Department, remembers this image being taken by the company photographer, though he doesn't remember the date. He recalled that security was very tight in that department. Based on documentation from MTC records, this was likely in the December 1959–January 1960 time frame. *Ted Ornas Collection/Betsy Blume*

▲ One of the first five prototype Scout chassis: a 4x2 mounting the Austin-built A-55 engine with a T-13 (Warner T-90) transmission. The slant-four was not yet fully developed and many of the other outside-sourced engines had been rejected due to costs. Note the exhaust manifold dumps on the driver's side and the pipes run all the way to the rear. *WHS 110532*

▲ One of the first five prototype chassis, this time a 4x4, again mounting the A-55 engine and mated to the T-90 and Model 18 transfer case. *WHS 111741*

A significant amount of testing was done on the engine, both ahead of the AM-80 and for the new 4x4. The reservations over its low output were recorded in the MTC minutes, but they went through the motions anyway. While the first prototype chassis were being assembled, an A-55 engine was installed in a B-100 truck lightened to reflect the weight of the new vehicle. The results were less than stellar when compared to a Jeep, which IH had bought for evaluation. Fuel economy was better, but acceleration was far inferior.

By September 1959, a new engine possibility entered the fray, one that could be produced in-house, but it couldn't be ready for quite some time. Some thought was given to using the A-55 for the initial launch, but the idea wasn't given much support.

Five prototype chassis were scheduled for completion early in 1960 (originally slated as three 4x2s and two 4x4s), and one was completed by February with an A-55 engine. By April, two Scout 4x4 chassis were being tested with A-55 engines. The reports stated, "It appears that the Austin engine may offer adequate power for some off-highway usage when the front axle drive is activated through the low speed ratio in the transfer case. However, it is apparently inadequate for normal traffic conditions in cities and on highways." That put paid to the idea of using the Austin engine.

In the end, the developmental delays for the new slant-four engine were less significant than expected and resulted in the all parts coming together at more or less the same time.

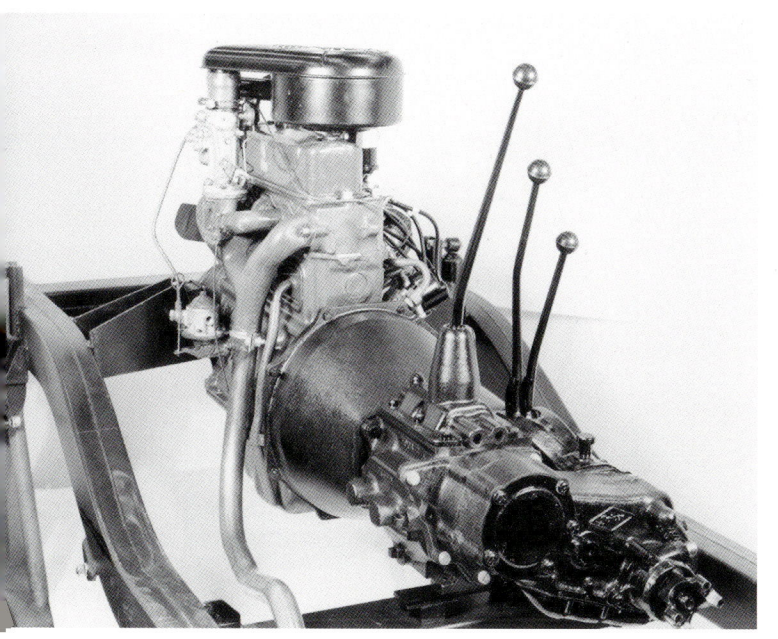

▲ A closeup of the A-55 engine in a test-fit situation. It's been mated to a T-14 transmission (Warner T-90) and a TC-144 (Spicer 18) transfer case. The 90.8-ci (1.5-liter) OHV four made only 52 horsepower in this single-carb configuration—about 55 percent of what the 152-ci (2.5-liter) slant-four made. It didn't take many miles of testing to determine this wasn't going to work. *WHS 110183*

▲ One of the five prototype Scouts is marked "E-922." There's photographic evidence of four prototypes with these kinds of numbers: E-922, E-923, E-924, and E-926. Most do not have the IH emblems on the grilles, though one is seen with and without. E-922 is a 4x2 configured as a Cab-Top. The color looks to be Blue Metallic (6282). The exhaust exiting on the right seems to indicate this one has a prototype 4-152 engine. *WHS 110470*

LAST-MINUTE CHANGES

As the Scout debut and the start of production loomed, last-minute changes were incorporated. This is not unusual with the introduction of a new vehicle, but they provide a sense of what might have been (or explain an early oddball Scout with strange features.)

Just a few months before production began, two bodies were planned: one for the 4x2 and another for the 4x4. The differences were in the floor area and the extra clearance needed for the transfer case. In July 1960, a standardized body was selected to simplify production. The floor of the 4x2 wouldn't be quite as roomy, but that was deemed acceptable. When the numbers were crunched, however, the retooling costs didn't justify the changes and the plan was scrapped. The issue would come up again in 1961, well into the production-Scout era.

For the introduction period, the Styling Department proposed a special seat upholstery called "Comanche"—a simulated white and tan calfskin. The Sales Department thought it would increase sales out West but might be a turnoff to easterners, so it was nixed.

By July, there was still no decision on the signature full-length top (at least three styles had been looked at). By August 18, however, the now-familiar design was approved and tooling rushed ahead.

About the time the full-length top was approved, the rear seat was made part of the options package. It was proposed as a rearward-facing seat that backed up against the permanently fixed bulkhead. But a fixed bulkhead would make access difficult and affect cargo space, so the option was dropped, though it would reappear later.

Another last-minute addition, optional dual fuel tanks, wasn't approved until September 1960, less

▲ E-923 was a Yellow (4154) Travel-Top 4x2. Because there is no sign of an exhaust pipe on the passenger side, it's likely this prototype mounts an A-55 engine. Note the small size of the side window—obviously a work in progress. *WHS 110454*

▲ It looks like E-924 was sent west to be thrashed at the Arizona proving grounds. It's a Cab-Top 4x4 mounting 6.00-16 non-directional tires. It doesn't appear to have an exhaust on the right, indicating A-55 power. *WHS 111605*

▶ A number of images show E-926 in various stages of "undress" for a photo demonstration on how to remove the top and doors. The colors on the 1960 transparency have shifted, but it looks to be painted Tan (1230) and is shown as a stripped-down 4x4 roadster with the other top options in the background, along with all the vehicles the new Scout could replace. Black and white images of this Scout show it without the grille badge. A version of this layout appeared in early promotional materials. E-926 has a right-exit exhaust, so is likely 4-152 powered. *WHS 110460*

than two months before the start of full production. Initially, the vehicle was to draw from both tanks, and a fuel-level sending unit was to be fitted to only one tank.

Even before production, it was noted that interior noise was objectionable due to resonance in the body sheet metal, especially in the 4x2 Scout. In November 1960 a fix was discussed but not until literally the same moment that production started was it decided to spray a product called VibraDamp on the inner roof panels, parts of the firewall, floors, and transmission tunnel.

Initial body colors were to be red, green, blue, yellow, and tan with the tops painted white. Even before production started, orders started coming in from dealers requesting Scouts painted all one color at extra cost. There was also a large number of requests for all-white Scouts, so white was made an option, as were tops painted the body color.

THE NAME OF THE GAME

As with any legendary vehicle, the source of the Scout's name is a subject of great interest. As it turns out, "Scout" was on the A-list right from the beginning, even though "Spartan" was seen on some of the concept drawings for the plastic body. Inconveniently, it turned out a Canadian manufacturer of tracked

▲ At least some of the prototype chassis, including this one, were 4-152-powered initially. The exhaust differs greatly from the production models in that it goes all the way to the rear. It's very likely this chassis is under one of the complete prototypes shown nearby. *WHS 110182*

▲ E-926, E-922, and E-923 (from bottom to top). E-926 is not wearing an IH badge in the grille as in other pics. In those days, vehicle manufacturers weren't paranoid about letting prototypes loose. Some of the more crude or battered ones were scrapped, but it is said that if an employee expressed an interest in a prototype, "arrangements" could be made. It's possible some of the first five Scouts transitioned to the outside world.
WHS 111617

▲ This interesting shot shows an early 4x2 with a short soft top. It's dolled-up with some chrome bits and pieces, but the most interesting part is the Comanche badge on the rear body. Some clues suggest the idea the truck had an upgraded interior for the Western markets that was rejected by planners. The full-length exhaust indicates a 4-152, but it exits at the right rear corner, something seen only on prototypes. The California dealer plate is the yellow and black 1956–1962 type, which lends further credence to this being a prototype. The only two known "Comanches" are the engine (which had not been so-named at this point in time) and the proposed interior with simulated calf hide. Could this be the first Scout Doll-Up?
WHS 110472

utility vehicles owned the trademark rights to "Scout," so while the IH legal department looked into the matter, "Scouter" was used as a fill-in and appeared on a number of the metal-body concept drawings and even on the first full-sized clay models shown in 1959.

The name problem was solved early in 1960 when the holder of the trademark, Bruce Nodwell, agreed to sell it to IH for $25,000 ($5,000 a letter, as Nodwell later liked to put it). Nodwell had invented and was producing a line of successful tracked cargo vehicles used to explore the muskeg- and snow-covered north. While he had used the Scout name for the early models, his very capable rigs were more commonly called Nodwells. Selling the name put a little extra money in his pocket with no serious loss to his firm.

Product Identification Committee Decision 454, dated March 1, 1960, announced the Patent Department had cleared the name for use. From then

▲ The original Scout was an earlier version of this tracked carrier used in the frozen north of Canada and the Arctic. Today, these rigs are best known as Nodwells, after their designer. *U.S. Defense Department*

on, the Scout name was permanent, but some other nomenclature remained to be determined. It was customary at IH to have a numerical component of the name designation to identify load capacity. Since the rated payload was 800 pounds, the designator became 80. Further identification was required for the drive system, so a Scout 80 with two-wheel drive was a plain old Scout 80 while the four-wheel-drive model was a Scout 80 (4x4), the "4x4" and parenthesis being part of the original nomenclature.

Scout enthusiasts are also familiar with the names given to Scout components. The names that made the final list for the 4-152 engine were Ranger, Commando, and Comanche. Contenders for the name of the full-length top were Sport-Top, Enclosall, and Travel-Top. The latter was chosen, of course, but Sport-Top was later used for the soft-top line.

▲ Bruce Nodwell, right, and George Kirkham at a show in 2004. Nodwell held the Scout trademark but was willing to part with it. Nodwell was a big deal in Canada, having been given the Officer of the Order of Canada in 1970, the country's highest civilian honor, for his contributions to Canadian industry. *George Kirkham*

TED ORNAS:
THE SCOUT'S DRIVING FORCE

Every great achievement has a driving force, a *prime mover*. The International Scout was a great achievement and Theodore "Ted" Ornas was the prime mover behind it. That is not to say he was solely responsible (he was always careful to spread the credit), but Ornas was the filter and director for the styling, and that's where it started. His persistence, even as executive interest waned, is why the Scout got going at the timely moment it did.

On both professional and personal levels, there was much more to Ted Ornas than the Scout. Dick Hatch, a designer hired by Ornas in 1967 who eventually ran the department, in 2009 conducted the last interviews with Ornas before his death in March 2009 at age 91. A few years before, Howard Pletcher produced a video interview of Ted Ornas, and together the two give a good picture of how Ornas reflected on his career. Pletcher and Hatch have graciously allowed us to draw from those interviews. In addition, Ornas wrote a family history that Betsy Blume, Ted's granddaughter, has allowed us to draw from.

Theodore Ornas was born July 29, 1917 in Cleveland, Ohio, to Theodore and Mary Ornas. He cited no particular childhood motivation to become an industrial designer and didn't even have a particular interest in art. After graduating high school and a short period working at the Cleveland Woolen Mill, he enrolled in the Cleveland Institute of Art in 1935, where he majored in industrial design. Ornas defrayed some of his costs by getting what was called a "working scholarship," working as a school janitor in exchange for half the $300-per-semester tuition.

With virtually no art experience, Ornas struggled through school, especially during the first year. He did manage to find part-time work with a local industrial designer, but it was without pay and done for experience. In his last year at school, 1939, he was selected for a stylist training program at General Motors, where he

▲ Ted Ornas was 43 and in his creative prime when this photo was taken in 1969. The Scout—his "baby"—had proven a hit and the new iteration would do even better. Ornas' stature in the International Harvester organization was pretty high at this stage. WHS 110319

▲ Ornas working on one of Leo Windecker's fiberglass aircraft designs in the early 1980s. Windecker had debuted a line of fiberglass aircraft in the 1960s, culminating with the Windecker Eagle AC-7, which in 1969 became the first powered aircraft built of composites to earn FAA certification. Undercapitalized, Windecker's venture had failed by the time he worked on the International SSV project. Ornas participated in an attempt to revive it in the early 1980s. *Ted Ornas Collection/Betsy Blume*

worked for $75, then $85 per month. A self-described "know-it-all," he ignored dire advice and left the GM program to work at George Walker Design doing a bit of just about everything, including some work on IHC projects, until 1945.

That year, Ornas left Walker and went to Donald Deskey and Associates in New York, where he worked on everything from architecture to yacht design. When that business got shaky, he left to form his own company with a workmate from Deskey, Dwight LeBarre. The pair moved back to Detroit and set up an office in the Ford building. Styling jobs were few, but they survived for a year on the small jobs they could generate. In 1947, they were within a week of shutting down when the universe suddenly smiled on them.

Ornas & LeBarre got a new lease on life when International Harvester called about designing a new line of medium- and heavy-duty trucks. Desperate

▲ A few lucky people have had Ted Ornas inspect their Scouts. One of the last was Rick Thompson, owner of a red '62 pictured in this book. When Thompson paid him a visit in Fort Wayne in 2006, Ornas was getting on but still living at home. He remarked that Thompson's Scout was certainly in the "eye candy" category. When asked to sign the inside of the glovebox door, Ornas was gracious enough to do so. Today Thompson's Scout is among a rare handful so adorned.

but capable men, they succeeded in getting the job. Despite a work schedule that caused health issues, Ornas & LeBarre managed to impress IHC with designs that included full-sized models. Along the way, Ornas & LeBarre also did work for Crosley and Nash, as well as the beginning of an IH project for a low-cost "austerity truck."

In 1951, International started courting Ornas for a position, and since his partnership with LeBarre was deteriorating, Ornas took the job. International wanted an in-house designer at Fort Wayne, but in the typical schizophrenic IH way, after the courtship, Ornas was all but ignored when he got there. His new "styling department" was little more than a hastily converted corner of an auditorium with no windows. Ornas's driving personality, anchored by that youthful know-it-all attitude, had been annealed and re-tempered in the forge called business to be at just the right amount of hardness to move him and his department higher in the corporate food chain. By the time the Scout project began in 1958, Ornas had acquired better quarters, staff, and stature in the company for having made the right calls in a number of critical developments.

Ornas was involved in the design of the L- and R-Line trucks, mostly as a contractor. As head of styling, he oversaw the development of the 1955 S-Line trucks, which improved the aging International Light Line by leaps and bounds. Even more telling were the A-Line upgrades that followed in 1957. Ornas was directly responsible for a couple of light-duty milestones, the first three-door SUV, the 1957 A-Line Travelall, and the first four-door SUV: the 1961 C-Line Travelall. The Light Line pickups benefitted from better looks and a more roomy cab on his watch. The legendary Loadstar, which debuted in 1962, was one of his proudest achievements and a medium-duty icon.

Ornas shepherded the new Scout II to production in 1971 and the S-Line medium and heavy trucks in 1977, but as International's financial travails and corporate chaos reached epic proportions in the late '70s, he began to think about retirement. It isn't clear how much the impending sale of the Scout line had to do with it, but Ted Ornas finally put in his papers in the fall and retired from International in December 1980 at age 63.

But that retirement did not end his career as an industrial designer. Most notably, he was involved with Leo Windecker, the composite materials pioneer who had helped with the Scout SSV development (see Chapter 6), to develop several aircraft projects. None of them bore much fruit, and by the time the '80s had progressed far, Ornas had enrolled in a local college to take art classes. He remained active in the Fort Wayne art scene for many years.

In the late '80s and the '90s, his pivotal role in the development of the Scout became generally known among a growing crowd of collectors and enthusiasts. As that status grew, Ornas admitted to being completely mystified but always expressed appreciation for the admiration of his work. On some occasions, owners would seek him out at home and arrive unannounced in a Scout. By all reports, he was gracious and forgiving of the intrusions. He was on hand at various times to dedicate or "consecrate" Scout exhibits, such as at the Auburn Cord Duesenberg Automobile Museum and the National Auto & Truck Museum, located next door to each other in Auburn, Indiana. He also wrote short history pieces for various Scout enthusiast publications and was always available for interviews. His donation of parts of his archive has also enhanced the research facilities of at least a couple of museums.

Inevitably, time would take its toll, and in the mid-2000s Ornas moved into an assisted-living facility with his beloved wife, Esther. Age would slowly claim him and on March 14, 2009, Theodore Ornas passed away at age 91. Esther followed just a few months later on August 6 at the same age.

Granddaughter Betsy Blume, daughter of the Ornases' oldest child, Donna, became the keeper of the family IH torch and opened up what remains of her father's personal archive to assist in the creation of this book. "I was just thinking the other day, if my grandfather was still alive, what would be the one thing he might want to say in the book," she commented. "I don't know if there is a way to add anything of a personal nature to the book from my grandfather's own words, but if there is room or an appropriate place for it, I had this quote in mind. Howard [Pletcher] asked my grandfather, 'Any final words for all the IH collectors out there that enjoy your products so much?' My grandfather answered, 'Well, I want to thank 'em for keeping them alive in their various clubs. I've seen some of the models that these people have refinished and they're works of art, and when I see them, I get a little lump in my throat.' I can only add, me too, Grandpa. Me too."

Chapter 2

SCOUT 80:
HUMBLE BEGINNINGS

By August 1960, the Scout design was largely worked out, tooling was under construction, and the line was being built in the recently purchased U.S. Rubber building near the Fort Wayne truck factory. Full production was originally slated to begin in early November 1960, but delays in getting body dies held things up until the last one arrived on November 27. The scheduled start of production was moved back to December 12, but in fact, the first production 80s came off the line on December 1, and by the projected December 12 date, some 30 had been built. Fort Wayne Works estimated they could ramp up to about 70 per day by January 15, 1961. Early units were tested and inspected by a special task force to ensure the new rigs were hitting the mark.

Press releases announcing the Scout were sent out on November 21, 1960, though industry insiders had long known it was being developed. More releases trickled out into January 1961, with the big one with lots of photos released on January 17, 1961. Legendary auto writer Tom McCahill of *Mechanix Illustrated* must have been one of the first to get a Scout test rig because the magazine published one of the first Scout Road tests in April. (Accounting for the usual magazine lead time, he must have had a Scout in January.) Known to be acerbic as well as a Jeep fan, McCahill was at or near the top of the auto writer pack in those days, so it's not hard to see why IH would have accommodated him. A good or bad review from McCahill could make or break a new car.

◀ The Panel-Top was approved in February 1962 as a special feature, meaning it didn't appear in the price book but was available by special order. A March 1963 MTC document stated 55 Panel-Tops had been sold for 1962, and as of January 23, 1963, 22 more had been sold. As result, IH decided to make the Panel-Top a regular production feature and put it into the price book (16738) at $119 list (the same cost as ordering a Travel-Top, incidentally). When the windows were enlarged and the roof height was raised on the Travel-Top for 1963, the Panel-Top followed suit. *WHS 110186*

Though McCahill wished for bigger tires, his review was positive overall and no doubt a major relief to IH.

Sales requested about 100 early-production Scouts to begin their marketing campaign. Some were slated for the press pool fleets and some for shows. Some were used as models for sales brochures and press photos. As far as we can tell, nearly all these Scouts were eventually sold into private hands, most likely sooner rather than later. The earliest production Scouts have turned up all over the country. It's probably safe to speculate that many of the first 100 serial-numbered Scouts were in the batch used by Sales.

The production Scout 80s arrived at dealer showrooms in January 1961, but in small numbers.

▶ *Meet the New Scout!* Dated December 1960, this brochure appears to be the first one produced for the Scout and could be found at dealers in early 1961. The small folder contained an actual black-and-white snapshot.

▼ The first production Scout, FC501 has survived. Sort of. Here's what's left of it. Built on December 1, 1960, the 4x2 Cab-Top was likely used for promotions, as were most of the first 100 built. Eventually, it was sold and ended up in Southern Ohio. Legendary Scout expert and parts guru, Phil Coonrod, found it in the early 80s, looking not much different than now, except that it had a rear axle. He knew what it was and packed it away as a retirement project. Coonrod set his business up in Colorado but has since been gearing up an outlet in Fort Wayne, where he once lived and worked at the Scout Plant. If the Harvester Homecoming organization is successful in setting up a museum, a restored FC501 will be a centerpiece. Apparently the front axle and transmission are original, but the engine, though a very early production unit, is not the one listed on the LST.

▲ FC522, the earliest known complete and restored Scout, belongs to Mike and Debra Ismail of International Harvester Only, a Scout Light Line Dealer in California; it's the 22nd Scout built (numbering began at 501). Configured as a 4x2 pickup, it rolled off the line at Fort Wayne on December 11, 1960. It's a basic single–fuel tank model without options. Mike was as faithful as he could be in the restoration. It bears the early IH tailgate and still mounts the original engine with a 953 serial number. As this book was going to print, there were plans to display it in the Petersen Auto Museum. About the time the 2nd Edition was being completed, FC520 was discovered complete and in good condition.

IH had predicted the compact 4x2 pickup would be the sales leader and the early production schedule reflected that prediction. It was soon clear the 4x4 was the home run and the Travel-Top station wagon was the preferred body. The sales breakdowns revealed a 33/67 percent ratio of 4x2 to 4x4 for 1961 and a 12/88 percent averaged ratio for 1962–1965 80s. Most private individuals were willing to pop for the extra $400 to get a 4x4, and most 4x2s ended up selling commercially or for fleet sales.

Sales were impressive by IH standards and a backlog of orders lasted for the first six months of 1961. That didn't stop IH from celebrating the 10,000th

▲ In June 1961, IH decided to offer 9.00-13 sand tires as an option. The tires were installed at the TSPC with hub spacers to prevent tire rubbing on hard turns and modifications to the kingpin bushings to prevent wheel shimmy. The initial list price of these tires was to be $99.50 for the front and $128 for the rear. WHS 111944

▲ Big friend, little friend. An early Scout 80 Roadster 4x4 timidly sits next to one of International's biggest and gnarliest 6x6s: the DF-808. Built in 1961, this was one of six similar trucks built for Dowell-Schlaumberger for use in the Sahara oilfields. It was powered by a Detroit 8V71 and mounted 14x20 Michelin-X Sand Sahara tires.

Scout in May. By then, production had ramped up to 164 per day. Despite a relatively short production year, 28,031 Scouts of all types had been sold.

CHANGES

Production backlogs in the early days kept model changes to a minimum, but a number of deficiencies eventually necessitated changes large and small during the first production year. Among the least known were front-axle changes to the Dana 27 axles. Both front and rear got larger outer pinion bearings and the axle shafts were slightly enlarged in diameter. These changes took place at approximately serial number FC15544 during the month of April but the old axles could be seen into May production until the stock was exhausted. The changes were from Dana, not from IH, and resulted in the "A" designation found on many Model 27 axles of the era. At the time it appeared in the Scout, the Model 27 was a relatively new product.

The two most common early Scout complaints had to do with water and noise. The noise issues related to sheet-metal harmonics and the sensation of being in a tin can. These problems would plague early Scouts despite the addition of a spray-on sound deadener called VibraDamp. Water leaks were also a continual problem and were first addressed with improvements in sealing that were implemented from June to October 1961.

Another early complaint regarded heat on the passenger floor from the muffler located directly below.

▲ In October 1962, a Consumer Relations Department News release stated that the Department of Defense had recently place an order for 500 Scouts. Given the date on this image (1965) and the 1965 Army registration number, this isn't one of them, but it does give an idea what a militarized Scout 80 looked like. It wasn't unusual to find these on American military bases all over the world. They weren't tactical rigs, but they were used in rough terrain and inclement weather. They were bottom-line rigs with options the military considered important. The optional brush guard (part number 244084 R91) was one of those, as were the front skid plate, a pintle hitch, and tow hooks. *George Kirkham*

▲ A 1961 factory-sponsored jaunt to the Canyonlands in Utah yielded a large number of advertising photos. Scouts were photographed in various locales, both to show off the vehicles and to acquire artwork for advertising. This Marlin Blue Metallic (610) Scout is running in Roadster mode with a few too many people aboard. Other shots show it with the Travel-Top and doors mounted. *WHS 111048*

On July 26, 1961, IH announced a heat shield was being incorporated into production to reduce floor heat. That same announcement (Sales Engineering Newsletter 212) stated the outside mirror bracket was being moved from the door hinge pin to the body hinge bracket so if the door was removed there was still a place to mount the mirror.

1962

One of the biggest early changes was purely cosmetic. At the time of the Scout debut, the Motor Truck Division was looking to separate themselves from the agricultural side of the company. Using "International" instead of "IH" was one way to do that. Sales requested that the stamped IH in the tailgate be replaced by International badging. The tailgate was altered to show an embossed "International" (painted to stand out above the body color) with "Scout" embossed in script below it. It came up in discussion several times and the tooling expenses were finally approved in October 1961 to be implemented starting in February 1962.

An improved one-piece sliding window assembly (the first type was three pieces) appeared early in 1962, greatly simplifying removal of the window assembly from the door. A front skidplate was made optional in May 1962 and the center skidplate was enlarged. The fuel filler neck was upgraded to include an integral

▲ This Hi Sport-Top became available sometime in early 1962 and offered up to 52 inches of height in the back for tall "cargo"—such as people seated on the wheel boxes, military personnel-carrier style. Auxiliary seat pads for the seat boxes (244239R91) cost $4.90 each. The Hi Sport-Top (244202R91) listed at $179. The Deluxe Wheel Discs on this rig were a recent addition to the "Necessories" catalog. They appeared in 1962 and were available for 15- and 16-inch wheels. The fronts were pierced for the locking hub and they cost $4.95 each. WHS 111946

▶ One of the up to 2,000 specially built "walk-thru" units offered to dealers in July 1962. They featured special Freedman seats, front and rear, in this black-and-white pattern with pleating in the inserts. The seat back on the passenger bucket tilted forward, a feature that was canceled for 1963 when the full-tilt seat was be adopted. International referred to this setup as the "interim walk-thru." These seats were also interim and were replaced late in 1962.
WHS 110338

▲ *Two lives, one Scout.* IH collector Herb Huddle has a 1962 4x2 Cab-Top with an interesting history. It was built in April 1962 with several others for the Valley Forge Military Academy in Pennsylvania. The Cab-Tops were ordered in a semigloss Military Green (5538, a custom color) with an "omit door glass" order. Later, special bumpers and a pintle hitch were fitted. The Cab-Tops were used to tow World War II–vintage M1 Pack 75 howitzers, which were used for cadet training and ceremonial purposes. The Cab-Tops operated with Roadsters in the same paint scheme. The accompanying picture is from *International Trails*, Vol. 33, No. 6 (December 1963), which had an article on the school and showed most of the Scouts. From this and the other images, it looks like the school had four Cab-Tops and five Roadsters. Interestingly, the Roadsters had the interim walk-thru conversion and the interim black-and-white seats. They were sold sometime in the 1980s; Huddle acquired his in the 1990s almost untouched.

SCOUT 80: HUMBLE BEGINNINGS

▶ On July 18, 1962, the 50,000th Scout rolled off the Fort Wayne line. International announced it on October 5 and it's likely this shot was staged sometime after the actual build date. FC50001 was a 1962 model-year 4x2 in Metallic Green (5359) and one of the 2,000 or so with the interim walk-thru conversion. It's shown here with the 1963 two-tone gray seats rather than the black seats available at the time it was constructed. The interim conversion lasted until the removable bulkhead was integrated into production in February 1963. The hubcaps are a mystery—a design that looks suspiciously like those used on the later Ford Bronco. Even odder, another shot of the same vehicle shows it with standard 4x2 dog dish–style hubcaps. WHS 110264

vent and a non-vented cap to prevent fuel leakage down the side of the Scout when parked on hills. More body-seal improvements were implemented, including the lower windshield frame, the seal between the top and the body, and the rear liftgate.

In March 1962, the International Sales Department requested a Travel-Top Scout with a walk-through interior feature to make them true station wagons. By then, everyone realized the disadvantage of a pickup body with a permanent bulkhead. A minimum-tooling approach for a removable bulkhead was approved on April 4, and by May it was decided that 1,000 specially modified Travel-Top Scouts without the bulkhead would be offered, using front bucket seats and a new rear seat option. Later, the number was upped to 2,000. This interim development was made to satisfy the market while International developed a better and more efficient method of delivering a walk-through option.

After the various districts had a look at the conversion, International announced on July 30 that walk-through Scouts could be ordered using option code 16024 ($73). This included bucket seats designed and built by Freedman Seating of Chicago and a spare tire mount attached to the interior of the tailgate. This was the first time buckets had been offered for Scout. A rear seat (also from Freedman) was optional with code 16823 ($44) and the outside spare tire mount with code 01578 ($9). Not long after, a dealer kit was made available so Scouts could be field-converted to the walk-through configuration (part number 244248R91, $29.60). The factory conversion and kit were completely different than the later removable bulkhead setup. The Interim Bulkhead appeared in production in early July, at about FC49491, and continued to be seen well into the '63 model year production as late as FC68XXX.

A 4.88:1 axle ratio was made optional early in the year for all 4x4s and 4x2s, but it couldn't be combined with the optional Powr-Lok limited slip. It was not listed on the PL-384 price lists, but according to a February 15, 1962, letter to district managers, it could be special-ordered. The option was later offered with a limited slip differential.

Another axle-related improvement was the addition of the RA-9 and RA-23 (Dana 44) heavy-duty rear axle as a Scout option. This upped the gross

▲ A great many go-to-work items were optional for Scouts in the early days. Some were farm-related, such as this Utemco three-point hydraulic lift first announced by International on January 26, 1962. Most commonly, the kit used a belt-driven hydraulic pump, but a PTO-driven setup was available if a rear PTO wasn't needed (this one has a rear PTO). The core was Utemco's Code 1200 heavy-duty rear bumper, to which the hitch's lower links attached. The upper links were powered by a 3-inch cylinder with a 5.4-inch stroke, and it could apply 2,000 pounds of upward force and 1,500 pounds down. The engine-driven pump produced 1,500 psi and delivered 5.5 gallons per minute of 10-grade hydraulic oil. This Scout also has the Utemco Code 38959 splined PTO, which produced a standard 540-rpm agricultural speed on a standard 10-spline shaft.

axle-weight rating (GAWR) from 2,300 pounds on the standard RA4/14 (Dana 27A) axle to 3,500 pounds for the new axle. The stated reasons were parity with Jeep and failures of the smaller axle when used with optional equipment. In October 1961, after engineering had test-fitted and tested the installation, MTC approved the addition. The option included heavier rear springs and larger rear brakes. This became option code 14009 for the open RA-9 axle ($50, circa 1965) and 14023 for the Powr-Lok-equipped RA-23 ($85, circa 1965).

The Panel-Top Scout was announced on April 19, 1962, as a $119 option over the standard Cab-Top. It was basically a Travel-Top, but the side window openings were not pierced. An indent in the shape of the original opening made a place for company logos or lettering. The option code was 16738.

In September 1960, before production began, Foreign Operations had requested a right-hand-drive (RHD) model for export. Executives asked for development and production costs, as well as sales predictions.

SCOUT 80: HUMBLE BEGINNINGS

◀ Starting on February 2, 1962, Scout tailgates began transitioning from the IH logo to this one. International retained these through the 800 era. This rig also wears Ramsey's two-piece rear bumper and drawbar with a ball hitch. WHS 111926

Initial forecasts had been listed at 1,000 units (400 4x2s and 600 4x4s), but those numbers were later raised to 3,000 total units. Foremost among the RHD markets was Australia, where IH had substantial connections. England was a possibility, but inroads against the "home team" (Land Rover) in that market were difficult. Africa was another possibility, but again Land Rover had longstanding connections there and the Japanese were making sales headway as well.

On August 29, 1961, Sales Letter 113 announced that production of RHD Scouts would start on or around September 1. The option code was 10525 and cost $14. The announcement was just in time for the 1962 model year and RHD Scouts would be available for every year of Scout production to follow. It didn't work out quite as planned, however. With shipping costs, rapidly rising import duties, a strong U.S. dollar, and protectionist policies overseas, IH couldn't sell Scouts at *competitive* prices. The biggest inroads were made in Australia, where the Scout was well liked throughout its production. Still, RHD proved most useful in the States, where there were extensive sales to the Postal Service. Most were 4x2 outfits and many were special orders to a specified pattern.

1963

There were many changes for 1963. Roll-up windows appeared as an option in late August 1962, just as the '63 models started rolling off the line (option code 16676, $35). The roll-up windows were an instant hit and for 1963; only 8.2 percent of Scouts were ordered

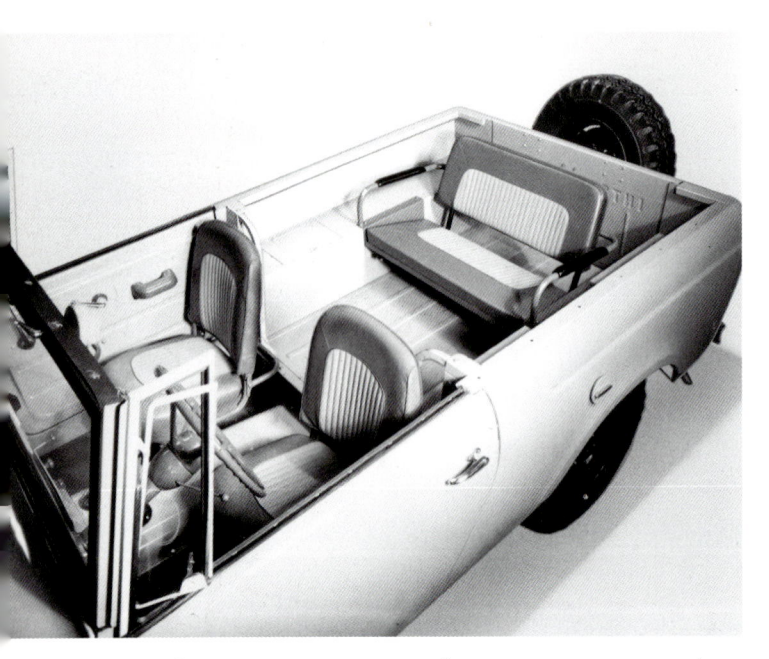

▲ These two-tone gray seats began appearing as early as October 1962, replacing the interim black-and-white seats in both the standard walk-thru and some of the interim walk-thrus. They became a key feature of the '63 models. The entire passenger seat tilted forward rather than just the seat back. The rear seat was positioned a little farther back than the interim seats. WHS 110090

▲ The standardized walk-thru setup used throughout the Scout 80 and the Scout 800 eras. The right image shows the interim walk-thru conversion built through January 1963. Scouts with the standardized setup began volume production in the January–early February timeline. Because it was a bolt-in job, it made production and later conversion much simpler for all involved. *WHS 111749*

▶ Rust was an immediate issue for Scouts in salty states. One could say that never changed, but 1963 was the first time International addressed it seriously. This illustration shows some of the steps taken to push back the specter of rust a little. Some dealers were also set up to do undercoating. *WHS 110091*

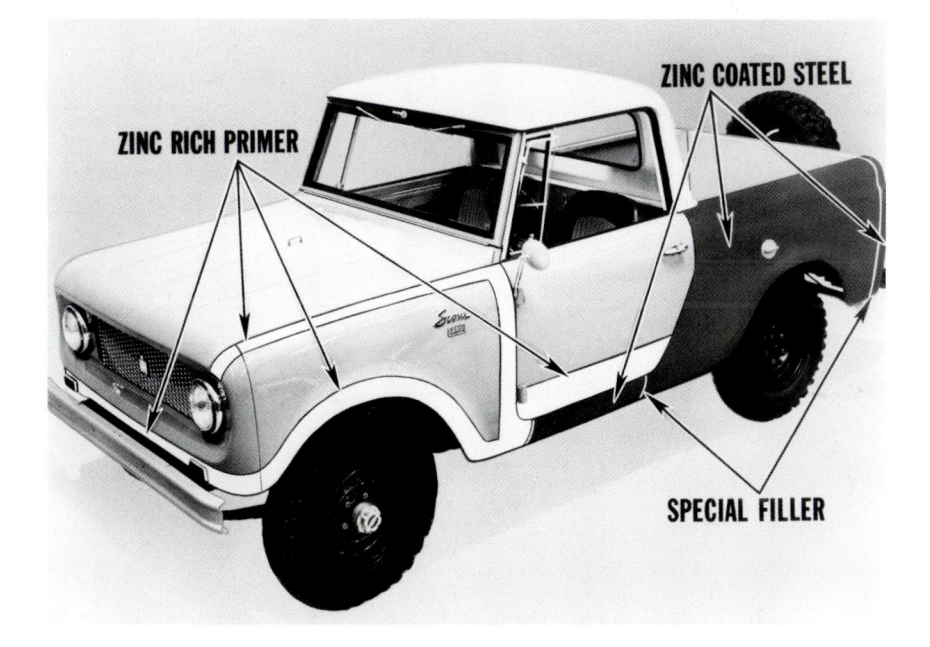

SCOUT 80: HUMBLE BEGINNINGS 41

▲ Scouts were synonymous with snowplowing from Day One and various snowplows, most commonly Meyer or Western, were approved for dealer or factory installation. For 1963, International decided to brand their own plows. Announced on September 16, 1963, the snowplows were built at IH's Canton, Illinois, works, where most of IH's tractor loaders, plows, plow blades, cultivators, disc harrows, and such were made. Here, a 1963 Scout 4x4 Cab-Top in Light Yellow (4285) poses in the studio for a shot that appeared in the first brochure. This is no doubt the 6.5-foot model 48057, which cost $321 in '63. The later model 48067 plow had a hydraulic angling feature and cost $471. An angling device was available to retrofit the standard unit. *WHS 111178, WHS 111176*

with the standard sliding windows. (For 1964, the number dropped to 5.2 percent and no doubt was lower in 1965.) All this led to IH making the roll-up windows standard in the 800 models.

The roof height of both the Travel-Top and Cab-Top were raised 1 1/2 inches for better head clearance. The Travel-Top also got larger side windows (13x34 inches, up from 11x34). The drip rails on both tops were improved and new soft tops appeared with removable sides and one with a raised roof height. Also in 1963, International instituted the first rust prevention program for the Scout. New seat upholstery appeared as well, the buckets having a two-tone pattern with dark gray bolsters and pleated light gray centers. The passenger bucket seat now pivoted at the front for rear seat access rather than just the seat back leaning forward. The rear seat design was changed and it was relocated to provide more legroom. The seatbelt anchors were built-in and reinforced rather than added simply by drilling holes. The heater, defroster, and fresh-air systems were improved and simplified. A new spare-tire mount was added to the tailgate, which tilted back so the liftgate could be opened.

Mechanically, there was a great deal to talk about. The brakes were larger and self-energizing to improve feel and performance. The front Dana 27A axle's tubes got thicker walls (5/16 inch versus 3/16 inch), and the dual fuel-tank selector switch was relocated outboard of the driver's seat. The engine was given a PCV (positive crankcase ventilation) valve to replace the road draft tube.

▲ Rick Thompson's restored '62 has been a consistent show winner since the eight-year restoration was completed in 2006. Produced in late February 1962, it still wears the old IH tailgate. The bumpers are painted black by owner preference, but they should be the standard Silver Grey. Note the muffler, which was still the short style that exited in front of the tire. The rear bumper (part number 244100R91) was a style announced in January 1961, in what seems to be the first Scout "Necessories" catalog. It listed for $32.30. The Warn manual locking hubs (872153R91) cost $67.50. This truck was used as a plow rig, so was ordered with the six-spring heavy-duty clutch and front and rear Powr-Lok limited slips.

▲ Another of the notable changes for 1963 was a new exhaust routed out to the driver's rear corner. This eliminated complaints of passenger floor heat and exhaust fumes, and made for a quieter Scout. The parts department continued carrying both the new and the old systems. WHS 111921

On the electrical front, a new 32-amp alternator replaced the 25-amp generator, and a 52-amp alternator was optional. The wiring, particularly where it routed through the firewall, was upgraded to prevent shorts. Turn signals were made standard and the dimmer switch was improved. Finally, the ignition system was radio-suppressed to eliminate ignition buzz on the AM band.

Three new colors appeared for 1963: Light Green (5551), Light Yellow (4285), and a new shade of Red (2150). These replaced Harvester Red (201), Green Metallic (5186), and Yellow (4154).

1964

Very little changed for 1964, but notably the 100,000th Scout was built on July 13, 1964 (a 1965 model year vehicle). The 4-152T debuted late in the November–December period (again for the 1965 model year).

1965

For 1965, the Cab-Top was painted the body color unless special ordered in white or another color. The Travel-Top remained Whitecap White (902). Four new colors appeared: Champagne Mist (1501), Aspen Green (5683), Apache Gold (4298), and Moonstone

▲ Danny Glover's (not *that* Danny Glover) restoration of a '64 Travel-Top highlights a good reproduction of the two-tone gray seats. This Scout has about all the comfort and style options available at the time, except for a radio. These include the walk-thru option, bucket front seats, rear seat, armrests, and sun visors on the inside. On the outside are chrome bumpers, wheel covers, and an outside spare-tire carrier.

▲ In July 1962, International publicly announced the availability of a fire conversion for the Scout 80, preceded by a sales letter on June 8. The first-response firefighting unit was intended primarily for factories or large facilities where fire or explosions were possible. It had a water tank of 50- to 75-gallon capacity to feed the one-inch hose and to refill backpack sprayers. An 8 1/4-horsepower gasoline engine drove a water pump that supplied up to 110 psi pressure to the 200-foot reel with a fog nozzle. The unit was equipped with six portable extinguishers, a rack-mounted 12-foot aluminum ladder, a pair of fire axes, brooms, shovels, and water buckets. A smaller water pump, a suction hose for refilling the tank, spotlights, floodlights, and a loudspeaker system were all optional. The Scout Cab-Top was largely standard and mounted 7.60-15 six-ply tires, here on what appear to be split-ring wheels. The heavy-duty RA-9/23 3,500-pound rear axle was recommended. The Scout shown is a '62 model. WHS 111273

◄ Back in the day it was popular to convert the short-wheelbase SUVs into fire trucks. Small, agile, and maneuverable, they were perfect first-response units in factories, on small airfields, and as brush rigs. Nearly every sort and brand of short-wheelbase 4x4 was converted at some point and Scout was no exception. UTEMCO, a division of Hendrickson Tandem Corporation, was International's choice for firefighting conversions and they supplied a lot of specialty equipment for the Scout. Beyond the fact this is a '61 or '62 model, little is known about it, but it appears to be configured as a brush rig. WHS 111934

46 INTERNATIONAL SCOUT ENCYCLOPEDIA

▲ For 1965, International still highlighted soft-top Scouts in advertising, even though orders were a tiny percentage of sales. This shot shows a 4x2 Light Yellow (4285) Scout with walk-thru, bucket seats, rear seat, roll-up windows option, whitewall tires, and wheel covers. Option 16704, omit standard top, was employed here along with seating option 16709—front bucket seats less bulkhead. The top was probably available as a TSPC deal (though not listed in the '65 price book), but more likely installed at the dealer. A 4x2 as a noncommercial light runabout was uncommon. *WHS 110452*

▲ A '65 Scout doing what Scouts do best: churning through rough terrain. This Cab-Top is marked as a State of Colorado Highway Department vehicle and is chained-up for the mud and the snow. International made Scout sales to many state agencies. Colorado was one of the best sales markets in the country. *WHS 111056*

Blue (6606). These colors would last well into the 800 era, but ostensibly they were created for the Champagne Series Scouts, replacing Beige (1384), Light Green (5551), Vegas Blue Metallic (6456), and Blue Gray (8371).

An era began in the 1965 model year with the introduction of the Red Carpet and Champagne Series Scouts (see Chapter 5). As SUVs (and trucks) grew more mainstream, the public required a greater level of style, comfort, and convenience. This was the era when the upscale push took hold at International.

Production of the Scout 80 ceased on July 14, 1965, to make way for the updated Scout 800s that started up shortly thereafter. According to the Line Setting Ticket (LST), the last Scout 80 produced (FC124894) was a Red (2150) Cab-Top 4x4 destined for North Dakota.

SCOUT 80: HUMBLE BEGINNINGS   47

SCOUT 80 DATA

MODEL YEARS 1961–1965

INTRODUCTIONS
1961: November 21, 1960
1962: November 1, 1961
1963: November 1, 1962
1964: November 1, 1963
1965: November 1, 1964

PRODUCTION DATES
1961: December 12, 1960-October 31, 1961
1962: November 1, 1961-October 31, 1962
1963: November 1, 1962-October 31, 1963
1964: November 1, 1963-October 30, 1964
1965: November 2, 1963-October 29, 1965

SERIAL NUMBER PREFIXES
FC: Fort Wayne, Indiana, plant
 (501 starting sequential number)
CW: Chatham, Ontario (5501 starting
 sequential number)
SC: Australia
Note: Sometimes seen in the early Scout 80 era, an "A" suffix indicates the GVW class of the vehicle, in this case 6,000 pounds and under.

SERIAL NUMBER RANGES
Includes canceled or forwarded serial numbers
1961: FC501–FC28865
1962: FC28865–58142
1963: FC58143–FC83951
1964: FC83952–FC107760
1965: FC107761–FC124894

MODEL CODES
The model code for the Scout was seven digits long but did not appear on the serial number plate. It could be found on the LST and sometimes on invoices or sales materials. The model code was not as useful in indicating characteristics as it would be starting in the 800 era. General Letter MT337, dated April 15, 1963, announced that the model codes had changed to coincide with the many upgrades for the 1963 Scouts. The first digit was changed from 4 to 7 and the fifth digit went from 6 to 0. The effective date was February 1, 1963, so early 1963 models built before still used the 4-series code. The serial number in this era always started with FC, the F for Fort Wayne assembly and C for the C-line, the assembly line on which it was produced. The sequential numbers in the serial number ran continuously from 1961 to 1966 and started at 501 for the first Scout built in 1960.

Digit	1	2	3	4	5	6	Description
	4						Scout 80 to 2/1/63
	7						Scout 80 from 2/1/63
		1					Scout 80 4x2
		8					Scout 80 4x4
			0				Scout 80 placeholder digit
			1				Scout 80 Red Carpet
				8			Scout 80
					0		Scout 80, removable bulkhead
					6		Scout 80, fixed bulkhead
						3	Front-end sheet metal only
						5	All

TOTAL SCOUT MODELS BUILT

Model-year production, not including canceled or forwarded serial numbers

1961: 28,031
1962: 28,975
1963: 26,360
1964: 23,827
1965: ~24,200*

* Includes some Scout 800 production

SCOUT 80 4X2 VERSUS SCOUT 4X4 MODELS

Fiscal-year production

1961: 6,527 / 20,101
1962: 2,847 / 23,801
1963: 3,166 / 21,201
1964: 2,408 / 19,456
1965: 2,880 / 20,216†

† Includes some 1966 Scout 800 production

SCOUT 80 SPECIFICATIONS

1961–1965, unless otherwise specified

Axles, Front

4x2: FA-4, dead, forged-steel I-beam
4x4: FA-14, live, Spicer 27 or 27A, closed knuckle
Optional 4x4: FA-24, live, Spicer 27A, w/ Powr-Lok (late 1961–1965)

Axles, Rear

Standard 4x2 and 4x4: RA-4, Spicer 27 or 27A, semi-float
Optional 4x2 and 4x4: RA-14, Spicer 27 or 27A, semi-float, w/ Powr-Lok
Optional 4x4: RA-9, Spicer 44, semi-float (1962–1965)
Optional 4x4: RA-23, Spicer 44, semi-float, w/ Powr-Lok (1962–1965)

Axles, Ratios

Standard: 4.27:1 (1961–1965)
Optional: 4.88:1 (1963–1965); 3.73:1 (1964–1965)

Brakes

Front 4x2: 9x2 in.
Rear 4x2: 9x2 in.
Total Lining Area 4x2: 139.4 sq. in.
Front 4x4: 9x1.75 in. (1961–1962)
Front 4x4: 10x2 in. duo-servo (1963–1965)
Rear 4x4: 9x2 in.
Rear 4x4 w/ RA-9 or 23 Axle: 11x1.75 in. (1962–1965)
Total Lining Area 4x4: 127.5 sq. in. (1961–1962)
Total Lining Area 4x4: 165.7 sq. in. (1961–1962)
Total Lining Area 4x4 w/ RA-9 or 23 axle: 156.3 sq. in. (1962–1963)
Total Lining Area 4x4 w/ RA-9 or 23 axle: 156.3 sq. in. (1962–1963)

CAPACITIES

GVW 4x4: 3,900 lbs. (standard)
GVW 4x2: 3,200 lbs. (standard)
FA-14 Front-Axle Capacity 4x4: 2,000 lbs.
FA-4 Front-Axle Capacity 4x2: 2,000 lbs.
RA-4 Rear-Axle Capacity 4x4: 2,300 lbs. (standard)
RA-9/23 Rear-Axle Capacity 4x4: 3,500 lbs. (optional 1962–1965)
RA-4 Rear-Axle Capacity 4x2: 2,300 lbs.
Main Fuel Tank: 11 gal.
Optional Auxiliary Fuel Tank: 11 gal.

CLUTCH

Standard 4-152: 10 in., 6-spring, 91.8 sq. in.
Standard 4-152T: 11 in., 9-spring, 123.7 sq. in. (1965)

DIMENSIONS

Wheelbase: 100 in.
Overall Length: 154 in.
Overall Width: 68.6 in.
Overall Height, 4x4 Travel-Top, Cab-Top: 68 in.
　(67 in. 4x2)
Approach Angle 4x4: 47 deg. (6.00-16 NDT tires)
Departure Angle 4x4: 35 deg. (6.00-16 NDT tires)
Ramp-Breakover Angle 4x4: 155 deg. (6.00-16
　NDT tires)
Approach Angle 4x2: 44 deg. (6.50-15 tires)
Departure Angle 4x2: 34 deg. (6.50-15 tires)
Ramp-Breakover Angle 4x2: 155 deg. (6.50-15 tires)
Ground Clearance to Chassis 4x2: 9 in. (6.50-15 tires)
Ground Clearance to Chassis 4x4: 10.3 in.
　(6.00-16 NDT tires)
Front-Axle Ground Clearance 4x2: 10.25 in.
　(7.35-15 tires)
Front-Axle Ground Clearance 4x4: 9 in.
　(6.00-16 tires)
Rear-Axle Ground Clearance 4x4: 9.75 in.
　(6.00-16 tires; RA-4, RA-14 axles)
Rear-Axle Ground Clearance 4x4: 7.81 in.
　(6.00-16 tires; RA-9, RA-23 axles)

ELECTRICAL

System Voltage: 12V
Generator Capacity: 25A (1961)
Generator Capacity: 30A (1962)
Alternator Capacity: 37A (1963–1965)

ENGINE 4-152, 1961–1965

Type: 4-cyl. slant-four OHV, IH 4-152
Displacement: 152 ci
Bore and Stroke: 3.88x3.22 in.
Compression Ratio: 8.19:1
Block and Head Material: cast iron
Number of Main Bearings: 5
Weight: 505 lbs. complete/dry
Gross HP Rating: 93.4 hp @ 4,400 rpm
Net HP Rating: 86 hp @ 4,400 rpm
Gross Torque Rating: 142.7 lb-ft @ 2,400 rpm
Net Torque Rating: 137.3 lb-ft @ 2,400 rpm

ENGINE 4-152T, 1965

Type: 4-cyl. slant-four OHV, IH 4-152
Displacement: 152 ci
Bore and Stroke: 3.88x3.22 in.
Compression Ratio: 8.19:1
Block and Head Material: cast iron
Number of Main Bearings: 5
Weight: 530 lbs. complete/dry
Gross HP Rating: 111.3 hp @ 4,000 rpm
Net HP Rating: 99.2 hp @ 3,600 rpm
Gross Torque Rating: 166.5 lb-ft @ 3,200 rpm
Net Torque Rating: 151.2 lb-ft @ 3,200 rpm

PERFORMANCE

Turning Radius 4x4: 21.5 ft.
Turning Radius 4x2: 19 ft.
0–60 Scout 80 4-152: 20.1 sec. (*Car Life*, 1961)
Quarter Mile Scout 80 4-152: 21.7 sec., 61 mph
Top Speed: 78 mph (*Car Life*, 1961)

SUSPENSION AND STEERING

Front Springs Length/Width 4x2/4x4: 40x1.75 in.
Front Springs No. Leaves 4x2: 6
Front Springs No. Leaves 4x4: 7
Standard Front Springs Rate 4x2: 198 lbs.-in.
Standard Front Springs Rate 4x4: 224 lbs.-in.
Standard Front Springs Capacity 4x2: 1,000 lbs.
Standard Front Springs Capacity 4x4: 1,025–1,030 lbs.
Steering Type 4x2/4x4: Manual, S-14 (Ross SE-54)
Steering Ratio: 24:1
Steering Wheel Diameter 4x2/4x4: 17 in.
Rear Springs Length/Width 4x2/4x4: 46x1.75 in.

Standard Rear Springs No. Leaves 4x2: 4
Standard Rear Springs No. Leaves 4x4: 5
Optional Rear Springs No. Leaves 4x4: 7 (1962–1965)
Standard Rear Springs Rate 4x2: 210 lbs.-in.
Standard Rear Springs Rate 4x4: 265 lbs.-in.
Standard Rear Springs Capacity 4x2: 1,150 lbs.
Standard Rear Springs Capacity 4x4: 1,275 lbs.
Optional Rear Springs Rate 4x2/4x4: 510 lbs.-in. (1962–1965)
Optional Rear Springs Capacity 4x2/4x4: 2,120 lbs. (1962–1965)

TIRES

Tire manufacturers varied. Common brands were General, Firestone, Goodyear, U.S. Royal, and Goodrich. Common sizes shown per data book; others available.

1961 4x2: 6.50-15, standard (4-ply)
6.70-15, optional (4- or 6-ply)
4x4: 6.00-16, standard (non-directional 4-ply)
6.00-16, optional (non-directional 6-ply)
6.00-16, optional (Firestone Town & Country 6-ply)
9.00-13, optional (sand tires)

1962 4x2: 6.50-15, standard (4-ply)
6.70-15, optional (6-ply)
4x4: 6.00-16, standard (non-directional 4-ply)
6.00-16, optional (non-directional 6-ply)
6.00-16, optional (Firestone Town & Country 6-ply)
9.00-13, optional (sand tires)

1963 4x2: 6.50-15, standard (4-ply)
6.70-15, optional (6-ply)
4x4: 6.00-16, standard (non-directional 4-ply)
6.00-16, optional (non-directional 6-ply)
6.00-16, optional (Firestone Town & Country 6-ply)
9.00-13, optional (sand tires)

1964 4x2: 6.50-15, standard (4-ply)
6.70-15, optional (6-ply)
4x4: 6.00-16, standard (non-directional 4-ply)
6.00-16, optional (non-directional 6-ply)
6.00-16, optional (Firestone Town & Country 6-ply)
9.00-13, optional (sand tires)

1965 4x2: 7.35-15, standard (Firestone Deluxe Champion, 4-ply)
7.75-15, optional (Firestone Deluxe Champion, 4-ply)
4x4: 6.00-16, standard (non-directional, Firestone 4- or 6-ply)
6.70-16, optional (Firestone Town & Country 4-ply, Super All-Traction 6-ply, General Tire Super All-Grip 6-ply)
7.60-15, optional (Goodyear Suburbanite WSW 4-ply, Red Carpet and Champagne editions)

TRANSMISSION AND TRANSFER CASE

4x2: 3-speed manual, T-13 (Warner T-90)
4x4: 3-speed manual, T-14 (Warner T-90)
Shift Type: floor
Ratios 4x2/4x4: 1st - 3.39:1, 2nd - 1.85:1, 3rd - 1.00:1, R - 4.531:1
Transfer Case: 2-speed, TC-144 (Spicer 18)
Transfer Case Ratios: Low - 2.46:1, High - 1.00:1

SCOUT 80 STANDARD COLORS

Color	Code	'61	'62	'63	'64	'65
Whitecap White	902	X	X	X	X	X
Harvester Red	201	X	X			
Green Metallic	5359	X	X	X		
Blue Metallic	6282	X	X			
Yellow	4154	X	X			
Tan	1230	X	X	X		
Vegas Blue Metallic	6456				X	
Beige	1384			X	X	
Blue Gray	8371				X	
Light Green	5551			X	X	
Red	2150			X	X	X
Light Yellow	4285			X	X	X
Champagne Mist	1501					X
Aspen Green	5683					X
Apache Gold	4298					X
Moonstone Blue	6606					X

COMANCHE:
THE ENGINE THAT MADE THE SCOUT

Though the Scout went from concept to production in an amazingly short time, the engine proved problematic. Four-cylinder engines were standard fare for the compact 4x4 market of the day, but IH built nothing suitable in-house. The engine could make or break the product and everyone knew it. The AM-80 Metro-Mite commercial van program came to the rescue with a possible contender: the Austin-sourced A-55 engine, but there were reservations over its lack of power relative to the market.

The 1.5-liter (90.88-ci), three–main bearing, OHV A-55 engine in the AM-80 was rated for 51 horsepower and 81 lb-ft of torque. Its origins were in Great Britain, debuting as the 1,200cc A40 Devon engine in 1947. In addition to the AM-80, the A-55 was also seen in the Nash Metropolitan beginning in 1954 in both 1,200cc and 1,500cc configurations. It also powered the 1955–1962 MGA sports car in both 1,500cc and 1,600cc twin-carburetor form. Upgraded to 1.8 liters, a five-main lower end, and nearly 100 horsepower, descendents of the A-55 powered the legendary 1962–1980 MGB sports car and other British-built cars into the early 1980s.

Later stellar history aside, the A-55 engine in 1959 wasn't the optimal choice on many levels, output being the main reason and a lack of "truck-like durability" not far behind. Unfortunately, readily available alternatives at the right price were few. In their search for a suitable AM-80 powerplant, and later for a Scout engine, IH looked at a long list of candidates, including industrial fours from Wisconsin, Hercules, and Continental. Little documentation remains as to exactly why these engines were not selected, but it was likely cost and packaging issues. IH also looked at Hercules, Perkins, Land Rover, and Standard diesels, but they were deemed too costly. For the Metro-Mite, it had come down to the 1.6-liter Volvo B-16A and the 1.5-liter Austin A-55.

When it came to the Scout, the A-55 was the readily available contender, but Austin's 2.2-liter A-70 (used in the Gipsy 4x4) and an overhead-valve four from Continental were also considered, as was, believe it or not, the Willys F134 engine. A B100 test chassis loaded to the predicted weight of the new vehicle was test-fitted with both the A-55 and the F134 engines. Predictably, the 90-ci A-55 engine was the fuel-economy winner, but the 134-ci Willys won the power contest. Because the new IH 4x4 was aimed squarely at taking market share from Jeep, the F134 was a politically incorrect choice.

It wasn't until late 1959 that a new contender came to light. MTC report 589, published on September 4, 1959, indicated the engine gurus were considering a new four-cylinder engine developed by IH. When pressed for an in-house alternative, the engineering department took cues from another manufacturer

SCOUT 80: HUMBLE BEGINNINGS 53

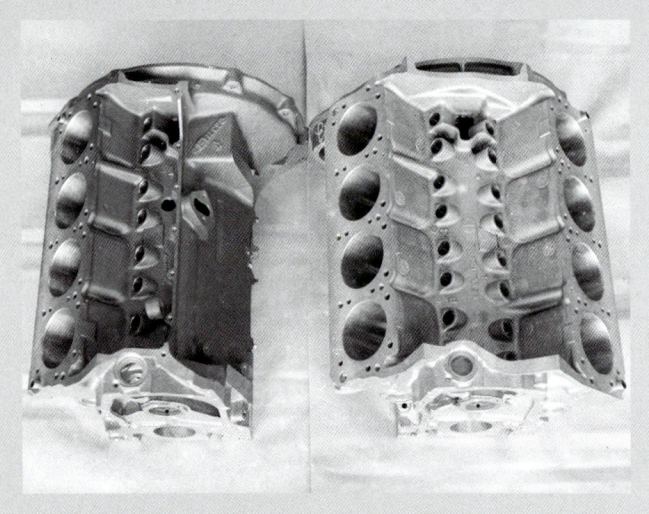

▲ The 4-152 and V-304 shared the same tooling and you can see the similarities in the block design. The indexing lugs on the four are where the two bosses on the right have been machined smooth. During the final machining, these lugs were removed on a spur line used for just for the fours. WHS 111619

▲ An early 4-152 engine as installed in a Scout 80 is shown with a hydraulic pump for a snowplow. Early upgrades included a PCV system and a more efficient replaceable cellulose air filter element to replace the oil bath. WHS 111505

with a new four-cylinder engine nearing completion. Pontiac was putting the finishing touches on their innovative Trophy 4 for the upcoming 1961 Tempest. They lobbed off the left bank of a 389-ci V-8 and ended up with a 195-ci slant-four. The standard one-barrel Trophy 4 produced 110 horsepower in production. (A high-compression variant made 120 horsepower, and a souped-up, hot-cam, high-compression four-barrel made 166 horsepower.) Built from 1961 to 1963, the Trophy 4–powered Tempest sold well and the engine didn't have problems beyond an inherent vibration (a trait the IH engine would share). The Pontiac idea was well publicized. It looked like a perfect answer to International's engine problem.

The first discussions were about a 133-ci four taken from the recently introduced V-266. Codenamed X-133, this engine was estimated to produce about 75 horsepower, putting it on par with Jeep's F134 in power and displacement. The MTC gave engineering the go-ahead and it was predicted it would be available for testing by February 1960. By this time, the overall design of the Scout was well on its way and five prototype test chassis were slated for completion by December 1959. They were planned for A-55 engines, but the chassis engineers left plenty of room under the hood for a larger engine.

Slant-four development progressed to the point where all of IH's small-series V-8 designs were considered, since the engineering was almost the same in each case. This included the V-304 and the V-345 then in production, but also the V-392 still on the drawing board. Each of the four-cylinder designs wore an X-number. The X-152 (87 horsepower estimate) was made from the V-304. The X-173 (100 horsepower estimate) was half a V-345, and the X-196 (115 horsepower estimate) was half a V-392. It isn't clear from the remaining records whether an X-173 test engine was actually built, but a 4-196-ci later appeared when the much-delayed V-392 debuted for 1966.

▲ Engines arrived at Fort Wayne from the Indy Engine Plant without ancillaries. A special area was used to add components such as carburetors, fuel pumps, generators, and pulleys. In later years, this was a basement room, but that does not appear to be the case in these 1960 images, taken when the line was being tested and calibrated.
WHS 116363, WHS 116358

The proposed engines were divided into two categories: short blocks (X-133 and X-152) and long blocks (X-173 and X-192). Short and long were defined by stroke length and the resulting block-deck height configurations. The smaller 266- and 304-ci V-8s had a 3.218-inch stroke while the 345- and 392-ci V-8s used a 3.656-inch stroke and, hence, taller blocks.

The advantages to these four-cylinder engines were patently obvious. First, the blocks could be machined on the existing long or short V-8 tooling with minimal adaptation. The four-cylinder blocks were cast with a vestigial left bank that was used to index the block for machining on V-8 tooling (the boring/honing tools on that side could be shut off when a four ran through). After block-machining was done, the four-cylinder went off on a spur line where

SCOUT 80: HUMBLE BEGINNINGS   55

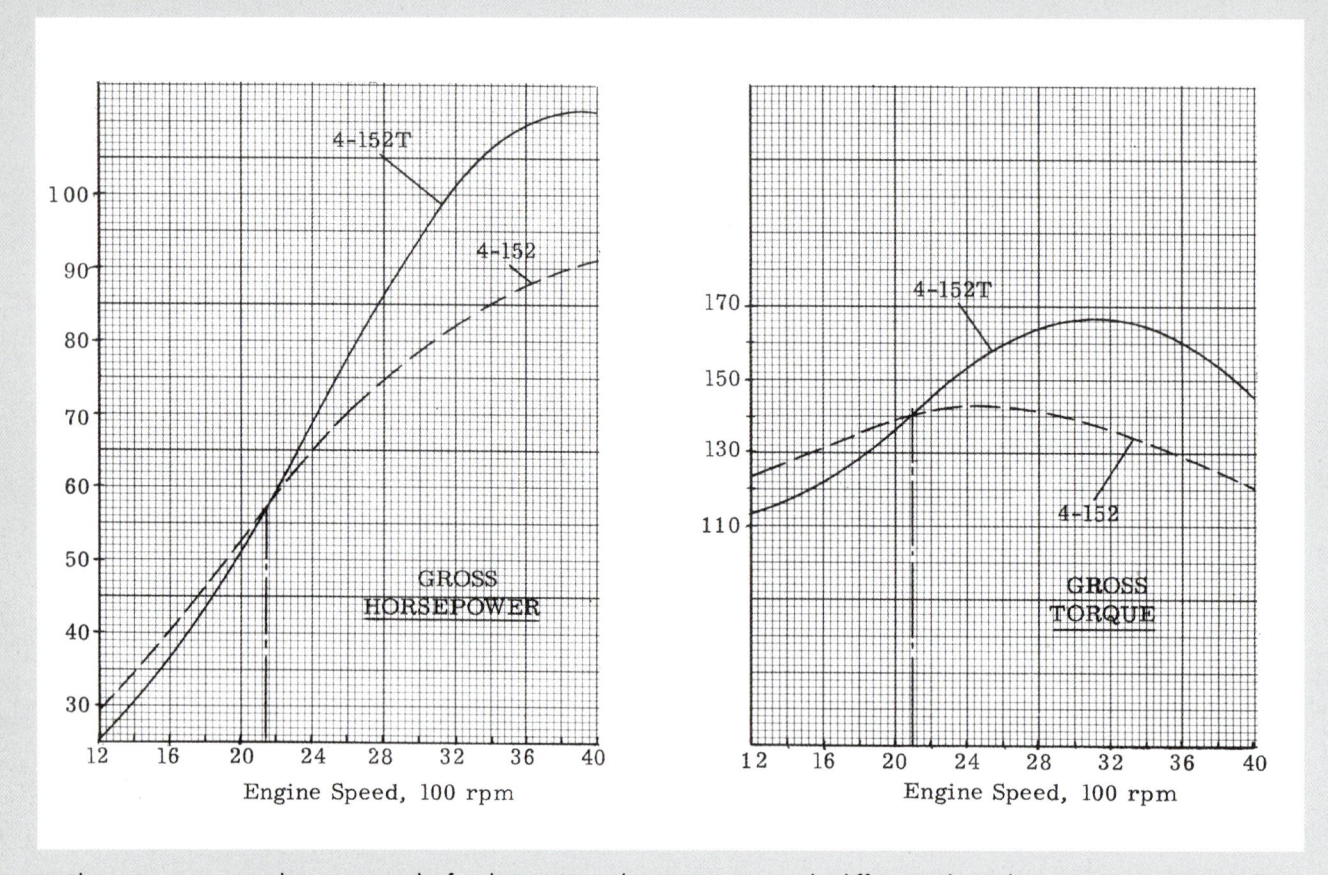

▲ The gross power and torque graphs for the 4-152 and 4-152T were vastly different. The turbo engine is more "peaky," meaning its power and torque sweet spots occurred over a narrower rpm range than the naturally aspirated 4-152, as well as at a higher rpm. The main purpose of the 4-152T was to improve on-road performance, especially at high altitude. All the literature warned that fuel economy would suffer. To avoid detonation, the carburetor was jetted extra rich, but evidently not enough—melted piston crowns were common.

the indexing lugs on the left side were machined off and a few other unique operations carried out.

The engines used mostly the same internal parts (pistons, connecting rods, pushrods, lifters, rocker arms, valves, to name just a few), and much of the external stuff was identical too, such as cylinder heads and right exhaust manifolds. New crankshafts, camshafts (same lobe profile design as the V-8), intake manifolds, and small parts were required, but they were all at least partially based on existing parts and tooling. While the IH Indianapolis Engine Plant had the ability to produce the special crankshafts needed for the four, their production schedule was maxed out so they contracted with Atlas Crankshaft Company of Ohio to produce them. The final cherry on top? Development was predicted to be very fast and just in time for the Scout debut. The Engine Plant was instructed to build two X-133 and three X-152 prototype engines for tests.

Development was held up briefly in October when the MTC expressed concern about how well the engine would fit in the newly designed Scout and the current Metro-Mite chassis. Installations of mockup engines in a prototype Scout chassis and the AM-80 were acceptable with minimum alterations required. It was noted running engines might not be ready in time for installation in the first five driving prototype Scouts, but the firewalls would be manufactured to accept the new engines at a later time. In actual fact, it appears test engines were available at the start.

By January 1960, to consolidate resources, it was decided to focus on the short-block designs. After testing, it was determined the operating fuel economies of the 133- and 152-ci engines were so close that the MTC decided to go for the higher-power unit. That was a good move—the 90-plus-horsepower, 152-ci four offered some competitive advantages against Jeep's 75-horsepower F134. Final MTC approval for the X-152 was given on May 17, 1960.

By September 1960, the slant-four's model designation was approved as the "4-152." The sales department was given the task of coming up with an appropriately whiz-bang marketing name. By November 1960, the field was whittled down to three choices: Ranger, Comanche, and Commando. The MTC approved all three and let sales decide which to choose. That decision came in December 1960 and the Comanche name was trademarked.

In August 1961, as Scout production became a less hectic undertaking, the Motor Truck Committee discussed a cost-reduction program for the 4-152. It used many of the same components as the medium-duty rated V-304, but because the Scout was not expected to be worked like a medium truck, IH thought they could save some money by using lighter-duty parts. The parts to be de-rated included the pistons, hardened exhaust valves, hardened valve-seat inserts, expensive steel-backed aluminum bearings, and the air filter. By December, it was determined a lower-cost engine would actually cost more due to added production complexity. A lower-cost bearing set had been approved and some 10,000 sets purchased. Plans were to install these bearings until stocks were exhausted, but the bearings proved defective and were scrapped.

MORE POWER!

By 1963, complaints about lack of power were coming in from people who carried or towed loads with their Scouts, especially on hilly terrain at highway speeds. This was partly a gearing problem and Engineering began looking at a close-ratio four-speed as the answer, among other ways to increase engine power. Sales was rabid on the topic and wanted something on hand for the start of the 1964 model year. That was not to happen, but a number of innovative ideas were discussed.

The obvious first thought was a six-cylinder engine. The installation of a 114-horsepower BG220 IH six had come up many times, including at the very start of the Scout development. The main roadblocks had been the multitude of Scout design alterations this would require. Such concerns had not changed by 1963—it was estimated the necessary body and chassis design changes would require some 4,000 man hours. Since the bigger engine would be offered as an option, it would also be necessary to have both four- and six-cylinder chassis available. By the late summer of 1963, a BG-220 had been installed in a 4x4 Scout for testing. Upgraded with high-altitude pistons, this Scout was shipped to Denver for more tests in high-altitude, mountainous terrain, and to demonstrate the idea to customers and gauge reactions.

Another idea for boosting power was to fuel-inject the 4-152. Designed by the Marvel-Schebler division of BorgWarner, the Bendix system would provide an estimated power increase of about 14 horsepower for a package wholesale price to IH of $220 with a sale of 3,000 systems annually. The Bendix electronic fuel-injection system and already been used on a few cars, most notably as an option in Chrysler products such as the legendary 300D of 1958. A system was installed in a Scout for further tests and went to Denver, Phoenix, and Los Angeles with the six-cylinder Scout for evaluation.

▲ The basic design was the same in the naturally aspirated and turbocharged engines, but the induction systems were vastly different. The turbo engine developed 5- to 6-psi boost, which about doubled the airflow. The turbo's intake manifold was the same, but the exhaust was new and directed the flow up and into the TRW turbo; the outflow dropped into a larger 2 ½-inch system. The air filter was designed for increased flow (basically it was an air cleaner from a six-cylinder truck application), and the Holley 1904 carb, used in a draw-through configuration, was re-jetted and calibrated for the increased airflow. The distributor was modified with a boost-actuated retard mechanism to avoid detonation.

The reactions from the Denver crowd were surprising. The BG-220 and the fuel-injected units were largely panned in favor of a V-266 installation. Barring that, the next preference was for the more powerful 140-horsepower BG-241 inline six instead of the BG-220. The V-8 idea was dismissed because of the time and effort required and the intent to offer a V-8 in a redesigned Scout just a few years hence. The BG-241 idea required the same effort as the BG220. The long-block four reentered the discussion, since the 4-196 was still on the drawing board but a long ways off.

While all this was going on, Product Engineering had been working with Thompson Products to develop a turbocharger installation for the 4-152. Thompson had developed the system for the Corvair Spyder. Estimated power for a turbocharged 4-152 was in the 112- to 115-horsepower range. By September of 1963, the six-cylinder and fuel-injected ideas had been officially rejected and the turbo engine had moved to the top of the list. Qualification tests were on track for completion of the by the end of December. Strident calls from Sales revived the BG-220 idea briefly, but by May 1964, the turbo 4-152 was approved and scheduled for an April 1965 introduction as part of an interim improvement program to "doll up" the Scout prior to the introduction of the Scout 90 (known as the Scout 800 in production).

High-altitude tests showed the turbo Comanche developed no less than 108 horsepower in the Mile-High City, where the naturally aspirated Comanche was making 79 horsepower on a good day. At sea level, the turbo 4-152 developed around 115 horsepower (111 was the rated power in production). Sales estimated a minimum of 1,000 additional sales of Scouts fitted with the turbocharged engine. Because the chassis modifications required to fit the turbo engine were relatively minor, the project moved rapidly and the engines went into production in December 1964 as the 4-152T. It became a $280 option for the Scout 80 in the 1965 model year and an option in the 800 model when it debuted later that year.

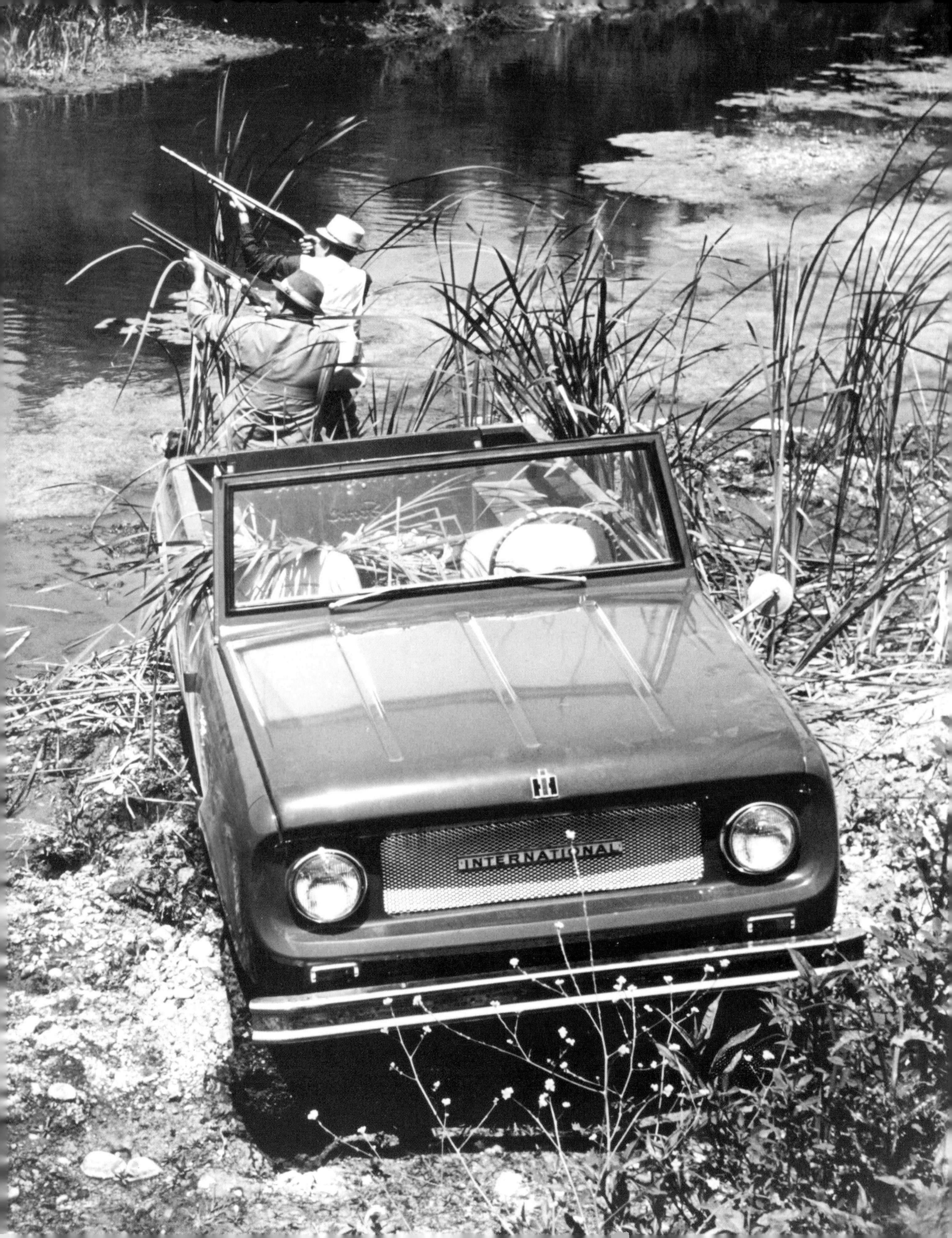
INTERNATIONAL

Chapter 3

SCOUT 800, 800A, AND 800B: MOVING UPMARKET

The Scout 80 introduction in 1961 was a marketing coup by a company not known for marketing coups. The Scout 80 may have gone more mainstream than International expected, but the market soon began outpacing the Scout. International Harvester wasn't built to compete in the Detroit-style marketing wars that began in the mid-1950s, where constant changes and one-upmanship ruled the product planning departments. Traditionally, IH played to a different marketplace, where stability and product stamina sold trucks. In the 1950s and 1960s, light trucks began playing to a more mainstream customer than ever before. There was money to be made as the markets expanded, and if IH wanted a share, they had to alter their marketing and product-planning strategies. That was tough in a company more focused on producing agricultural products than trucks. Plus, in the IH corporate structure, the truck divisions probably had less autonomy than they needed.

Suddenly, a Jeep or a Scout was not just a tool for geologists, surveyors, park rangers, and oilfield workers. It was a second or third "car" that did triple duty as a daily-use vehicle, suburban hauler, and outdoor recreational vehicle. The Scout had the work and outdoor duty down pat. It was the daily driver part where it didn't do well, and people demanded more and more comfort and refinement from all brands in that vehicle class.

Not long after the Scout 80 debut, IH planners began thinking about the next iteration. Early planning documents indicate a rather leisurely

◀ The '66 800 was available as a roadster for about $200 less than a base Travel-Top. The buyer ordered either a full-length or short soft-top from the accessory catalog, or no top at all. Because all the tops were interchangeable to a great degree, an owner wanting a soft top was better advised to simply order a soft top kit and change out the tops as desired. Well into the 800 era, the old-style sliding windows were still an option on the Utility model. Option 16774, "Sliding Window Glass in Lieu of Roll-Down Windows," resulted in a $39 credit.

approach, but the response to the Scout took IH by surprise and it forced them to work harder to retain their newly claimed market share.

The first time a new model discussion appeared in the documents was on March 30, 1961. S. Colacuori in Product Development proposed expanding the Scout line to include an enlarged half-ton version with a wheelbase lengthened one foot to 112 inches and a 1,000-pound payload. The concept was named the Scout 90 (the 90 to reflect a payload class of approximately 900 pounds). They wondered, however, if it would interfere with sales of half-ton pickups in the lightest half-ton class (namely the C-100).

Though it was no doubt discussed and recorded in earlier documents, a major Scout upgrade pops up again in several MTC documents from March 1964. By then, top-secret planning was underway for a totally new Scout some years hence called the X-Scout (which would become the Scout II), but the need for an interim model was becoming apparent. Paramount was a Scout with increased engine power and a quieter and more weather-tight body. To that end, the installation of a BG-220 six (and other engines) was discussed, as was a fixed windshield, the fold-down windshield being one of the major sources of wind and water leaks. As an alternative to engine upgrades,

▲ Though the 800 shared its overall lines and look with the 80 model, the new grille and badging set it apart instantly. But the differences were more than skin deep—the 800 bodies were more weathertight and quieter to boot. WHS 111160

▲ The 800 Travel-Top made a good plow rig and International wasted no time shooting pictures in the winter of 1965–1966. Meyer was already fast friends with IH and moved quickly to have an 800 plow kit available. There were 6.6- and 7-foot blades, as well as a lot of options on the lift and tilt apparatus, from full hydraulic running off an engine-driven pump to electrohydraulic. WHS 111275

▲ The Cab-Top translated nicely to the 800 line, and with the fixed windshield, standard roll-up windows, and better sealing, it was a nice upgrade. This '66 is resplendent in Red and has the Custom trim. In this era, the two basic trim levels were Utility and Custom. Utility was bare-bones, but Custom offered chrome front bumpers, outside mirrors, hubcaps, and Deluxe interior with floor mats, armrests, and headliner. It cost $214 more than the Utility package. *WHS 111656*

a couple of four-speed transmissions were also considered, including a close-ratio unit that optimized drivability with low-power engines, and one with an overdrive top gear. Finally, a Dana 20 transfer case was discussed because its straight-through, centered rear output made for a *much* quieter unit than the offset rear-output Dana 18, even if it did eliminate a PTO/overdrive point. A much more efficient and attractive dash was also discussed as a vital necessity.

As the mid-1960s neared, introduction of the X-Scout kept moving further back and the need for a bridge vehicle could no longer be ignored. Planners moved forward with ideas and the project soon became known as the "Interim Scout." By June 1964, the Interim Scout had also become known as the Scout 90, still a designator that made some sense. By this time various approvals were made regarding interior and exterior refinements. The dash was one early achievement, with the stylish new layout using the individual gauges seen in the C-Line light trucks. Six new grille/front-end designs were also considered. By September, tentative plans were in place to begin production of the Scout 90 in May 1965. By November 1964, the new grille design with its "International" badge was approved.

By February 1965, the Scout 90 project name was gone, replaced by Scout 800. By then most aspects of the design were approved, including standard roll-up windows with vent windows (sliders were optional), more sound-deadening material, plus a fixed windshield. An optional four-speed, close-ratio manual transmission option was near completion.

Most of the bases were covered with regards to civilizing the Scout, but the gorilla in the room was the engine. Among the more numerous complaints about the Scout was a lack of get-up-and-go. More than that, when it was loaded to the gills, it was totally overwhelmed. Remember, this was the beginning of the horsepower race and the muscle car era—the "need-for-speed" was invading all parts of the American automotive scene.

The Sales Department had been bemoaning the lack of an engine upgrade almost from the day the Scout had debuted. Development and integration costs had prevented much work in this area, but events of the mid-'60s soon heightened awareness that the competition was pulling ahead. Within a couple of years of the Scout intro, Ford was pursuing a rival entrant in the same category. It seemed clear to everyone that the Scout was the inspiration for Ford's sporty new bobtail 4x4, which provoked more than a little pride at IH, as well as anguish. As details slipped out, IH learned this new Ford 4x4 would have a standard six and an optional V-8. That rig would be called the Bronco, and many discussions would follow at IH on how to counter it.

If the new Ford wasn't enough to inspire corporate paranoia and powerplant envy at IH, arch-nemesis Jeep was introducing a 225-ci, 160-horsepower V-6 engine option in the CJs *and* a sporty new sport-utility vehicle powered by that same engine and called the Jeepster Commando. *And* there was news of a pickup-based General Motors utility due for a late-'60s intro (the Blazer) *and* a similar Chrysler project that had just started (the Dodge Ramcharger/Plymouth Trail Duster).

By March 1965, IH was well along with designing the totally new X-Scout for a late-1960s intro. A V-8 option was planned and International didn't want to blunt the significance of that by giving the interim rig the option first. IH believed a new Turbocharged 152 four was going to be enough to hold the line, but the opposite soon became apparent: The 4-152T was problematic, a customer-complaint generator, and still just a four.

Some hope came with the news that the V-392-ci V-8 project was nearing completion, which meant the long-awaited 4-196 four-cylinder could proceed to production. But, again, the 4-196 was still a just a four—a bigger one, and more reliable, sure, but still just a four nonetheless. Sales was harping for a six or a V-8 to keep up with the Joneses, but management still resisted.

By May 1965, the production target date for the 800 was July 7, with a tentative goal of building some preproduction units in early to mid-June. That came to pass when the very first line-built 800 (serial number FC150000) was roll-tested on June 14, followed closely by five others: FC150001, 150002, 150003, 150006, and 150007 (-004 and -005 are missing and apparently stricken from the serial number list). A few more followed in June and July, mixed in with the last of the 80 models. Most of these first few 800s were test and engineering units. The first ones with shipping addresses outside IH test facilities were in the FC150019 serial number range and up. (Scout 800 FC150000, incidentally, was a Doll-Up and a 4x2.)

A note about the serial number jump is in order. Some of the first 800 Scouts had serial numbers in line with current Scout 80 production, but likely it was

▲ Here is a '67 Custom Travel-Top with a V-8 (note the "V" on front fender) in Apache Gold (4298). It's mounting the luggage carrier (244 689 591), which was normally a dealer-installed item at this time and could be found in the International "Necessories" catalog. In 1966, this carrier cost $74.50 plus installation. *WHS 111099*

deemed prudent to differentiate the new models from the old, so while the last 80s were in the FC124000 range, the 800 line started at FC150000. On the pre-production 800s, the LST shows a typed 120000 number scratched out and a 150000 number handwritten over it.

A stylish high-end Scout to be called the Sportop (not to be confused with the Sport-Top convertibles) was intended to go into production at the same time as the rest of the 800s, but it didn't fare so well on the new schedule due to delays in getting special parts (see Chapter 5). The Scout 800 line would ramp up to full production in late July 1965 and be announced shortly thereafter (August 4). It took some time for 800 production to ramp up, but IH had prepared by putting the 80 line on overdrive before it was shut down. Sales then offered generous incentives to help the dealers sell them.

MEET THE SCOUT 800

A great many ideas had been discussed with regard to making the 800 a memorable new entrant. Some of those fell by the wayside, but when the Scout 800 made its public debut in August 1965, it was clear IH had a lot to crow about.

Mechanically, there were many upgrades. The 93-gross-horsepower 4-152 soldiered on as the base engine, with the 111-horsepower 4-152T as the top option. The standard T-14 manual three-speed carried on as the base transmission for the 4x4, and the T-13 did the same in the 4x2. The big news, however, was the availability of a four-speed close-ratio box. Initially, this was listed in specs as the T-21, but that IH designation also applied to some medium-duty five-speeds, so it was soon changed to T-45 (T-44 for the 4x2), a close-ratio T-18C transmission (4.02:1 first gear) that gave the low-powered fours better gear spacing to work with.

▲ During the 800 era, International began a more serious push into the RV market. The "Campermobiles" ad campaign highlighted the slide-in and chassis-mounted campers available from IH dealers. This shows a 10.8-foot Dreamer camper on a '66 Scout 800—a four-cylinder, no less. This Apache Gold (4298) Cab-Top is a Custom model 4x4 and it appears the camper accents were painted to match the Scout. This Scout appears in some advertising and brochures from 1966, but the Dreamer decal on the front is conspicuously missing. International also used other manufacturers to build campers that were rebadged as International Campermobiles. *WHS 111320*

The Dana 20 "silent" transfer case was a big deal in making the 4x4 Scout a better and quieter daily driver. In two-wheel drive, the rear axle was driven directly via the centered rear output. The offset Model 18 transfer case from the Scout 80 had the rear output offset to the right so even in two-wheel drive a significant portion of the geartrain in the unit was operating, resulting in more noise and parasitical

▲ Most people regarded the 800 dash as a big improvement over the 80. The instruments were exactly the same as those used in the C-Line light trucks. The 1966 model year did not yet have the padded dash that became standard in 1967. This Utility 800 is partially upgraded with deluxe pieces, such as the floor mats and seats. The Deluxe interior included padded interior door panels, the better seats (shown here in the Champagne color), and headliner.

drag. The additional benefit to the centered output was that it made for a better driveshaft angle, resulting in less vibration and longer U-joint life. IH made a serious effort at ride quality improvements by offering lighter standard spring rates, better shock tuning, and better riding two-ply (four-ply rated) tires. The net result was a Scout that rode better than the previous models by a goodly margin.

Cosmetically and ergonomically, the 800 made great strides. The dash was a great highlight, along with a revised instrument panel using individual gauges. Two-speed wipers were a big plus. The heater and defroster was improved, and more sound-deadening upholstery was added, including a standard headliner. Seats were also improved in various ways, the standard seats benefiting the most. The tailgate was redesigned for one-handed opening, and the door handles went from twist type to push button along with more effective door latches and better window glass regulators.

1966

Because the 800s were offered so early in 1965, some people think they were '65 models. Not so. They were always intended to be '66 models. It took a while to revamp the assembly line for building the new rigs, which is why the 80 line was shut down in early in

▶ Late in 1967, International was very happy to win a Postal Service contract for 7,920 right-hand-drive, two-wheel-drive Scouts, 6,745 with automatic transmissions (T-28/BorgWarner 8), and four-cylinder engines. Development of these rigs had started in early 1966. This prototype was built in that time frame, presented to Postal Service officials, and favorably received. Production units had windows on both sides and the hubcaps were not those typically seen on production units. The Postal Service specified Powr-Lok limited-slips in the rear and a number of interior changes to suit the postal life. The development of the automatic for the postal Scouts led directly to the introduction of an automatic transmission option in left-hand-drive Scouts for 1969. These units had a special serial VIN range outside the normal G-number range, running from G710001 through G716745. The VIN prefix uses a 3 as the third digit to indicate a Postal Scout.

July to reset. Normally, a model-year changeover would occur in late July or August in a matter of days or a week, depending on the number of changes to the line. The differences between the 80 and 800 were considerable (more than what appears on the surface), requiring extra time. The first 800s to run through tested the line, trained the workers to the new units, and produced test models and vehicles for the press fleet and advertising. The end result was a model line that came to dealers a bit earlier than normal, but not by much.

IH wanted their flagship model, the Sportop, ready in time for the 1966 debut. As it was, they publically announced the Sportop on February 3, and increasing numbers of Sportops became available for delivery to dealerships (see Chapter 5). The Custom model filled the gap as the top-line Scout and was a package equivalent to the Scout 80 Champagne Edition.

The first big news for 1966 was the August 18 announcement to dealers that the 4-196 would be an available option, which was followed up by an August 29 general public announcement. Installations of the new engine had begun August 17 at a rate of 10 per day. Few were in time to be in '66 models.

On October 7, 1966, not long after the end of model-year production, it was announced to dealers that the troublesome 4-152T turbo was being deleted from the options list, but that the factory still had a number of engines available for special order. Some were sold by special order for the 1967 model year (We found '67 LST showing at least a few were assembled in '67.) IH was not sorry to see the 4-152T go bye-bye.

As with the previous generation, both 4x4 and 4x2 models were offered. A few changes on the 4x2 models are worthy of note. First, they eliminated the wimpy four-lug wheels, allowing IH to add the 3,500-pound Dana 44 rear axle to the options list. Heavy-duty springs were also optional, as were higher-rated tires. While IH did not list a higher optional GVW for the 4x2 with these upgraded parts as it did with the 4x4s, it stands to reason that with the optional parts, it was, in fact, higher.

Initially, the most of the Scout 80 colors, including those special colors for the Champagne Series Scout,

▲ For the 1967 model year, a padded dash and sun visors were made standard as part of the safety upgrades mandated for 1968. This '67 Cab-Top Utility has few options, not even a radio.

carried over, but later new colors were added. The Panel-Top carried over to the 800 as well, but would soon fade away.

1967

In October 1966, IH announced to dealers that the 1967 models would have a greater array of standard features. Among these were dual-circuit brakes, a padded dash, padded sun visors, inside and outside mirrors, seatbelts, glare-reducing wiper arms, windshield washers, backup lights, hazard lights, and 7.35x15 tires (street tread). They didn't mention that some of these items would soon be mandated by federal law.

The big news on February 20 was the availability of the 155-horsepower, 227 lb-ft V-266 V-8 (dealers had been apprised on February 9). IH finally got sorely needed "power parity" with the competition and the bragging rights needed to stay a contender in the SUV marketing wars. Getting the V-8 to production was a long haul. Beyond a batch of 50 specially built units completed in January, availability of the V-266 was slated for April, but some reports show units trickling into dealers in March.

The V-8 was not an easy shoe-in. For various reasons, the engine had to be mounted forward of the four-cylinder location, so the mounts, the first welded crossmember, suspension, location of the body on the chassis, steering, and eventually the front axle had to be changed from the four-cylinder configuration. IH had hoped to make the V-8 option available with the introduction of the new 1967 models in November or even December, but the front-end issues plagued engineers well into

▲ In the latter part of 1968, International was working up a new Postal Service model to pitch to the government. No doubt influenced by the sliding doors of Jeep's DJ post office rigs, International developed one of their own in the hopes of a new contract for up to 11,000 units. Next to nothing is known about these rigs, though it is known the burlier T-39 automatic was slated for use. This one was shot in November 1968 at the Styling lab in Fort Wayne. International did not win the contract, but did build postal-spec 800As and 800Bs. They also briefly considered offering this setup as a commercial option but rejected the idea due to costs and an unclear market potential.

December and the V-8s didn't start rolling down the line until January 1967.

Along with the new engines, the axles got some upgrades. These were mostly made necessary by the V-8. The 2,300-pound-rated Dana 27A (RA4/14) rear axles were unsuitable for the V-8 so the 3,500-pound Dana 44 (RA-9/23) was made standard with the V-266. These were not the D-44s that had previously been optional, but a wider version (by about 2.5 inches) also used in the Metro-Mite van. The wider, stronger axle was needed to match the wider front axle required of the V-8. Rather than complicate the assembly line with two rear axles, the bigger units were fitted across the board. Similarly, the tried-and-true FA-14/24 (Dana 27AF) up front was inadequate for the V-8-powered rigs. The mounting of the V-8 engine and resulting steering changes necessitated a wider axle (1.562 wider track), so the stronger FA-11 (Dana 30) was used, but only in the V-8 rigs. This axle had a slightly beefier center section and thicker tubes, but the outer ends were the same as the D-27AF.

With all the necessary differences, it was decided to make the V-8 a separate model rather than just an option. For that reason, the model code had a 2 in the fifth position and the sales-data sheets listed the four- and eight-cylinders as separate models. When the sixes debuted later in the 800 run, they also were listed separately, but their chassis and suspension were very much like the V-8's.

A little tidbit regarding developing a V-8 option is that an outside engine was considered. AMC had begun offering a 343-ci version of their new lightweight V-8 and it attracted the attention of IH marketing and engineering planners. The 343 two-barrel was rated at 200 horsepower, about 54 more than the IH 266. On top of that, it was 138 pounds lighter (556.5 versus 693.6). Rough number-crunching showed costs were about equal (both tooling and purchase costs), but the kicker was the potential loss of IH parts sales, which was the stated reason for not proceeding.

One of the few downsides to the Dana 20 (TC-145) transfer case was that it eliminated the PTO port and thus the possibility of adding an overdrive or a PTO for winches or other tools. The V-8 further complicated winch-mounting because the front crossmember was moved forward and there was no kit available for it. Both issues were solved by Ramsey, which developed the STC-2 PTO that replaced the bottom cover of the transfer case and engaged the T-case gears directly. They also developed a winch kit for the V-8, which IH offered both as a factory and a dealer installation. The PTO winch was in its sunset years at this point, however, with electrical systems and batteries improving enough to allow for lighter, easier-to-install, and more efficient electric winches.

1968

The big news for '68 started with a single-stick version of the Dana 20 (TC-145) transfer case that was announced on July 29, 1967, and had been incorporated into production a short time before. This version had been discussed in March 1967 and it was noted that IH had enough twin-stick units in stock to carry production of the four-cylinder models to September 1967 and the V-8s at least to May. Yet, for whatever reasons, it took more than a year to implement the change.

SCOUT 800 LST ANOMALIES

A very frustrating notation on 1966 model-year 800 LSTs are a Q or, very rarely, an R suffix after the 800 on the first line (e.g., Scout800Q). It is unclear what these suffixes signify, but their meaning is hotly debated. A survey of several LSTs revealed one of these suffixes assigned to many early Scout 800s, usually spaced well apart, but sometimes closely spaced or even consecutive. There is little evidence to support any of the popular theories.

THE SCOUT 800 CUSTOM

Especially in the early days of 1966 production, International highlighted the 800 Custom as the top-drawer Scout. These rigs carried on the tradition of the Champagne Series (CS) Scout 80 Doll-Up. The 800 Custom was very much the same as the CS, having almost identical champagne-colored upholstery, door panels, and carpet. The headliner was different but in a similar style. Because the 800 dash was different, it didn't use the same dash pad as the 80 CS, but the accoutrements matched closely. The non-Custom Scouts were listed separately as Utility models, making the Custom a "special" in its own right.

The model code (see "Scout 800 Data," page 84) had an 8 as the sixth digit, indicating a Doll-Up interior. The Scout 800 Custom Doll-Ups have the model code 780908 (same as the Champagne) and the LST stamped "DOLL-UP-SCOUT" at the top. Also, the accessory codes list the top-line interior options, notably the component code 16023, versus 16022 for the Utility version and 16025 for the Cab-Top.

▲ Developments leading up to the '69 800A models included a fiberglass top. Styling worked up some renderings and in September a basic design was approved for a mockup. After the mockups were reviewed, the project was approved in principle, and by November, this is what was emerged. Testing reported problems with roof-seam bonding and the liftgate hinge attaching points. Production problems with the vendor would have delayed the top well into the '69 model year, and when costs were reviewed, the project was canceled at a January 23, 1968, meeting. This was a precursor to the fiberglass top on the 1976–1980 Terra and Travelers.

Surveys of the early 800 LSTs show a *lot* of these Custom 800 Doll-Ups produced early on. An unofficial count yielded 25–30 percent Custom Doll-Ups in this period (perhaps as many as 2,500 units). They were noted in a serial number range of FC150000 to FC158000, after which they began to dial back in number, indicating production roughly into early November 1965. Based on the evidence, educated speculation is that increased Doll-Up production was intended to make up for the long delays in getting the Sport Scouts (see relevant sections of "Scout 800 Data," pages 84–89) into volume production. The Sportops were planned as the upper-end Scouts, a category that had proven to be a good seller.

Some owners of early 800 Custom Doll-Up models have taken to calling them the Champagne Series. On the surface, that may seem a fair representation, but it is not an accurate one. Examination of period 800 literature yields no references to them as Champagnes, only as the 800 Custom, which is the most accurate term.

IH was really pushing the 4-152T turbo engine in the early days of the 800 and, in concept, the turbo

▲ This August 1967 rendering by Kikuyo Hayashi depicts one possible Scout facelift for 1969. A little bit of the Scout II can be seen in the front end. In fact, something similar was used for the Australian assembled Scouts, so this may have been a design study for that market. Hayashi was one of International's more gifted designers and was also known in the early 1980s for his work on covers of science fiction books and magazines.

▶ The mocked-up front end from September 1967 based on the Hayashi rendering in the nearby image. Note the emblems. There were three variations of this look, each with the same front end but with the badging positioned differently.

72 INTERNATIONAL SCOUT ENCYCLOPEDIA

▲ By May 1968 the details were pretty well in place for the '69 model year, as seen in this preproduction 800A four-cylinder in the Copper Metallic (2282) that was to debut for '69. It was a base model without the Custom trim option and is the earliest image found that depicts the final 1969 design changes. Note the reflectors. This series of shots also showed new designs that radically improved the baseline seats.

▶ The facelifted 800A models debuted with a few basic styling changes and the lineup was broadened with special models. Instead of separate Utility and Custom models, "Custom" became an options package (code 16023 for the Travel-Top, $245) and included color-keyed floor mats, headliner, chromed outside mirrors and front bumper, door trim panels, armrests, lighter, and chrome hubcaps. The four-cylinder and V-8 models remained separate, as did the PT-6 engine when it debuted later in 1969. This is a Travel-Top V-8 in Lime Green Metallic (5857) with the step rear bumper and luggage rack. *WHS 110863*

was a good fit with an upscale Scout. A fair number of early 800 Customs were built with the 4-152T. A look at the LST will determine if an early model Doll-Up 800 was originally powered by the problematic turbo engine. Many were reverted to natural aspiration in the intervening decades.

SCOUT 800S WITH FOLD-DOWN WINDSHIELDS

A persistent urban legend in the Scout community regards 800s with fold-down windshields. Some scoff, some believe, but most have never seen one or seen proof they existed. Well, for the record, they did exist. It was a

▲ Early in 1968, legendary car customizer George Barris was nearing the peak of his stature and was approached about designing a custom trim package for the Scout 800. Barris Kustom City also built a custom show vehicle with simulated snakeskin paint and interior that debuted at a December 1968 custom car show in Las Vegas. It continued to make the rounds of car shows into 1969. The Scout and a resulting kit were both called *Python*. The kit re-created certain design elements from the the show car, including body appliqué, a hood with a scoop, and special wheels. In February 1969, the SPC rejected the trim package based on cosmetics and economics. The *Python* was sold to an IH draftsman in the 1970s and was used as a daily driver. It was last seen around 1980.

little-known option, but it appears on LSTs as "fold down windshield" with component code 16536. The option was available all the way through the end of the 800B in 1971 and a surprising number of them were built. More difficult to find are mentions of this option in sales literature. It may have once existed on some obscure salesmen options sheets, but because it did exist as a component code, IH had the means to produce it according to some specific plans. Since the first edition of this book, quite a number of 800, 800A, and 800B survivors with fold-down windshields have been found and verified.

INTERIM, INTERIM: THE 1969–1970 800A

Imagine how execs and product planners at International felt about delays on the X-Scout rollout. While the Scout 800 rollout had gone reasonably well and IH was holding its own in the market, it was clear some extra pizzazz was needed to help carry the 1969 Scout to the new-model goal line. A number of improvements had been developed and it was thought a little rollout hoopla might help. Along with that came a new designation, 800A.

The 800A got a bit of a facelift, most notably in the grille. New features, mandated by DOT regulations, were round side reflectors. (These would morph into rectangular lights on near-future models.) Sliding rear side windows were added as a factory option, though dealer-installed sliders had been available for some time. A few minor details and color changes also were notable in the three-model lineup. The

▲ Sometimes called *Python II*, this 800A was done up for dealer shows about the same time the 800A was introduced. Few particulars are known other than that it was done in-house. It appeared in a number of press and marketing images in 1969 and 1970.

Cab-Top became the Utility Pickup. The Travel-Top station wagon remained the most popular model, but a Roadster was now available with a sporty new Whitco Safari Top, though few were sold that way from the factory.

The engine lineup changed greatly with the 800A. It had long been assumed that the 4-152 would eventually be phased out as the standard engine. That moment came on November 14, 1968, after which the 4-152 was no longer installed and the 4-196 became standard. Along the same lines, the 193-horsepower V-304 replaced the V-266 as the top dog in the lineup. That left a middle spot open and in waltzed an inline six that IH dubbed the "PT-6" (Power Thrift-6), later known as the 6-232. It was none other than AMC's 232-ci inline, a modern lightweight that cranked out 145 gross horsepower and 215 lb-ft. For a time, it was available only in the Aristocrat, which debuted in April (see Chapter 5), but was made generally available after a June 9, 1969 announcement, becoming a very popular option. This trio of engines really covered the bases and made Scout the Bronco's equal. It couldn't quite match up to the new Blazer, though, which had a potent 350-ci V-8 as an option. Then again, the Blazer didn't have a super-economy four to delight the truly thrifty.

Among the big news for the 800A model (announced on January 20) was the availability of an automatic transmission, specifically the BorgWarner Model 11 (IH T-39). The installation was based on the 1967 development work done for a right-hand-drive automatic-transmission Postal Scout using the BorgWarner Model 8 automatic (IH T-28). Because the Models 8 and 11 transmissions were so similar, many of the lessons learned with the T-28 translated to the T-39, making its approval a no-brainer. The T-39 was offered behind the 6-232 (PT-6) and the V-304.

▲ In an attempt to add some sportiness, the roadster was given a little pizzazz when Whitco debuted a new top dubbed the Safari Top. This snazzy roadster in Plum Metallic (729), complete with chrome wheels and red-stripe tires, made the publicity rounds for both Whitco and International. The top was available as a dealer-installed accessory or via the aftermarket. It boasted 84 square inches more viewing area than the previous top. *WHS 110862*

▲ The 200,000th Scout rolled off the line on November 27, 1968. The 1969 800A just happened to be sold to a Fort Wayne employee, Mrs. June Pontello, who was there to accept the keys from the manager of the Fort Wayne sales branch, production superintendent, and works manager. It was eight years, almost to the day, from when the first Scout was produced. *WHS 110254*

▲ One big deal with the intro of the 800A was the availability of an automatic transmission. The BorgWarner Model 11 (IH T-39) was shifted via a console shifter with the transfer case lever integrated into the assembly. The T-39 was available with a choice of the new V-304 or the PT-6 six-cylinder.

Early in 1969, Dana informed IH they would stop producing the two-piece rear-axle shafts that had long been used for Scouts and other 4x4s. The Scout was the last holdout of this old design, which was the only option before upset axle-forging. The new one-piece flanged shafts would be stronger and more reliable, as well as cheaper. Implementation began in March 1969, but the old stock was used up first.

Along similar lines, the introduction of the PT-6 engine created a second application requiring the wider Dana 30 axle, leaving the Dana 27 for four-cylinder models only. The Scout Product Committee (SPC) crunched the numbers and decided in February to homologate the line and use the wider Dana 30 axle for all applications. This change more or less coincided with the introduction of the Aristocrat in April, but the system was purged of Dana 27As first, hence there may be four-cylinder Scouts built after April that don't have the Dana 30.

◀ This is a 1970 four-cylinder 4x2 Travel-Top with a lot of high-end options. How can you tell a four from a V-8 or a six? The V-8 models had the "V" emblem on the fender, of course, but there are were other "tells" that indicated a six-cylinder unit as well as a V-8. A body lift on the V-8 and the sixes is most visible at the front, with a gap between the top of the bumper and the lower edge of the grille panel. Also, the rear wheel appears to be set back in the wheelwell on the sixes and V-8s but is more centered with on the four-cylinders. Take note of the hubcaps, something seen mainly in 1970, mixed in with the hubcaps commonly seen on the 1969s 800As as well as the 800Bs of 1971. Strangely, the same hubcap was used on 1969 and up D-Line half-ton pickups. Note also the round reflectors. This particular Scout is seen in a number of '70- era brochures and Data Sheets and was obviously taken from early production. *WHS 111339*

On the documentation front, IH announced that beginning with the 800A, model codes would no longer reflect the Deluxe/Custom trim levels; rather, those trim levels became options. As a result, model codes were somewhat homogenized. This is only important when comparing early and late option codes.

The idea of an air-conditioning option for the 800 line had long been discussed, and a system was finally developed in time for the 800A, but it was determined to be unsatisfactory. That development was curtailed for the 800A because the upcoming X-Scout was higher on the agenda and no further efforts were to be wasted on the 800 lines. The same situation was encountered with power steering for the 800A.

In February 1968, the gear ratios in the T-44/T-45 (T-18C) four-speed transmission were tightened. Customers were finding the gap between first and second

▲ A winch-equipped '69 800A 4x4 Cab-Top four-cylinder in Medium Blue Metallic (6756). A variety of winches were available in the 800A era. The factory would install a Ramsey X46R PTO on the line, but dealers offered many other factory-approved choices. Shown here is a Warn Model 6000 (Warn part number M6-2651, IH part number 2651) for four-cylinder application. This was an early version of the legendary Warn 8274 winch. On the four-cylinder applications, it fit between the bumper and the front crossmember; the six and the V-8 had that crossmember farther forward, so the kit had the winch in front of the grille. The 1969 list price from International was $290 for the winch and $30 for the mount. The winch was rated for 6,000 pounds and the drum held 150 feet of 5/16-inch cable. The winch gear ratio was 136:1 and the no-load line speed was 60 feet per minute (2000-pound load was 28 feet per minute). The kit features controls both at the winch and inside the cab. This factory image most likely depicts a test-fit to gain factory approval of the kit. *WHS 111264*

▲ Early SSII? Research at the Wisconsin Historical Society uncovered a lot of great factory images, many of which had no context or accompanying information. This is a doorless, roll bar–equipped 1969 or early-1970 800A with sand tires. *WHS 111676*

▶ The pickup and 4x2 Scouts were still in production in 1970, the last year for the 800A. They were certainly low volume, this particular combo being among the lowest. This Gold Metallic (4357) pickup is well equipped with chrome hubcaps (10534, $24), and chrome front and rear bumpers (01600, $15, and 01611, $26, respectively). It's long been incorrectly stated that one defining difference between the 800A and 800B was the rectangular side lights of the '71 800B models versus the round reflectors of the 800A. In fact, it's less clear cut. While there are '70 model 800As with round reflectors, many, if not most, have the rectangular lights. Other visual cues must be used to tell them apart. The approximate start for the addition of side marker lamps was at VIN G378002 in October of 1969. *WHS 110870*

objectionable. Both transmissions had a 4.02:1 first gear and a 1.97:1 second; customers wanted the ratios to be closer and second-gear starts possible, so a 2.41:1 second gear was instituted starting with the 1969 800A models.

The 1970 model year 800A had a few notable difference from the 1969s. One was the addition of the DOT-mandated side marker lights. They were federally mandated on vehicles built after January 1, 1970. The changeover began October 1, 1969, at approximately VIN G378002, so a fair number of early 800As still had reflectors.

The 1970 model year also brought the SR-2 Doll-Up (see Chapter 5).

INTERIM, INTERIM, INTERIM: THE 1971 800B

By the end of 1970, the X-Scout was well on its way, but brewing labor troubles would delay it even more. IH execs agreed to another minor refresh and a couple of new special models, but asked that any serious new developments have application to the X-Scout. Proposed changes were few, but some of the most

▲ The 800Bs were the last facelift for the 800 line and production started on August 17, 1970. Changes on the outside were few, mainly the grille and new colors. The IH emblem moved from the hood to the right grille, replacing the International badge. The '71 800B had the shortest production run of any Scout. Even in the '71 800B era, International dragged out the roadster every once in a while just to show Scouts were for fun-loving people. Not many were actually sold like this. This looks to be a Red (2150) 4x4 in Custom trim. *WHS 8755*

interesting and cherished Scout specials would come from this era.

On June 22, 1970, IH announced in sales letter MT1102 that production of the 1971 800B model would commence on August 17 with retail deliveries beginning on or after September 30 when the public model announcement would be made.

The cosmetic changes for the 800B were few. It got a front-end revision that included removal of the "IH" from the hood and the offset "International" badge from the grille. A new "IH" badge was added to the driver's side of the grille and the headlight bezels went from black to bright. An "800B" replaced the "800A" on the right dash panel. Some new color choices were offered as well.

Mechanically, changes were also few. A few new emission controls were added, and later in 800B production the Dana 44 limited-slip rear-axle option changed over from a Powr-Lok to the less expensive (and less effective) Trac-Lok. There were also some upgrades to the TC-145 (Dana 20) transfer case to address noise issues.

The most exciting part of the 800B line was the addition of two special models: the Comanche and the

▲ Frost Green Metallic (5865) was a new color for 1971, and this Custom 800B 4x4 V-8 Traveltop wears it well. Frost Green survived into the Scout II line and was a popular color. WHS 9301

Sno-Star. Derived from an earlier idea, the Comanche was given a Southwest motif and new colors (see "Scout 800A and 800B Data," pages 90–95). The Sno-Star originally had been slated as an addition to the 800A line in its second year, but because of a few production issues, it didn't come until the 800B line.

The 800B remained in production from August 17, 1970, to March 8, 1971, making it the

▶ The 800B interior didn't change much from the 800A, except for the "800B" on the right dash. This is the Custom interior (16023) that included the better seat upholstery, vinyl floor mats, door panels, full headliner, dual armrests, lighter, and bucket seats. *WHS 110858*

▲ Sales to the U.S. military continued through the 800 era. Surviving LSTs show varied equipment, from relatively standard 800 Traveltops like this, to PTO winch-equipped Cab-Tops with lots of extras. Most were painted in the appropriate color for the service: gray for the Navy and various shades of green for the Army. Most often, the green was glossy for non-tactical vehicles. A few LSTs specified red, obviously for base fire departments. Many were equipped with brush guards and tow hooks, like this 800 model. *George Kirkham*

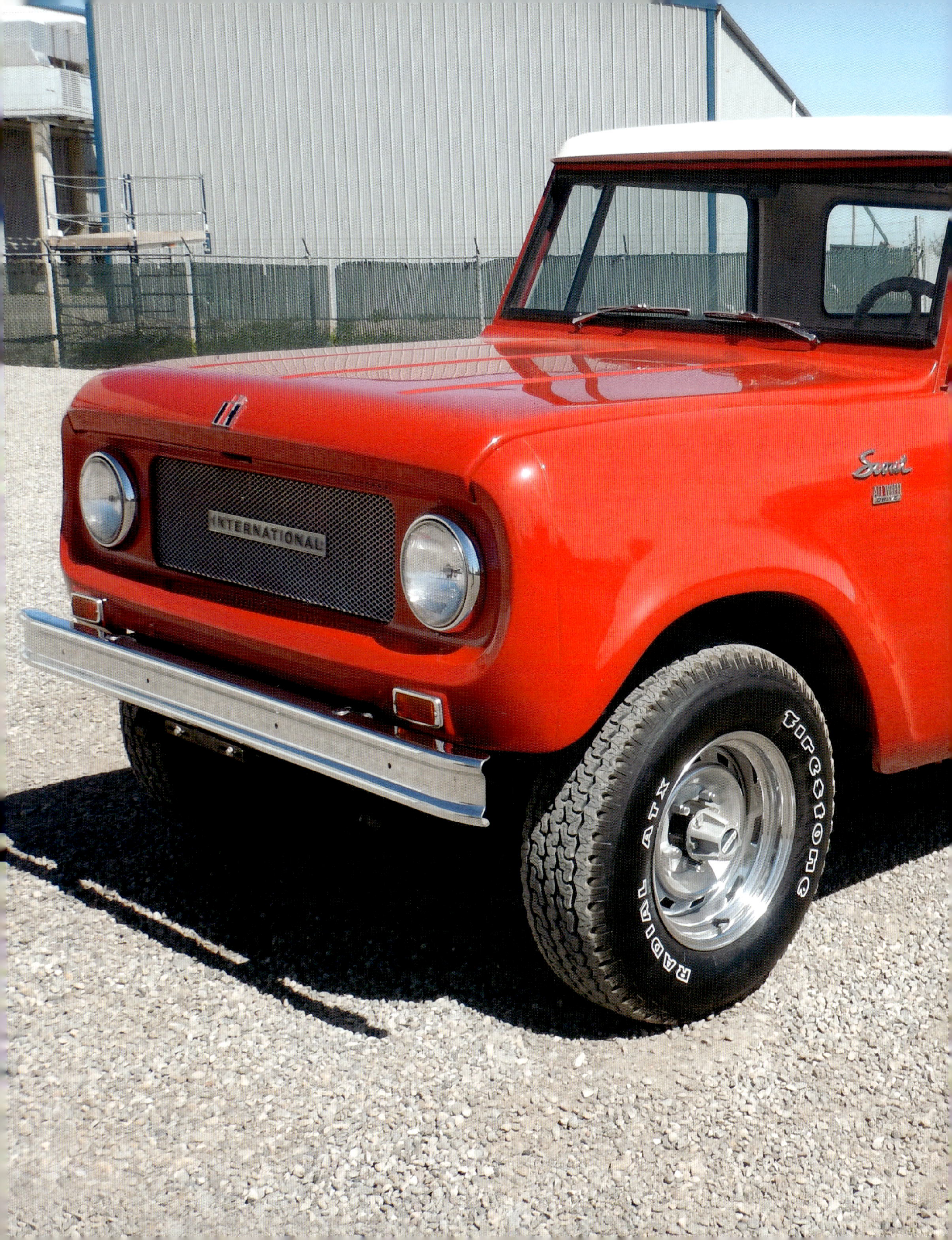
INTERNATIONAL
Scout
Firestone
RADIAL ATX

◀ Scouts were built at Chatham, Ontario, from March 1962 to November 1968 in the 80 and 800 lines. The Canada-built Scouts were assembled from components shipped from Fort Wayne, only about 240 miles away. They were largely the same as the U.S.-built Scouts but differed in minor details due to local parts content. Local content was maximized on the small parts to aid the Canadian economy. Also, a separate dealer network existed north of the border, so the range of options and features differed as well. George Kirkham's 1967 800 was assembled in Chatham in November 1966. It was originally configured as a Cab-Top with a Doll-Up interior (16026) but is now a Travel-Top. It's equipped with a 4-152 engine, T-14 transmission, and 4.27 axle ratios with front *and* rear Powr-Loks (02024 and 14023, respectively). It was originally sold with 6.00-16 Firestone Town & Country tires on the standard steel rims.
George Kirkham

shortest-production Scout model. It also has the distinction of being the last evolution of the 800 series. After the 800B was phased out, tooling for the 800B was shipped to Brown Corporation, of Ionia, Michigan, to begin the process of building continued service parts for older Scouts.

There are some indications the 800B line came close to living on. A March 5, 1970, discussion by the SPC mentioned that some in upper management expressed interest in continuing the 800 model after the Scout II debut as a low-cost, austere Scout for off-road applications, export, and government contracts. Marketing was firmly against the idea, however, and thought maintaining 800 production after the new Scout debuted would be detrimental to sales. As an alternative, Marketing suggested a very austere short-wheelbase chassis (85 inches) using Scout II components. Styling explored the idea and a couple of prototypes were built (see Chapter 6).

SCOUT 800 DATA

MODEL YEARS 1966–1968

INTRODUCTIONS
1966: August 4, 1965
1967: ~November 1, 1966
1968: ~November 1, 1967

PRODUCTION DATES
1966: October 29, 1965–October 31, 1966
1967: November 1, 1966–October 31, 1967
1968: October 31, 1967-November 1, 1968

SERIAL NUMBER/VIN INFORMATION
FC: Early models with serial numbers (150000 starting sequential number)
Model code plus G: Later models (165294 starting sequential number)*
Model code plus C: Canadian models (000501 starting sequential number)†

* Starting with the 1966 model year, IH went to a 13-digit VIN and the sequential numbers of all the Fort Wayne products were mixed together. This makes it very difficult to search out specific models. The number sequence at Fort Wayne began at G165000. Research shows the first Scout was G165294, necessitating scrolling through 294 LSTs to find it. Interestingly, the first vehicle in the G165000 range was a Fleetstar 2000D semi-truck. Scout serial numbers were often issued in blocks, so finding one usually yields others. Still, there are singles and big gaps between the batches. Searching for specific Scouts in the G-number era (1965–1972) is a test of researcher sanity.

† Information on Chatham VINs is limited, but Howard Pletcher reports finding information that C000501 was built on February 28, 1966.

SERIAL NUMBER AND VIN RANGES
1965–1966: FC150000–160573 ‡
1966: G165294–G222729 §
1967: G222730–G280875 §
1967: G710001–G716745 **
1968: G280876–G329141 §

‡ The total range of 800s built using the old FC serial number system. Some documents show the changeover to the 13-digit VIN on December 12, 1965; others show it beginning on October 1, 1965, but there may be a few FC-numbered Scouts that came later.

§ The total range of vehicles built at Fort Wayne for the model year indicated. Includes Scouts and other trucks built at Fort Wayne for that model year. It was beyond our ability to accurately scour out Scouts from the total.

** The special VIN range for the 1967 Postal Scout contract that produced 6,745 right-hand-drive (RHD), 4x2, T-28 automatic (BorgWarner 8) Scouts into April 1968.

MODEL CODES
For the 1966 model year, IH changed the VIN and model codes to create a 13-digit VIN. They used a six-digit model code, eliminating the somewhat redundant 2 (since all Scouts were 100-inch wheelbase), though it's still seen on some parts of the LST. The model code then became the VIN prefix followed by the sequential numbers that began with a G to indicate Fort Wayne assembly, or a C to indicate Chatham assembly. The sequential number ran consecutively year after year thru 1972, when the system was changed again.

Digit	1	2	3	4	5	6	Description
	7	1					Scout 800 4x2
	7	8					Scout 800 4x4
			0				Early 1966 Scout 800
			1				1966-on Scout 800
			3				Post Office Scout 800
				9			All 800
					0		Gas engine (4-cyl. all)
					1		Gas engine (seldom seen for Scout)
					2		Gas engine (typically V-8)
						1	Sportop, Convertible
						2	Sportop, Hardtop
						3	Cab/Cowl (half-Scout or Glassic)
						4	Roadster, standard
						5	Cab-Top, standard
						6	Cab-Top, w/ Doll-Up
						7	Travel-Top, standard
						8	Travel-Top, w/ Doll-Up
						9	Roadster, Custom Interior

TOTAL SCOUT 800 PRODUCTION

1966: ~7,000–10,573 ††
1966: 24,410
1967: 21,911
1968: 19,880

†† Estimated possible range of 800 production based on FC serial numbers issued before the changeover to the G numbers. Does not include canceled serial numbers, of which there were many.

SCOUT 800 SPECIAL MODEL PRODUCTION BY YEAR AND TYPE

1967 RHD automatic: 1,808 (1,680 Postal Service; 128 commercial)
1968 RHD automatic: 5,329 as of 9/1/68 (5,065 Postal Service; 264 commercial)

SCOUT 800 SPECIFICATIONS

Axles, Front

4x2: FA-4, dead, forged-steel I-beam
4x4: FA-14, live, Spicer 27AF, closed knuckle (1966–1968 4-cyl.)
Optional 4x4: FA-24, live, Spicer 27AF, w/ Powr-Lok (1966–1968, 4-cyl.)
4x4 V-8: FA-11, live, Spicer 30 hybrid (1967–1968)
4x4 V-8: FA-21, live, Spicer 30 hybrid, w/ Powr-Lok (1967–1968)

Axles, Rear

Standard 4x2 and 4x4: RA-4, Spicer 27A, semi-float (1966–1967)
Optional 4x2 and 4x4: RA-14, Spicer 27A, semi-float, w/ Powr-Lok (1966–1967)
Optional 4x4: RA-9, Spicer 44, semi-float (1966–1967)
Standard 4x4: RA-9, Spicer 44, semi-float (1967–1968)
Optional 4x4: RA-23, Spicer 44, semi-float, w/ Powr-Lok (1966–1968)

Axles, Ratios

4-cyl.: 4.27:1 (1966–1968, optional V-8)
Optional 4-cyl. and V-8: 4.88:1 (1966–1968)
V-8: 3.73:1 (1967–1968, optional 4-cyl. but only w/ 4-speed)

Brakes

Front 4x2: 9x2 in.
Rear 4x2: 9x2 in.
Total Lining Area 4x2: 157.4 sq. in.
Front 4x4: 10x2 in. duo-servo
Rear 4x4: 9x2 in.
Rear 4x4 w/ RA-9 or 23 Axle: 11x1.75 in.
Total Lining Area 4x4: 165.7 sq. in.
Total Lining Area 4x4 w/ RA-9 or 23 axle: 156.3 sq. in.

Capacities

GVW 4-cyl. 4x2: 3,200 lbs.

GVW 4-cyl. 4x4: 3,900 lbs.

Optional GVW 4-cyl. 4x4: 4,700 lbs.
   (w/ RA-9 or 23 axle)

GVW V-8 4x4: 4,200 lbs.

Optional GVW V-8: 4,700 lbs. (w/ heavy-duty
   springs)

FA-14 Front-Axle Capacity 4x4: 2,100 lbs.

FA-24 Front-Axle Capacity 4x4: 2,500 lbs.

FA-11 Front-Axle Capacity 4x4: 2,500 lbs.

FA-4 Front-Axle Capacity 4x2: 2,000 lbs.

RA-4 Rear-Axle Capacity 4x2/4x4: 2,300 lbs.
   (standard)

RA-9/23 Rear-Axle Capacity 4x4: 3,500 lbs.

Main Fuel Tank: 11 gal.

Optional Auxiliary Fuel Tank: 11 gal.

Clutch

Standard 4-152: 10 in., 6-spring, 91.8 sq. in.

Standard 4-152T, 4-196: 11 in., 9-spring, 123.7 sq. in.

Standard V-266 V-8: 10-in. heavy-duty

Dimensions

Wheelbase: 100 in.

Overall Length: 154 in.

Overall Width: 68.6 in.

Overall Height, 4x4 Travel-Top, Cap-Top: 69.5 in.
   (68 in. 4x2)

Approach Angle 4x4: 47 deg. (6.00-16 NDT tires)

Departure Angle 4x4: 35 deg. (6.00-16 NDT tires)

Ramp-Breakover Angle 4x4: 155 deg.
   (6.00-16 NDT tires)

Approach Angle 4x2: 41 deg. (7.35-15 tires)

Departure Angle 4x2: 34 deg. (7.35-15 tires)

Ground Clearance to Chassis 4x2: 10.4 in.
   (7.35-15 tires)

Ground Clearance to Chassis 4x4: 10.4 in.
   (6.00-16 NDT tires)

Front-Axle Ground Clearance 4x2: 10.25 in.
   (7.35-15 tires)

Front-Axle Ground Clearance 4x4: in. (6.00-16 tires)

Rear-Axle Ground Clearance 4x2: 7.75 in.
   (7.35-15 tires)

Rear-Axle Ground Clearance 4x4: 9.75 in.
   (6.00-16 tires; RA-4, RA-14, RA-24 axles)

Rear-Axle Ground Clearance 4x4: 7.81 in.
   (6.00-16 tires; RA-9, RA-23 axles)

Electrical

System Voltage: 12V

Standard Alternator Capacity: 32A

Optional Alternator Capacity: 52A

ENGINE 4-152, 1966–1968

Type: 4-cyl. slant-four OHV, IH 4-152

Displacement: 152 ci

Bore and Stroke: 3.88x3.22 in.

Compression Ratio: 8.19:1

Block and Head Material: cast iron

Number of Main Bearings: 5

Weight: 505 lbs. complete/dry

Gross HP Rating: 93.4 hp @ 4,400 rpm

Net HP Rating: 86 hp @ 4,400 rpm

Gross Torque Rating: 142.7 lb-ft @ 2,400 rpm

Net Torque Rating: 137.3 lb-ft @ 2,400 rpm

ENGINE 4-152T, 1966–1967

Type: 4-cyl. slant-four OHV, IH 4-152T

Displacement: 152 ci

Bore and Stroke: 3.88x3.22 in.

Compression Ratio: 8.19:1

Block and Head Material: cast iron

Number of Main Bearings: 5

Weight: 530 lbs. complete/dry

Gross HP Rating: 111.3 hp @ 4,000 rpm
Net HP Rating: 99.2 hp @ 3,600 rpm
Gross Torque Rating: 166.5 lb-ft @ 3,200 rpm
Net Torque Rating: 151.2 lb-ft @ 3,200 rpm

ENGINE 4-196, 1966–1968

Type: 4-cyl. slant-four OHV, IH 4-196
Displacement: 195.45 ci
Bore and Stroke: 4.135x3.656 in.
Compression Ratio: 8.10:1
Block and Head Material: cast iron
Number of Main Bearings: 5
Weight: 545 lbs. complete/dry
Gross HP Rating: 110.8 hp @ 4,000 rpm
Net HP Rating: 102.7 hp @ 3,800 rpm
Gross Torque Rating: 180.2 lb-ft @ 3,200 rpm
Net Torque Rating: 176.1 lb-ft @ 3,200 rpm

ENGINE V-266, 1967–1968

Type: V-8 OHV, IH V-266
Displacement: 265.76 ci
Bore and Stroke: 3.625x3.218 in.
Compression Ratio: 8.4:1
Block and Head Material: cast iron
Weight: 778 lbs. complete/dry
Gross HP Rating: 154.8 hp @ 4,400 rpm
Net HP Rating: 142.5 hp @ 4,400 rpm
Gross Torque Rating: 227.1 lb-ft @ 2,800 rpm
Net Torque Rating: 216.4 lb-ft @ 2,800 rpm

PERFORMANCE

Turning Radius 4-cyl. 4x4: 20.1 ft.
Turning Radius V-8 4x4: 20.8 ft.
Turning Radius 4x2: 18.8 ft.

SUSPENSION AND STEERING

Front Springs Length/Width 4x2/4x4: 40x1.75 in.
Front Springs No. Leaves 4x2: 6
Front Springs No. Leaves 4x4: 7
Optional Front Springs No. Leaves 4x4: 8
Standard Front Springs Rate 4x2: 198 lbs.-in.
Standard Front Springs Rate 4x4: 224 lbs.-in.
Optional Front Springs Rate 4x2: 224 lbs.-in.
Optional Front Springs Rate 4x4: 265 lbs.-in.
Standard Front Springs Capacity 4x2: 900 lbs.
Standard Front Springs Capacity 4x4: 1,005 lbs.
Standard Front Spring Capacity V-8 4x4: 1,095 lbs.
Optional Front Spring Capacity 4x2: 1,005 lbs.
Optional Front Spring Capacity 4x4: 1,180 lbs.
Steering Type 4x2/4x4 4-cyl.: Manual S-12 (Ross 24J)
Steering Type 4x4 V-8: Manual S-8 (Ross 24J)
Steering Ratio: 24:1
Steering Wheel Diameter 4x2/4x4: 17 in.

Rear Springs Length/Width 4x2/4x4: 46x1.75 in.
Standard Rear Springs No. Leaves 4x2: 4
Standard Rear Springs No. Leaves 4x4: 5
Optional Rear Springs No. Leaves 4x4: 6
Standard Rear Springs Rate 4x2: 210 lbs.-in.
Standard Rear Springs Rate 4x4: 265 lbs.-in.
Optional Rear Springs Rate 4x2: 265 lbs.-in.
Optional Rear Springs Rate 4x4 & V-8: 310-510 lbs-in.
Standard Rear Springs Capacity 4x2: 795 lbs.
Standard Rear Springs Capacity 4x4 4-cyl. and V-8: 1,040 lbs.
Optional Rear Spring Capacity V-8 4x4: 1,430 lbs.

TIRES

Tire manufacturers varied. Common brands were General, Firestone, Goodyear, and Goodrich. Common sizes shown per data book; others available.

1966 4x2: 7.35-15, standard (4-ply passenger)
        7.75-15, optional (4-ply passenger)
      4x4: 6.00-16, standard (non-directional 4-ply)
        6.00-16, optional (non-directional 6-ply)
        8.45-15, optional (passenger car or all-terrain)
        9.00-13, optional (sand tires)

1967 4x2 4-cyl.: 7.35-15, standard (4-ply passenger)
        7.75-15, optional (4-ply passenger)
      4x4 4-cyl.: 7.35-15, standard (4-ply passenger)
        7.35-15, standard (4-ply all-terrain)
        7.75-15, optional (4-ply passenger, or all-terrain)
        8.15-15, optional (4-ply passenger, or all-terrain)
        8.45-15, optional (4-ply passenger, or all-terrain)
        6.00-16, optional (4-ply, non-directional)
        6.00-16, optional (6-ply, non-directional)
      4x4 V-8: 7.75-15, standard (4-ply passenger)
        7.75-15, optional (4-ply all-terrain)
        8.15-15, optional (4-ply passenger, or all-terrain)
        8.45-15, optional (4-ply passenger, or all-terrain)
        6.00-16, optional (4-ply, non-directional)
        6.00-16, optional (6-ply, non-directional)

1968 4x2 4-cyl.: 7.35-15, standard (4-ply passenger)
        7.75-15, optional (4-ply passenger)
      4x4 4-cyl.: 7.35-15, standard (4-ply passenger)
        7.35-15, standard (4-ply all-terrain)
        7.75-15, optional (4-ply passenger, or all-terrain)
        8.15-15, optional (4-ply passenger, or all-terrain)
        8.45-15, optional (4-ply passenger, or all-terrain)
        6.00-16, optional (4-ply, non-directional)
        6.00-16, optional (6-ply, non-directional)
      4x4 V-8: 7.75-15, standard (4-ply passenger)
        7.75-15, optional (4-ply all-terrain)
        8.15-15, optional (4-ply passenger, or all-terrain)
        8.45-15, optional (4-ply passenger, or all-terrain)
        6.00-16, optional (4-ply, non-directional)
        6.00-16, optional (6-ply, non-directional)

TRANSMISSION AND TRANSFER CASE

Standard 4x2: 3-speed manual, T-13 (Warner T-90)

Standard 4x4: 3-speed manual, T-14 (Warner T-90)

Standard 4x4 V-8: 3-speed manual (Warner T-90)

Optional 4x2: 4-speed manual, T-44 (Warner T-18C)

Optional 4x4: 4-speed manual, T-45 (Warner T-18C)

Optional 4x4 V-8: 3-speed automatic, T-39 (BorgWarner M-11)

Shift Type: floor

Ratios 4x2/4x4 4-cyl., T-13/T-14: 1st - 3.39:1, 2nd - 1.85:1, 3rd - 1.00:1, R - 4.531:1

Ratios 4x4 V-8, T-14: 1st - 2.80:1, 2nd - 1.85:1, 3rd - 1.00:1, R - 4.53:1

Ratios 4x2/4x4, T-44/T-45: 1st - 4.02:1, 2nd - 1.97:1, 3rd - 1.41:1, 4th - 1.00:1, R - 4.73:1

Transfer Case: 2-speed, TC-145 (Spicer 20)

Transfer Case Ratios: Low - 2.03:1, High - 1.00:1

SCOUT 800 STANDARD COLORS

Color	Code	'66	'67	'68
Champagne Mist Metallic	1501	X ‡‡		
Aspen Green	5683	X	X	X
Apache Gold	4298	X	X	X
Moonstone Blue	6606	X §§		
Bahama Blue Metallic	6701	X	X	X
Tahitian Yellow	4853	X	X	X
Malibu Beige	1503	X	X	X
Light Yellow	4285	X ***		
Red	2150	X	X	X
Alpine White	9120	X	X	X
Medium Blue Metallic	6756			X
Silver Grey (bumpers)	884	X	X	X

‡‡ Champagne Mist Metallic replaced by Malibu Beige in late-1965 production for 1966 model year.

§§ Moonstone Blue replaced by Bahama Blue in late-1965 production for 1966 model year.

*** Light Yellow replaced by Tahitian Yellow in late-1965 production for 1966 model year.

SCOUT 800A AND 800B DATA

MODEL YEARS 1969–1970 (800A) AND 1971 (800B)

INTRODUCTIONS
1969: ~November 1968 (800A)
1970: ~November 1969 (800A)
1971: September 30, 1970 (800B)

PRODUCTION DATES
1969 800A: November 1, 1968–October 31, 1969
1970 800A: November 3, 1969–August 16, 1970
1971 800B: August 17, 1970–March 8, 1971

SERIAL NUMBER PREFIX
Model code, plus G (329142 starting sequential number, 1969 model year)

IH used a 13-digit sequential VIN and the sequential numbers of all the Fort Wayne products were mixed together. IH assigned blocks of VINs to various models, and without information on those assignments, it's very difficult to track individual models.

VIN RANGES
1969 800A: G329142–G384198 *
1970 800A: G384199–421500 * †
1971 800B: G421501–G445189 * ‡

* This is the total range of vehicles built at Fort Wayne for the model year indicated. It includes Scouts as well as other trucks built there for that model year.
† The beginning number is accurate for the start of the model year, but the ending is an estimate based on charts of VINs issued by month. It could be off by hundreds.
‡ The starting number is an estimate based on charts of VINs issued by month. The ending number is one of the last—if not *the* last—800B built in late March 1971 before Scout II production began in April. This was the last 800B LST found in researching the LST files.

MODEL CODES
For the 1966 model year, IH changed the VIN and model codes to make a 13-digit VIN. They used a six-digit model code, eliminating the somewhat redundant 2 (since all Scouts had a 100-inch wheelbase). The model code then became a VIN prefix followed by the sequential numbers that began with a G indicating Fort Wayne assembly or a C for Canada. The codes changed slightly between the 800 line and the 800A and 800B models. The sequential number ran consecutively year by year through 1972, when the system was changed again.

Digit	1	2	3	4	5	6	Description
	7	1					Scout 800A 4X2
	7	8					Scout 800A 4x4
	8						Scout 800B Models
			2				800A/B
			3				800B Comanche
			4				800A Aristocrat, 800B Sno-Star
					8		800A/B
						0	Gas engine, typically 4-196
						1	Gas engine, typically 6-232
						2	Gas engine, typically V-8
						6	SR-2, 6-232
						8	SR-2, Comanche V-304

Digit	1	2	3	4	5	6	Description
				4			Roadster
				5			Cab-Top
				7			Travel-Top

TOTAL SCOUT 800A/B MODELS BUILT
1969 800A: 17,926
1970 800A: 15,404
1971 800B: ~5,800 §

§ Estimated by subtracting Scout II production from 1971 model-year Scout total.

SCOUT 800A/B 4X2 VS SCOUT 800A 4X4 MODELS
1969–1970 800A 4x2: 5,794
1969–1970 800A 4x4: 20,649
1971 800B 4x2: unknown

SCOUT 800A/B SPECIFICATIONS

Axles, Front
4x2: FA-4, dead, forged-steel I-beam
4x4: FA-14, live, Spicer 27AF, closed knuckle (4-cyl. to April 1969)
4x4, PT-6, V-8: FA-11, live, Spicer 30 hybrid (also 4-cyl. from April 1969–1971)

Axles, Rear
Standard 4x4: RA-9, Spicer 44, semi-float
Optional 4x4: RA-23, Spicer 44, semi-float, w/ Powr-Lok

Axles, Ratios
4-cyl.: 4.27:1 (optional V-8, 3.73:1, 3.31:1 opt.)
6-cyl.: 3.73:1 (4.27:1, 3.31:1 opt.)
V-8: 3.73:1 (3.31, 4.27:1 opt.)

Brakes
Front 4x2: 9x2 in.
Rear 4x2: 9x2 in.
Total Lining Area 4x2: 157.4 sq. in.
Front 4x4: 10x2 in. duo-servo
Rear 4x4: 9x2 in.
Rear 4x4 w/ RA-9 or 23 axle: 11x1.75 in.
Total Lining Area 4x4: 165.7 sq. in.
Total Lining Area 4x4 w/ RA-9 or 23 axle: 156.3 sq. in.

Capacities
GVW 4x2: 3,200 lbs.
GVW 4-cyl. 4x4: 3,900 lbs.
Optional GVW 4-cyl. 4x4: 4,700 lbs. (w/ RA-9 or 23 axle)
GVW 6-cyl., V-8 4x4: 4,200 lbs.
Optional GVW 6-cyl., V-8: 4,700 lbs. (w/ heavy-duty springs)
FA-14 Front-Axle Capacity 4x4: 2,100 lbs.
FA-24 Front-Axle Capacity 4x4: 2,500 lbs.
FA-4 Front-Axle Capacity 4x2: 2,000 lbs.
RA-9 Rear-Axle Capacity 4x4: 3,500 lbs.
RA-23 Rear-Axle Capacity 4x2/4x4: 3,500 lbs.
Main Fuel Tank: 11 gal.
Optional Auxiliary Fuel Tank: 11 gal.

Clutch
Standard 4-152: 10 in., 6-spring, 91.8 sq. in.
Standard 4-152T: 11 in., 9-spring, 123.7 sq. in.
Standard V-266 V-8: 10 in., heavy-duty

Dimensions
Wheelbase: 100 in.
Overall Length: 154 in.
Overall Width: 68.6 in.
Overall Height, 4x4 Travel-Top, Cap-Top: 69.5 in. (68 in. 4x2)
Approach Angle 4x4: 44 deg. (E78-15 tires)
Departure Angle 4x4: 35 deg. (E78-15 tires)
Ramp-Breakover Angle 4x4: 155 deg. (E78-15 tires)
Approach Angle 4x2: 42 deg. (7.35-15 tires)
Departure Angle 4x2: 32 deg. (7.35-15 tires)
Ground Clearance to Chassis 4x2: 8.7 in. (7.35-15 tires)

Ground Clearance to Chassis 4x4: 9.4 in. (E78-15 tires)
Front-Axle Ground Clearance 4x2: 10.2 in.
   (7.35-15 tires)
Front-Axle Ground Clearance 4x4: 7.7 in.
   (E78-15 tires)
Rear-Axle Ground Clearance 4x2: 6.8 in.
   (7.35-15 tires)
Rear-Axle Ground Clearance 4x4: 6.8 in.
   (E78 tires; RA-9, RA-23 axles)

Electrical
System Voltage: 12V
Standard Alternator Capacity: 37A
Optional Alternator Capacity: 52A

ENGINE 4-196
Type: 4-cyl. slant-four OHV, IH 4-196
Displacement: 195.45 ci
Bore and Stroke: 4.135x3.656 in.
Compression Ratio: 8.10:1
Block and Head Material: cast iron
Number of Main Bearings: 5
Weight: 545 lbs. complete/dry
Gross HP Rating: 110.8 hp @ 4,000 rpm
Net HP Rating: 102.7 hp @ 3,800 rpm
Gross Torque Rating: 180.2 lb-ft @ 3,200 rpm
Net Torque Rating: 176.1 lb-ft @ 3,200 rpm

ENGINE 6-232 (PT-6)
Type: 6-cyl. OHV, IH 6-232
Displacement: 232 ci
Bore and Stroke: 3.75x3.50 in.
Compression Ratio: 8.50:1
Block and Head Material: cast iron
Number of Main Bearings: 7
Weight: 478 lbs. complete/dry
Gross HP Rating: 145 hp @ 4,300 rpm
Net HP Rating: 120 hp @ 4,000 rpm
Gross Torque Rating: 215 lb-ft @ 1,600 rpm
Net Torque Rating: 202.5 lb-ft @ 1,000 rpm

ENGINE V-304/V-304A
Type: V-8 OHV, IH V-304, V-304A
Displacement: 303.68 ci
Bore and Stroke: 3.875x3.218 in.
Compression Ratio: 8.19:1
Block and Head Material: cast iron
Weight: 700 lbs. complete/dry
Gross HP Rating: 193.1 hp @ 4,400 rpm
Net HP Rating: 180 hp @ 4,400 rpm
Gross Torque Rating: 272.5 lb-ft @ 2,400 rpm
Net Torque Rating: 262 lb-ft @ 2,800 rpm

PERFORMANCE
Turning Radius 4-cyl. 4x4: 20.1 ft.
Turning Radius V-8 4x4: 20.8 ft.
Turning Radius 4x2: 18.8 ft.

SUSPENSION AND STEERING
Front Springs Length/Width 4x2/4x4: 40x1.75 in.
Front Springs No. Leaves 4x2: 6
Front Springs No. Leaves 4x4: 7
Optional Front Springs No. Leaves 4x4: 8
Standard Front Springs Rate 4x2: 198 lbs.-in.
Standard Front Springs Rate 4x4: 224 lbs.-in.
Optional Front Springs Rate 4x2: 224 lbs.-in.
Opt Front Springs Rate 4x4: 265 lbs.-in.
Standard Front Springs Capacity 4x2: 900 lbs.
Standard Front Springs Capacity 4x4: 1,005 lbs.
Standard Front Spring Capacity V-8 and 6-cyl.
   4x4: 1,095 lbs.
Optional Front Spring Capacity 4x2: 1,005 lbs.
Optional Front Spring Capacity 4x4: 1,180 lbs.
Steering Type 4x2/4x4 4-cyl.: Manual S-12 (Ross 24J)

Steering Type 4x4 6-cyl. and V-8: Manual S-8 (Ross 24J)
Steering Ratio: 24:1
Steering Wheel Diameter 4x2/4x4: 17 in.

Rear Springs Length/Width 4x2/4x4: 46x1.75 in.
Standard Rear Springs No. Leaves 4x2: 4
Standard Rear Springs No. Leaves 4x4: 5
Optional Rear Springs No. Leaves 4x4: 6
Standard Rear Springs Rate 4x2: 210 lbs.-in.
Standard Rear Springs Rate 4x4: 265 lbs.-in.
Optional Rear Springs Rate 4x2: 265 lbs.-in.
Optional Rear Springs Rate 4x4 V-8 and 6-cyl.: 310–510 lbs.-in.
Standard Rear Springs Capacity 4x2: 795 lbs.
Standard Rear Springs Capacity 4x4 4-cyl. and V-8: 1,040 lbs.
Optional Rear Spring Capacity V-8 4x4: 1,430 lbs.

TIRES

Tire manufacturers varied. Common brands were General, Firestone, Goodyear, and Goodrich. Common sizes shown per data book; others available.

1969   4x2 4-cyl.: 7.35-15, standard (2-ply passenger)
                7.75-15, optional (2-ply passenger)
       4x2 6-cyl.: 7.35-15, standard (2-ply passenger)
                7.75-15, optional (2-ply passenger)
       4x4 4-cyl: 7.35-15, standard (2-ply passenger)
                7.75-15, optional (2- or 4-ply passenger, or all-terrain)
                8.25-15, optional (2- or 4-ply passenger, or all-terrain)
                8.55-15, optional (2- or 4-ply passenger, or all-terrain)
                6.00-16, optional (4-ply, non-directional)
       4x4 6-cyl.: 7.35-15, standard (2-ply passenger)
                7.35-15, standard (4-ply all-terrain)
                7.75-15, optional (4-ply passenger, or all-terrain)
                8.25-15, optional (4-ply passenger, or all-terrain)
                8.55-15, optional (4-ply passenger, or all-terrain)
                6.00-16, optional (4-ply, non-directional)
       4x4 V-8: 7.75-15, standard (4-ply passenger)
                7.35-15, standard (4-ply all-terrain)
                7.75-15, optional (4-ply passenger, or all-terrain)
                8.15-15, optional (4-ply passenger, or all-terrain)
                8.45-15, optional (4-ply passenger, or all-terrain)
                6.00-16, optional (4-ply, non-directional)

1970   4x2 4-cyl.: 7.35-15, standard (2-ply passenger)
                7.75-15, optional (2-ply passenger)
       4x2 6-cyl.: 7.35-15, standard (2-ply passenger)
                7.75-15, optional (2-ply passenger)
       4x4 4-cyl: 7.35-15, standard (2-ply passenger)

7.75-15, optional (2- or 4-ply passenger, or all-terrain)
8.25-15, optional (2- or 4-ply passenger, or all-terrain)
8.55-15, optional (2- or 4-ply passenger, or all-terrain)
6.00-16, optional (4-ply, non-directional)

4x4 6-cyl.: 7.35-15, standard (2-ply passenger)
7.35-15, standard (4-ply all-terrain)
7.75-15, optional (4-ply passenger, or all-terrain)
8.25-15, optional (4-ply passenger, or all-terrain)
8.55-15, optional (4-ply passenger, or all-terrain)
6.00-16, optional (4-ply, non-directional)

4x4 V-8: 7.75-15, standard (4-ply passenger)
7.35-15, standard (4-ply all-terrain)
7.75-15, optional (4-ply passenger, or all-terrain)
8.15-15, optional (4-ply passenger, or all-terrain)
8.45-15, optional (4-ply passenger, or all-terrain)
6.00-16, optional (4-ply, non-directional)

1971 4x2 4-cyl.: E78-15, standard (2-ply passenger)
F78-15, optional (2-ply passenger)

4x2 6-cyl.: E78-15, standard (2-ply passenger)
F78-15, optional (2-ply passenger)

4x4 4-cyl.: E78-15, standard (4-ply passenger)
F78-15, optional (4-ply all-terrain)
F78-15, optional (4-ply passenger, or all-terrain)
G78-15, optional (4-ply passenger, or all-terrain)
H78-15, optional (4-ply passenger, or all-terrain)

4x4 6-cyl.: F78-15, standard (4-ply all-terrain)
F78-15, optional (4-ply all-terrain)
G78-15, optional (4-ply passenger, or all-terrain)
H78-15, optional (4-ply passenger, or all-terrain)

4x4 V-8: H78-15, standard (4-ply passenger)
H78-15, optional (4-ply all-terrain)
G78-15, optional (4-ply passenger, or all-terrain)

TRANSMISSION AND TRANSFER CASE

Standard 4x2: 3-speed manual, T-13 (Warner T-90)
Standard 4x4: 3-speed manual, T-14 (Warner T-90)
Standard 4x4 PT-6 V-8: 3-speed manual, T-14 (Warner T-90)
Optional 4x2: 4-speed manual, T-44 (Warner T-18C)
Optional 4x4: 4-speed manual, T-45 (Warner T-18C)
Optional 4x4 6-cyl. and V-8: 3-speed automatic, T-39 (BorgWarner M-11)
Shift Type: floor
Ratios 4x2/4x4 4- and 6-cyl., T-13/T-14: 1st - 3.34:1, 2nd - 1.85:1, 3rd - 1.00:1, R - 4.531:1
Ratios 4x4 V-8, T-14: 1st - 2.80:1, 2nd - 1.85:1, 3rd - 1.00:1, R - 4.53:1
Ratios 4x2/4x4, T-44/T-45: 1st - 4.02:1, 2nd -1.97:1, 3rd - 1.41:1, 4th - 1.00:1, R - 4.73:1
Transfer Case: 2-speed, TC-145 (Spicer 20)
Transfer Case Ratios: Low - 2.03:1, High - 1.00:1

SCOUT 800A/B STANDARD COLORS

Color	Code	'69	'70	'71
Red	2150	X	X	
Dark Blue Metallic (Aristocrat)	6748	X		
Silver Metallic (Aristocrat)	8528	X		
Flame Red (SR-2 1970)	2289		X	X
Burnished Gold (SR-2 1970)	4393		X	
Plum Metallic	729	X	X	
Copper Metallic	2282	X	X	
Alpine White	9120	X	X	X
Lime Green Metallic	5857	X	X	X
Medium Blue Metallic	6756	X	X	X
Gold Metallic	4357	X	X	X
Prairie Gold (Comanche)	4398			X
Silver Grey (bumpers)	884	X	X	X

THE 800 ENGINE EVOLUTION

By the mid-1960s, forthcoming new-model competition from Ford, GM, and Dodge was inspiring anxiety at International. One big source of angst was International's power deficit. International's four-cylinder lineup couldn't hold market share against sixes and V-8s. Thinking IH market share was in a death spiral, Sales demanded a quick answer.

Ideas abounded, but development budgets were slim outside the X-Scout. The BG-220 six-cylinder idea was still being proposed with all its associated development costs, but that engine series was fading fast and didn't package all that well in the Scout. Several pages of MTC report 757 are taken up with analysis of the possibility of buying the 225-ci V-6 from GM,

▲ The 4-196E derived from the V-392E that finally debuted in 1966 after a long and difficult development process. Installation was relatively simple. In its early-800-era form, this engine cranked out 110.8 gross horsepower at 4,000 rpm and 180.2 lb-ft at 2,000 rpm. The 4-196E delivered the raw power and torque of a small six. The disadvantage was a general lack of smoothness versus a six-cylinder.

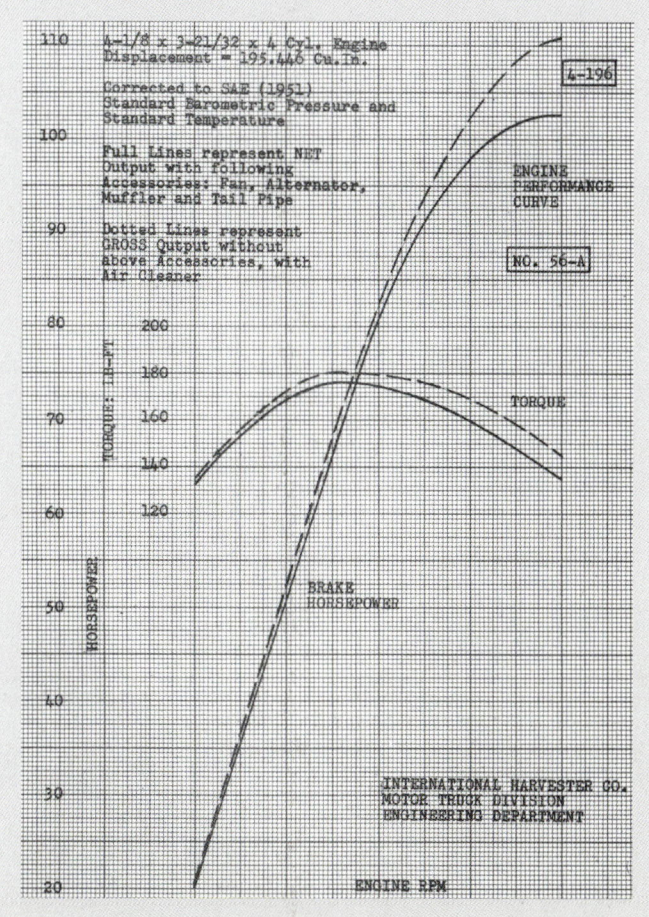

▲ The performance curves for an 800-era 4-196E. Compare this to the 4-152 and 4-152T graph found in the Chapter 2 engine section.

as well as discussion about a V-266 V-8 option, but consensus for a firm direction could not be reached. By the time 1966 rolled in, so had the Ford Bronco, with its standard six and optional V-8, and the V-6 Jeeps. IH Sales went into full panic mode—the V-266 idea was looking pretty good.

THE V-8 STORY

International product planning had wanted to hold back the V-8 card for the new X-Scout but finally realized it needed to be played right away. In February 1966, the MTC approved a project to test-fit a V-8 into the 800 and Engineering quickly produced a functioning test vehicle. It was inspected by Product Reliability in July and found to have many faults related to the axles, steering, and suspension. It would take long months of work and refinement to get a satisfactory installation, but the V-8 option was finally available in March 1967 with the desired effect on Scout sales.

The V-266 engine had debuted for the 1959 model year as one of three V-8s in the "small series" (the "SV" V-8s). Coming in 266-, 304-, and 345-ci displacements, these medium-duty powerplants were designed to replace a line of aging inline sixes. IH already had a large-series V-8 line for their medium-heavy and heavy-duty trucks, but the SV engines were also suitable for light and light-medium trucks. The light truck market was beginning to demand a V-8 and it was a bandwagon IH needed to be on.

The short-block V-266 had been intended mainly for the Light Line as a modern alternative to the BD-264 inline and first appeared in the B-Series for 1959. It was oversquare, meaning bore dimension was greater than the stroke. This was a new trend in engines that yielded very broad torque and power ranges. The V-266 was rated at 155 horsepower—a rating it would keep throughout its time in Scouts. Those numbers were on par with the IH 264-ci "big" six, but the V-8 had a wider operating rpm range and was easier to package in a light truck—and later the Scout. The 304-ci SV engine used a larger bore but the same stroke as the 266. Many have wondered why two such similar engines were produced, but it's clear the V-266 was intended as a Light Line option only. The long-block 345 had the same 3.875-inch bore as the 304 but a longer 3.656-inch stroke.

Unlike the passenger car engines used in most other SUVs, the IH small-series V-8s were built to heavy-duty specifications. They used timing gears rather than chains, had skirted Y-blocks that offered superior support for a forged-alloy crankshaft with steel-backed, copper-lead bearings. The cylinder

▲ When the V-8 option finally debuted in February 1967, it was a pretty big deal. The 155-horsepower V-8 gave the Scout some suds for work, play, or just keeping up with traffic. When fitted with a V-8, the Scout 800 was substantially different mechanically than a four-cylinder, so it was treated as a different model.

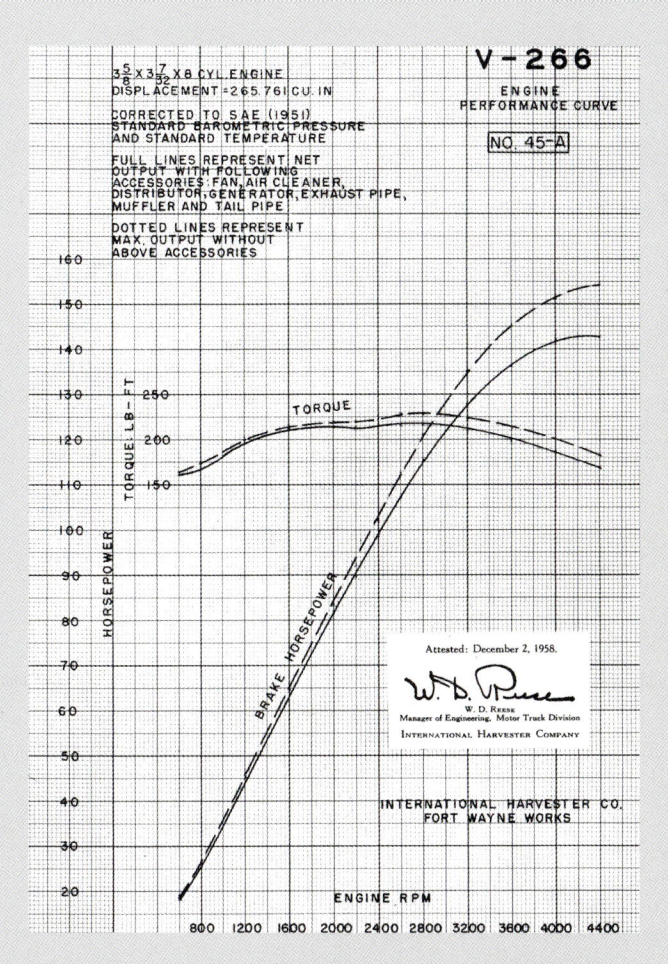

▲ The V-266 had a broad, flat torque curve but was short nearly 50 horsepower and 50 lb-ft from the Ford 289. It matched the Jeep V-6's initial power rating but was a few lb-ft short on torque. What it lacked in output, it made up for in durability, though, in light-duty use, that advantage was largely moot. This graph shows both gross (advertised) and net ratings.

▲ Among the biggest hurdles to installing the V-8 were the front axle and steering. The Dana 30 axle (FA-11) was born from this project. While it used the same axle ends as the previous Dana 27AF (FA-14), it was wider and had a higher GAWR (gross axle-weight rating—often spoken by engineers as "gower"). The chassis changed substantially with the V-8, with the front crossmember moving forward to just behind the bumper, versus about a foot back on a four-cylinder. Also, the steering box mounted outside of the chassis instead of inside and the steering geometry was different. Because the V-8 engine sat forward compared to the four, the mounts were placed differently. This is probably a prototype V-8 installation because it appears a lot of fabrication has been done and components relocated. *WHS 111277*

SCOUT 800, 800A, AND 800B: MOVING UPMARKET

heads used hardened-steel exhaust-seat inserts and Stellite valves with valve rotators. The pistons used a steel insert for the top ring and had big 1.06-inch wrist pins. In short, the V-266 was intended for hard, continuous use and, yes, you could call it overkill in a rig like the Scout.

By the end of 1967, discussions had turned to discontinuing the 266 V-8 and replacing it in Scouts with the V-304 V-8, mostly to consolidate engine production. The 304 was a relatively easy upgrade that appeared as an option in the 1969 model year, offering a nearly 40-horsepower boost in power and 45 lb-ft

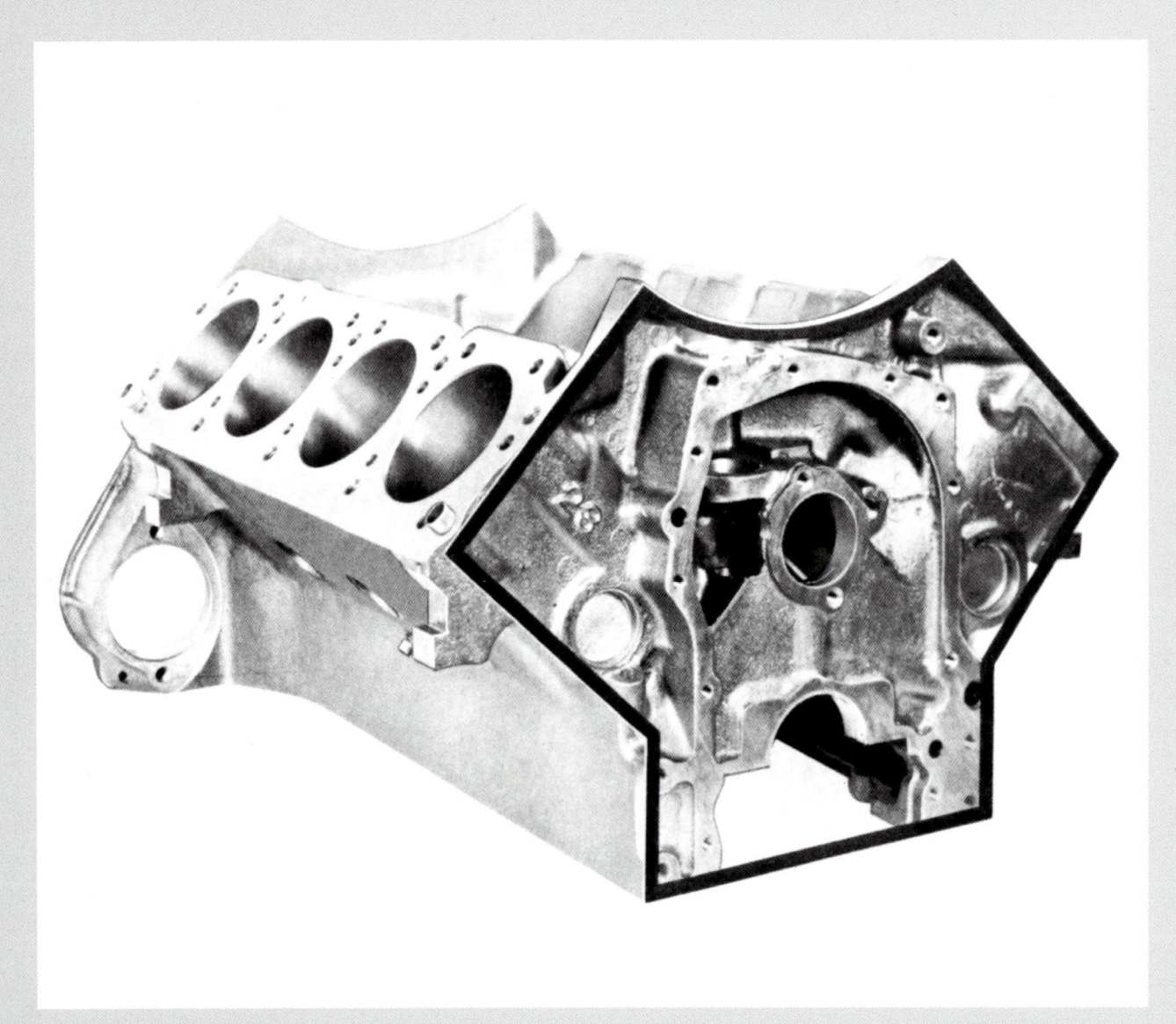

▲ The deep-skirt Y-block V-8 construction was a true heavy-duty powerplant. Was that a real technical advantage in the light-duty realm? Probably not, given that the longevity of a light-duty engine in light-duty use was perfectly adequate. International knew it too, which is why there were numerous discussions over engine cost-cutting for the light line. For much of the Scout era, it didn't make much sense from a production standpoint to produce different engines for light- and medium-duty. Later it did, but no matter what, these super-stout engines allowed International some bragging rights that might have swayed a few potential customers and made diehards of the existing ones.

▶ The AMC sixes were a lightweight casting with seven main bearings, a good-breathing cylinder head, great power output, and compact dimensions. The 232 made 145 horsepower and 215 lb-ft, which equaled the larger, heavier BG-241 but offered fuel economy very close to the 196-ci four. This 1968 image shows a PT-6 in IH Red and was one of a series of images used for brochures and advertising. *WHS 111630*

▲ The 6-232 and the Scout was a match made in heaven. The engine's compact dimensions and light weight made for a good fit in the 800A chassis and also the later Scout II. The sixes mounted to the chassis similarly to the V-8.

▶ The V-304 upgrade for 1969 offered a power boost for the Scout. This 1969 shot shows a V-304 engine. By this time, V-8s in the light line had mostly transitioned from oil bath to paper-element air filters and spin-on versus canister filters. In this era, the canister oil filter was still seen as stocks of old parts were used up. *WHS 111022*

▲ The V-304 package size was virtually identical to the V-266, eliminating fitment issues with the upgrade. This V-304 in a '69 Aristocrat does not have the emissions-control devices commonly found just a short time later on the V-304A engines.

▶ As late as 1969, International was still presenting an engine performance graph for the V-304 from 1960, but the "official" rated output had not changed. The 38-ci difference between the 266 and 304 doesn't explain the substantial difference in power and torque. Sources from the Indy plant claim the output of the early V-304 was "optimized" and the 140 horsepower (net) of the later 304s more accurately represents the V-304 than the 180 horsepower (net) figure from the 1960s.

of torque over the 266. It was in the same family as the V-266, and even shared many parts. The 304 continued to be the top-dog option in the 800A and 800B and even translated to the Scout II as a middle-option powerplant.

In 1970, International revisited their earlier idea of reducing manufacturing costs for light-duty engines by dialing back the heavy-duty meter a little. Starting with the V-304, which was being used less and less in medium-duty applications, some of the heavy-duty internal components were no longer needed, including the steel-insert pistons, which were replaced with more conventional and less expensive all-aluminum pistons. New cylinder heads used induction-hardened cast-iron valve seats instead of alloy steel inserts.

The Stellite-faced exhaust valves were replaced with a less expensive austenitic steel valve. The positive exhaust-valve rotators were replaced with "slow rotators." The intake valves also went from an alloy construction to a more conventional aluminized steel. Crankshaft bearings went from steel-backed tri-metal to tin-aluminum. The new crankshaft was quench-hardened rather than induction-hardened. In light-duty applications, these changes didn't give away much, but they did reduce the bragging rights a little. The new "cost-reduced" V-304 was designated V-304A. It began appearing in production in Mid-May of 1970, around the VIN range of G412593.

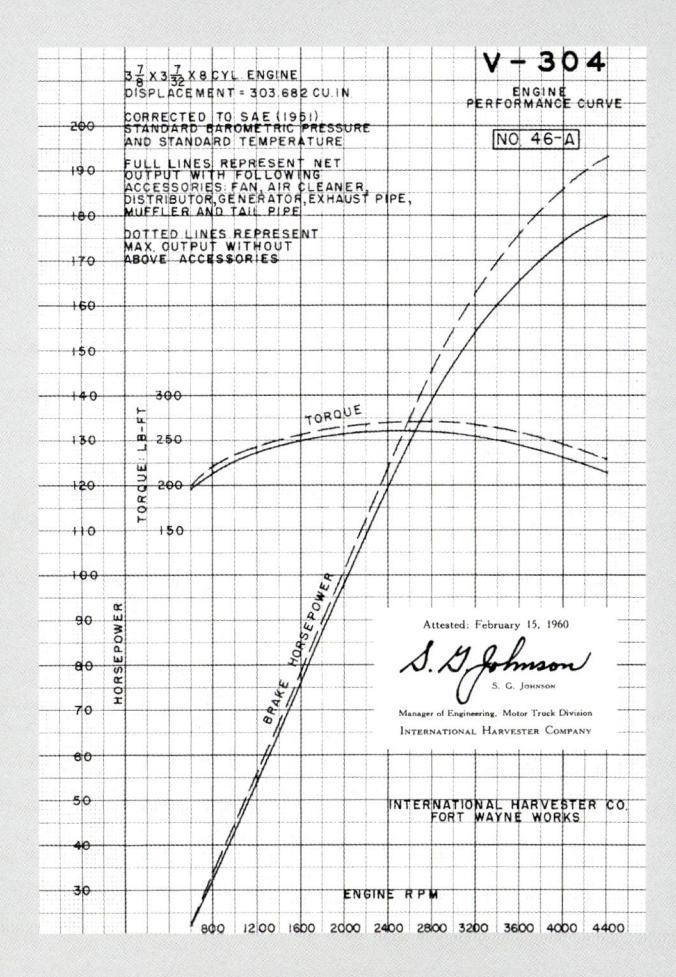

THE FOUR-CYLINDER STORY

The 4-196 program had many hiccups but by August 1966, ten 4-196 engines were being installed in Scouts each day. Far from being a bolt-in, the taller engine required some sheet metal revisions to the firewall and hood. Internally, the engine mimicked the V-392 with the same very large bores. Again, all the heavy-duty components transferred over from the V-8, including the Stellite faced valves, hard steel exhaust valve seats, and valve rotators. Typical of all the IH slant-fours and V-8s, timing gears were used instead of chains. The engine debuted at the beginning of the emission control era, but the original output was 111 horsepower and 180 lb-ft, which beat the 4-152T for torque and equaled it for power.

Late August 1966 also marked the end of the problematic 4-152T turbo four. By October, the engine had been removed from options lists and only a few leftover engines were available by special request. The worst of the bugs had been worked out of it by then, but it would be remembered as being two steps ahead of its time—and perhaps one step too far.

The 4-152 was dropped as the base engine after the 1968 model year. It would have died sooner, but IH was completing a large Postal Service order that required them as well as marketing the 4-152 as a stationary engine. The engine was phased out of production in November 1968, replaced by the 4-196E as the standard engine in the 800A models that debuted for 1969.

▲ The V-304A debuted in 1970 as a cost-saving measure. Power and torque output didn't change and there is little to tell the engines apart externally except the "A" stamped into the block. According to Terry Hankins, a former engineering manager at the engine plant in Indy, V-304A parts were often in short supply in the program's early days, so it's possible to find engines that have cost-reduced parts with heavy-duty components mixed in. Sometimes, even a complete heavy-duty engine was used. Later in the 1970s, the V-304 heavy-duty engines were used less and the V-304A more. Shown is a '71 800B with a modulated dry air filter and a hot-air duct from the exhaust manifold.

▲ Non-emissions contoured pistons are easy to distinguish by their unique crown design. To 1966, all Scout engines used this design. The contours of the cylinder head combustion chamber matched the piston.

AMC TO THE RESCUE

International had lusted after a light, modern six-cylinder to replace its aging BG Series 220 and 241 in light trucks. With the cost to develop their own six just for the Light Line prohibitive, they looked for an outside source. By October 1968, they had chosen the AMC 232 six and Engineering had signed off on a Scout installation. It became an option for the 1969 model year as the PT-6 and first appeared in the limited-production "Aristocrat" Doll-Up. Later known as the 6-232, the AMC 232, and its successor, the 258, would remain part of the Scout and light-truck programs for several years.

The 232-ci AMC sixes were relatively new designs, having debuted near the end of the 1964 AMC model year. In 1966 a short-stroke "economy" 199-ci version was built. For 1971, the 258-ci was introduced and coexisted with the 232 until it was dropped after 1979. The 232 was used in virtually all AMC car lines. After the AMC purchase of Jeep, it saw service in most Jeeps as well, evolving into the 4.0-liter in 1987 and surviving through 2002.

WHAT'S WITH THE "E"?: THE 4-152E, 4-196E, AND V-304E

Exhaust-emission controls became a major fact of life about the time the 4-196 debuted in 1966. Some built-in features allowed it to run a little cleaner, including familiar external equipment such as air-injection reactors, and carburetion and tuning alterations, but internal engine changes also helped. Ultimately, all light-duty IH engines received the same internal and external changes. On the inside, the traditional IH contoured combustion chamber changed along with the piston crown to increase squish area. As a result, the light-duty emissions engines all got an "E" designation, even if it was not always used on paperwork. In fact, aside from the service manuals, the designation often isn't seen

▲ The first Scout engine to get the E-suffix head and pistons was the 4-196. In short order, the 4-152 and V-304 also were modified in the same way (easy since the pistons and heads were the same). The 4-152E, V-304E, and V-345E got the flat-top pistons (right), and the 4-192E and V-392E got the notched pistons (the notch being necessary for clearance for the larger valves). The combustion chamber was completely redesigned to match.

anywhere except on the model stamp on the engine, which most often (but not always) had the "E" stamp.

The "E" pistons are readily identifiable because they are flat-tops or notched flat-tops, versus the non-emissions piston with a dished cup on one side of the crown and a raised part on the other. The combustion chamber shape and volume changed as well, so the cylinder heads were different castings. These pistons and combustion chambers were much easier to manufacture and resulted in some cost savings. Since the 4-196 (and the V-392 V-8 it was based upon) debuted at this time, it wasn't built any differently, but the 4-152 and V-304 had the changes implemented in the 800 era and an "E" added to their designation.

As for the V-266E, this engine had been on the chopping block for a long while. Although it was discontinued in the Scout for 1969, limited production continued until late 1969, but it's unclear how they were used. While many (if not most) V266E engines used the contoured pistons and combustion chambers, some late production units have been found with flat-tops.

International would go on to experiment a lot with combustion chamber designs and make one more revision before the Scout was discontinued.

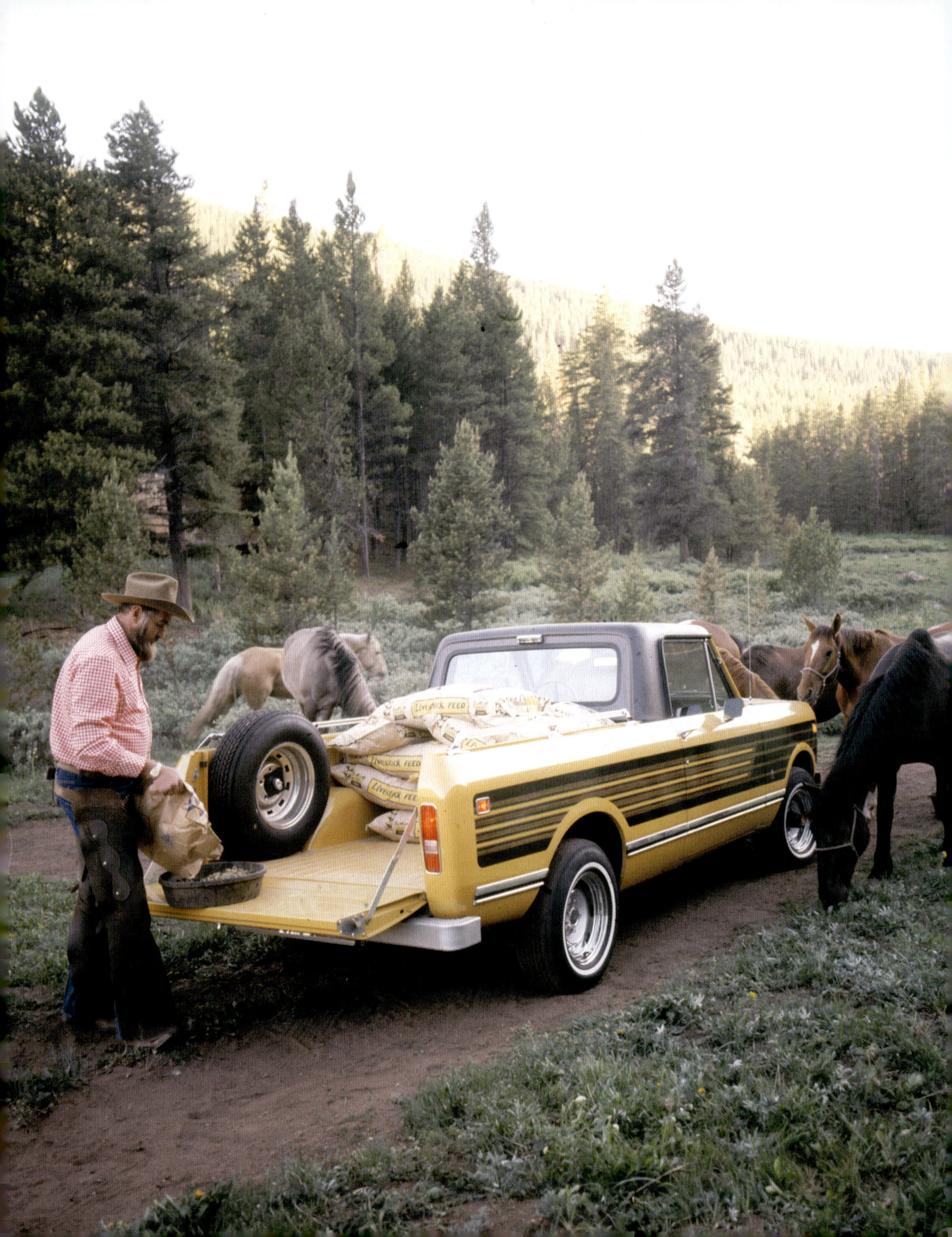
Livestock FEED

Chapter 4
SCOUT II, TRAVELER, AND TERRA:
ANYTHING LESS IS JUST A CAR

"It was the best of times, it was the worst of times." That quote from Charles Dickens' *A Tale of Two Cities* could apply to the Scout's last nine years. The Scout's initial success inspired the development of an upgraded model, but it was a long time coming. Approval was given to pursue a new Scout program on January 28, 1964, with a proposed production date of 1967 or 1968. It was later reported (MTC 794) that a more complete concept was approved by the Advanced Product Committee on November 11, 1964, based on a fiberglass prototype, again with a 1967–1968 timetable. These were the long-term goals, but long-term goals were interrupted and delayed by a succession of short-term improvements to the 800 models needed to hold the line against new competition in the SUV market.

MTC 794 listed 32 desirable new features:

1. Dimensions and wheelbase remain generally the same as existing Scout.
2. Distinctive styling readily identifiable in relation to current Scout.
3. Flush floor when doors are opened (no step-down). Doghouse eliminated except for six-cylinder engines.
4. Transmission tunnel height decreased.
5. Revision of driver and passenger doors to facilitate easier entry to the rear compartment and push-button door handles with a rotary lock.
6. Fold-up second seat.

◀ New appliqués appeared for 1978, including this pattern that came in white, black, and woodgrain, depending on the body color. The two-tone woodgrain pattern (10843, $169) looks great on this Rallye Gold (4415) '78 Terra with a Midnight Brown top, Deluxe exterior (16835, $124), and radial tire package with chrome Rallye wheels (10938, $388). WHS 111351

7. Lowering of rear passenger seat by addition of a floor well.
8. Relocate spare tire inside of body.
9. Body extended 10 inches to provide a load space of 40 1/3 inches behind second seat; 60 1/2 inches with seat folded.
10. Fixed position windshield.
11. Double wall tailgate construction. Improved tailgate latch.
12. Silent transfer case.
13. Single lever transfer case.
14. Synchronized-in-low transmission.

▶ These are very first corporeal iterations of the design that became known as the Scout II. They sported custom-made fiberglass bodies built at Creative Industries in Detroit before Ornas' design studio was given a permanent home at Fort Wayne in late 1964. Though there is no exact documentation, there are clearly two distinctly different bodies here, indicating at least two mockups were built from late November 1964 to January 1966. The front end and hoods are different, but more telling are the windshields and rooflines. The rear-window treatments are very different from the later version as well. One of these prototypes was approved by the Advanced Engineering Committee on November 11, 1964. The basic shape is evident on both and one is amazingly like the production model.

George Kirkham Collection

15. Belleville clutch pressure plate for reduced pedal effort.
16. Wide track axle with 40-degree wheel cut.
17. A 17-gallon gas tank in rear overhang of body.
18. Split brake system.
19. 4-196, V-266 engines available. Provision for six-cylinder engine.
20. Optional engines, including V-266.
21. Improved engine mountings.
22. Model 44 rear axle.
23. Improved brakes, front hubs, and drums.
24. Interior trim to comply with 1968 safety regulations.
25. Windshield wipers, heater-defroster, hazard lights, side marker lights, backup lights, padded sun visors, breakaway rearview mirror, outside rearview mirror, windshield washers to comply with new safety requirements.
26. Collapsible steering column.
27. Longer springs for a better ride.
28. Improved steering gear.
29. Aluminized exhaust system, muffler, and mounting.
31. Corrosion protection.
32. Travel-Top effects.

▲ The Cab-Top was part of the plan right from the beginning, and that's clear from this fiberglass prototype. It could be an adaptation of one of the Travel-Top prototypes or a third or fourth body. *George Kirkham Collection*

▲ What appears to be yet another fiberglass prototype shows a fair bit of evolution from the first versions. Note the differences in the roof and rear windows and the body lines along the side.

From the very start, the "Interim Scout Programs" seriously degraded the resources available to complete the all-new Scout program, which nonetheless continued at the most rapid pace possible. Reporting on the February 17 and 18, 1965, meetings, MTC 737 first identifies a working name for the new product: X-Scout. The X designation had recently come into favor among IH planners and was being used on other experimental and developmental projects. MTC 737 also announced that moving the X-Scout to Springfield was being given "serious" consideration.

Engineering had proposed an April 1968 production date and a June 1968 introduction. Knowing that GM was going to introduce the Blazer SUV around that time, Sales implored them to beat those dates. As a result, the MTC notes target dates of January 1966 for engineering release, a funding appropriation date of April 1966, a production start in August 1967, and an October 1967 introduction. The short term, however, got in the way of the long term, with product planners and engineers scrambling to keep the existing 800 selling against the new

▲ A color front view shows a front end similar to that on one of the earlier units and what became the final design.
George Kirkham Collection

▲ Body development was ongoing. In this shot, three members of International's highly prized clay model crew translate the lines onto a Cab-Top model. This image is undated but is likely from February or March 1965, not long after the design studio moved into a new 16,800-square-foot facility in Fort Wayne. Seated up front is Howard Arnold. The gent to the right is Franz Mueller, another of the Ornas inner circle, and the man on the far side Bob Zimmerman. These names have largely been lost to time, but they are very important to the creation and evolution of the Scout. Mueller and Zimmerman provided very useful input to this book.
WHS 110362

Ford and Jeep SUVs that were stealing big chunks of market share.

An altered and more complete X-Scout timetable is found in MTC 747, dated June 25, 1965, calling for an experimental release of October 1965, with an experimental vehicle due in January 1966. Fund appropriation was listed as due in March 1966 with a production release projected for April. Production was slated for November 1967 with a product announcement in December. In just over a month, the timing changed again with the experimental release moved back to January 1966, an experimental vehicle in April, appropriation in June, production release in July, production start in February 1968, and the announcement in March 1968. The timetable would change many more times in the next few years.

Actually, delays in the X-Scout program would prove beneficial. Upgrades to the interim Scouts and all their special variants paved the way for many common Scout II features. The V-8, the 4-196 four-cylinder, AMC sixes, a four-speed transmission, the automatic transmission, and the Dana 30 front axle were only a few developments completed concurrently for both the 800 line and the X-Scout.

Tooling costs were the often-reported reasons for delays in the X-Scout program; work done for the 800 models somewhat reduced those costs but not enough to enable IH to offer a new Scout at a competitive price. MTC documents reported that the design work on the X-Scout was nearly complete by July 1967. The various

▶ A view through the interior from the tailgate area seems to indicate it was more or less complete and leaves the possibility it may have been a running prototype. There is no documentation to substantiate this theory, and the recollections of the few surviving people who were there say all the fiberglass Scouts were non-driving. Another noteworthy item is the rather overlarge wheelwell on the right. These early prototypes had a side-mounted fuel tank (right-side fill) and the box on the inside is just part of the space made for it. A side fuel tank was obviously a bad idea, and when the production Scouts hit the road, they had chassis-mounted tanks.

▲ December 1967: all-metal prototypes. The rather gruesome-looking tailgate from the 1965 fiberglass Scout is gone, replaced by an elegant version that might remind one of the Traveler many years later. The rear wing windows are a nice touch and were given serious consideration. There are several other differences from the production model, first among them, the fuel filler door on the passenger side. This was eventually rejected for cost and complexity issues, but it sure is classier than the production setup.

▲ The other side of the preproduction Scout shows a Ramsey PTO winch fitted into what appears to be a custom-made mount. There are many interesting things to note here, including the internally splined front hubs and those big, tall seats up front. And don't forget the red undercarriage paint, something seen on several of the early test rigs. As for the front axle and hubs, perhaps International was testing a preproduction Scout Spicer 30 axle with the Ford knuckles and hubs. The 40-degree-steering-angle Dana 30 was derived from the Spicer open-knuckle design first used for the Ford Bronco and specified by Ford. Ford was one of the first companies to adopt the smooth, one-piece internally splined hubs as opposed to the two-piece bolted hubs favored by IH, Jeep, and others.

▶ This is more or less the production tailgate by early 1968, but notice the International badge above the license bracket. In production, the badge mounted up and to the right.
WHS 110360

SCOUT II, TRAVELER, AND TERRA: ANYTHING LESS IS JUST A CAR

▲ The X-Scout was much more evolved in July 1968. At this point, they are showing the difference between a plain Scout and one with a doll-up stripe and the 80/800 wheel covers used on 800s. Also, the International badge is inside the lower grille slot, something that marked all later prototypes. The badging on the side—including a Scout emblem behind the front marker light, the IH on the fender near the door, and "All-Wheel Drive" below—lasted almost to production when the Scout II badge at the rear fender replaced the one up front. The fuel filler still hasn't moved to the driver's side.

▶ This badging came very near to production, with the IH emblem on the front edge of the fender above a marker light that was lower than in production, a Scout plaque in the same general location as with the 80 and 800, and an All-Wheel Drive badge below.

players—New Product Development, Marketing, Engineering, and Manufacturing—were tasked with reducing costs by examining every aspect of the new Scout.

Items 16, 23, and 31 on the original feature list of 32 items were earmarked as high-cost. The front-axle upgrade was eventually deemed unnecessary, reducing the cost of brake changes. Corrosion resistance was not directly addressed in any of the uncovered product planning documents, but it is known the issue was eventually addressed.

By July 1968, IH appeared to have overcome many of the production obstacles and anticipated a December 1970 production start. From that point, the X-Scout stayed remarkably close to schedule. In January, February, and March 1969, it was down to deciding which options to offer and what identification badges to affix. One previously rejected idea was resurrected in March: the open-knuckle, 40-degree-steering-angle Dana 30 axle. Marketing became obsessed with this, and axle costs had come down a little, so it was approved.

Until May 8, 1969, "X-Scout" had been the working name for the new rig. On that day, the Scout Product Committee formally requested the model designation be changed to 810 but asked that it not appear on the vehicle. It's interesting to note that among the other proposed numeric designations, 900 was near the top of the list but rejected due to possible confusion with the model 900 pickup.

Also on May 8, IH made a fateful engine decision. Because the Chevy Blazer was being offered with a 350-ci V-8, it was proposed to offer the V-345 V-8 as the top-dog option for the 810 and to consider the V-392 for later. The V-345 became part of the Scout legend, but the V-392 didn't, mostly due to emissions and fuel-economy trends that would arise.

International had long been pursuing a fully synchronized three-speed manual transmission to replace the T-90, which was synchronized only in second and third gears. The proposed transmission was to be designated T-424 and was sufficiently stout to be mounted behind the V-345 (the T-90 was not). BorgWarner had the T-15 series worked up by then, but the expense to field it was deemed excessive and the idea was dropped. The project was revisited and eventually a fully synchronized three-speed (the Warner T-15A, IH T-332) was adopted for the Scout in 1974.

On December 3, 1970, the Scout 810 designation was officially changed to "Scout II." Management had expressed displeasure with 810 and suggested the need for a catchier name. Unfortunately little is known about alternatives that might have been floated. Interestingly, images exist of some prototypes with just "Scout" and no "II." The badging, which had already been more or less decided upon, was altered to reflect the new choice. Evidently, there was even a cost savings on badging, but the vehicle load capacity sticker on the glovebox door were marked "810." The 810 sticker appeared through most of 1971 model year production.

Research indicates the first eight prototype Scout IIs were built in November and December 1970, starting with the sequential number G425121 and running through G425128. The first assembly line–produced Scout II was a Yuma Yellow (4385) 4x4 Travel-Top with a 6-232, T-39 automatic, manual steering, 3.73:1 axle ratios, and a Trac-Lok rear axle. The LSTs for these units have a lot of handwritten notes on them and most of the new rigs went straight to Engineering for tests. Each was listed with a "Pilot Model" number. Approximately 25 preproduction pilot models were built shortly thereafter, starting with G442553. Some of the first batch went to Arizona for the Scout II unveiling. Full production started on April 19, 1971, with G447268. Some, but not all, pilot models and prototypes had dark-gray grilles, which sometimes can be used as a "tell" when viewing early Scout II images.

The new Scout II officially debuted at the San Marcos Hotel in Chandler, Arizona, on April 20–23. Production had recently started and 32 of the earliest production Scout IIs were there in various configurations for the press and interested parties to ogle. Product News G-11 dated May 24 highlighted the new "WOW Wagon" campaign slated to start June 24. It would be a long while before the dealerships had many to sell, however, and those on the lots were mainly there to inspire orders while the remaining 800B models were sold off. The new Scout IIs were rotated among automotive journalists as rapidly as possible, and the reviews were almost overwhelmingly positive.

International seemed poised for another coup.

BIG SHOES TO FILL: THE TERRA AND TRAVELER STORY

On May 5, 1975, the last International Light Line truck rolled out of the factory at Springfield, Ohio, ending almost 70 years of building light trucks. It wasn't really a surprise for anyone paying attention, but it was a sad day, nonetheless. Into the breach rode the

▲ This X-Scout-based prototype of a possible export pickup was photographed in September 1968. The body was closed off at the back and the rear chassis was ready for the installation of a separate pickup bed, possibly even one of the existing short-bed boxes from the light-truck line. The wheelbase was going to be 122 inches, according to the International image information. Because of their more compact size, Scouts were the only Light Line product with serious export potential. Howard Pletcher recalls a couple of stake-bed Scouts being sold to employees in the early '70s.

▲ By December 1968, the project was getting together and down to the finer touches. Here, the base exterior trim is compared to what would be the Deluxe with the bright trim, side molding, and hubcaps. The subtle detail here is the lack of a "II" with the "Scout" on the rear fender, though the Scout script is the same as production. Note that the front marker light has moved up a few inches, but "All-Wheel Drive" is still on the front fender. That would go away and the IH would move down on the fender. The fuel filler has finally moved to the driver's side. These changes came late. The decision to do so, and add a "II", didn't come until December 17, 1970.

▲ A glimpse of the Custom interior as envisioned in December 1968 shows most of what the production Scout had but also many subtle and not-so-subtle differences, such as the faux wood trim on the transmission tunnel cover.

Terra and Traveler for 1976. These Scouts, stretched to a 118-inch wheelbase, were nearly identical until either a bulkhead and cab or a full-length fiberglass top and hatchback were added. The Terra pickup had a nominal 6-foot bed, up to a 2,400-pound payload, and a GVW equivalent to a short half-ton pickup. The Traveler was a two-door SUV groomed by Marketing to replace the venerated Travelall as a people-mover or RV hauler.

One might think the whole thing was planned for 1976, but it was more a case of random events. The upcoming long-wheelbase Scouts added production volume to the Scout line, making them a little more profitable and helping International maintain a tiny share of the light-truck market.

The earliest evidence for the idea of long-wheelbase Scout appeared with an experimental 108-inch-wheelbase Scout 80 in 1961. It featured a full-length top and presented very much like an early version of a Traveler. Experiments in 1970–1971 to produce a 122-inch-wheelbase export pickup on the Scout II chassis were another step toward long-wheelbase vehicles, and photographs reveal an 18-inch stretch of a Scout II pickup completed in early 1972. Little more than a year later, in May 1973, the Styling Department made concept drawings of a proposed long-wheelbase Scout Hatchback. Finally, a lot of styling work was done to the pickup cab roof in December 1973 and to an extended-cab version nicknamed the "Greenhouse" Club Cab.

▲ As the Scout reached production status, a driving chassis was produced as a showpiece and training aid. It highlighted the Scout's extra-stout boxed chassis and beefy, truck-like underpinnings. It was first seen publicly at the Chandler, Arizona, Scout II press debut in April 1971.
WHS 110767

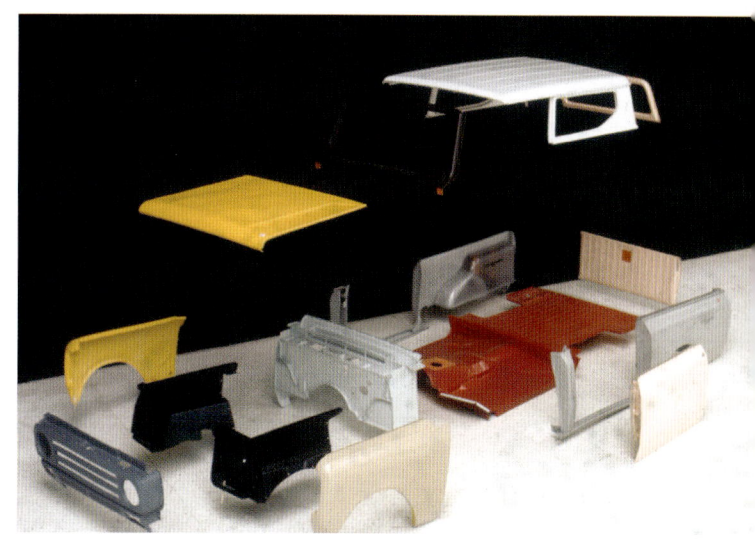

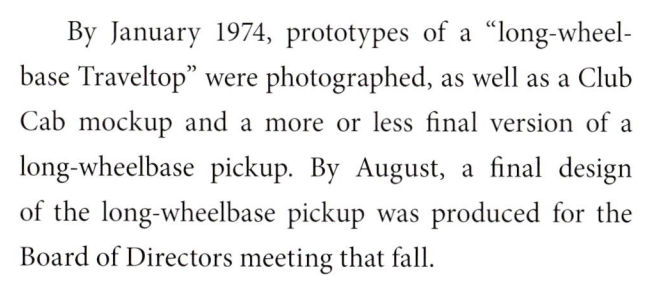

▲ The new Scout II body may have resembled the old Scouts superficially, but it was a completely new manufacturing ballgame. This exploded view shows the new body parts for the Scout II from about 1969.
WHS 110353

By January 1974, prototypes of a "long-wheelbase Traveltop" were photographed, as well as a Club Cab mockup and a more or less final version of a long-wheelbase pickup. By August, a final design of the long-wheelbase pickup was produced for the Board of Directors meeting that fall.

The top designs were the most difficult element in the stretched Scout idea. From an engineering standpoint, adding 18 inches to the Scout wheelbase was child's play. The long-wheelbase chassis had 18 inches added to the flat area near the middle in the form of a long sleeve. From there, it was simply a matter of a longer rear driveshaft, brake and fuel lines, and wiring harnesses. For the Terra, the optional Traveltop heavy-duty springs were fitted as standard (with no upgrade for 1976, at least). The Traveler used the same springs and spring options as the 100-inch Traveltop. The few extra pounds of metal in an 18-inch stretch amounted to nothing and, in the case of the Traveler, much of

that was saved with the use of a plastic top versus a metal one. The 6,200-pound GVW didn't change from the 1975 XLC Traveltop or Cab-Top to the 1976 Terra or Traveler.

Fiberglass tops, often referred to as "plastic" in International literature, had been on the development agenda for Scout since the first-generation 80 era as both a weight-saving device and a way to offer easy styling changes, such as a hatchback. Most early problems revolved around the Traveltop tailgate/hatchback. As early as the 800 era, International wanted a once-piece hatchback, but the hinge attachment was always cited as the problem (along with budgets, of course). It finally bore plastic fruit for the production Terra and Traveler, but even after those models were in production, IH was experimenting with metal tops. Various interviews with former IH employees reveal that certain elements in the company were unreasonably resistant to the plastic

▲ The Scout chassis was stouter than the average SUV, consisting of a heavy-gauge, fully boxed chassis. It measured 4x3 inches and was made up of 0.120-inch (nearly 1/8-inch) with three welded crossmembers. The section modulus (an engineering measurement of the cross section of a structural member) was 1.87 cubic inches. The springs were long—56 inches in the rear and 43 inches up front—a feature that improved ride quality at any load capacity. The 4x4 and 4x2 chassis were identical, save the location of the transmission crossmember. The 4x2 used a tubular front axle from Dana that featured the same steering knuckles as the 4x4 driven axle.

tops and kept their development at a slower pace than they might have been.

Many early brochure and publicity images show smooth, gel-coated tops, often in body colors, on both the Terra and Traveler. Almost immediately, the production tops for the Traveler were changed to white with a textured finish and shortly after augmented with an optional dark brown. The earliest PL-382 Price Books for 1976 show a non-textured smooth top for the Terra, which came standard in body color, though an option was listed for a two-tone combination under paint instruction code 201SP. Later literature shows smooth-texture tops in body color but indicates the white and brown tops were standard. Photographic evidence shows smooth-texture tops on apparently unaltered rigs that have survived to recent times, though they are rare.

The Terra and Traveler debuted in July 1975 as 1976 models but were still a long way from going on sale. Those early announcements were likely a means to counter the bad news about the light trucks and give loyal customers something to hang onto. International made a lot of marketing hay all the way to March 1976, when the units became widely available on dealership lots. Regular production of the long-wheelbase Scouts started in August 1975 with very heavy production volume starting in late September. The Scout II

100-inch-wheelbase Cab-Tops were discontinued as the Terra assumed the duties of the recently deceased full-sized pickup line still cooling in its grave.

All three Scout versions were built on the same line, spaced intermittently according to the production schedule. The LST dictated what accoutrements were added to make a Terra or a Traveler, and various mounting holes were drilled in the two individual models on the line as needed when those holes were not universal to both. The goal was to minimize the number of special parts required for each and it worked remarkably well. Any number of Terras later became Travelers and vice versa.

By the time the Terra debuted, Marketing had figured out how to place it. Initial advertising pitted the Terra pickup head to head against imports. Since the Terra was smaller than a full-size American pickup but bigger than imports, it was touted as the "just right" choice. The standard 4-196 four-cylinder gave them better fuel economy too. The new-for-1976 diesel option didn't hurt either, though it wasn't everyone's cup of tea. The Traveler was billed as a "lean, mean Suburban-beating machine" and the successor to the Travelall. The wide range of engines gave the marketing folks a lot of ammo.

1971

The Scout II 1971 model year was abbreviated with production starting so late. The hot and heavy part of the year began June 24, 1971, the official "WOW" date for the dealership launch. By that time the "WOW Wagon" ad campaign was in full swing. There were still a lot of 800Bs to sell and Scout II production bugs to work out, so 1971 production continued into November.

The '71 Scout IIs were available in seven colors with the ever-popular Travel-Top, which by this time was more often being written "Traveltop," headlining

▲ A preproduction Scout in the photo studio. It's highly likely this is 810 Pilot #2, which rolled off the line November 11, 1970, in a batch of about eight. It was Flame Red (2289) and had the Custom exterior trim and luggage rack. It was powered by a V-304A backed by the T-39 automatic and had 3.31:1 axle ratios. *WHS 111387*

▲ A bird's-eye view of the '71 Scout II showed a vastly improved interior. This is the Custom interior that featured nylon seat inserts. In 1971, the Custom interior was a $249 option and included carpets (passenger area only) and an insulated rear cargo mat. *WHS 111388*

▲ The Cab-Top Scout II for 1971 offered more cargo room than the previous generation, as well as all the other improvements that came with the new generation. Many Cab-Tops were sold as base or lightly optioned mods, but most of the high-end options were available. This Yuma Yellow 4x4 isn't one of those highly optioned units, but it does have a few practical additions, including a rear-step bumper, wheel covers, and front tow hooks. *WHS 111020*

▲ The "WOW Wagon" campaign started in the summer of 1971 and encompassed print, television, radio, and dealerships. Marketing the Scout II was a full-court press. Gilmore International was a rural Illinois dealer that appeared to have a nice display out front. *WHS 25636*

the choices. The Cab-Top was still available for those who needed a really short pickup. Both 4x4 and 4x2 Scout IIs were available, but the 4x2 was more homogenized, requiring fewer special parts and assembly. Both the 4x2 and the Cab-Top were slow sellers but apparently deemed necessary by IH management.

The initial price for a base 1971 4x4 Traveltop was $3,212 and the 4x4 Cab-Top was $3,063. The 4-196E was the standard engine for both, with the venerable T-13 (4x2) or T-14 three-speed manual behind it. Optional was the 6-232, a popular choice, with the V-304A as the middle option and V-345 as the high-power option. The T-44 (4x2) and T-45 close-ratio four-speed manual transmissions were options behind any engine, and the BorgWarner automatics served out their last tour of duty for IH as the automatic option this year. The T-39 was used behind the 6-232 and the V-304A. The stronger T-49 BorgWarner was used behind the 345 only. The T-49 was stronger because it had a 12-inch torque converter with a 29-spline input shaft (and a few other internal upgrades).

The Scout II chassis sat three inches lower than the 800 line for more easy access, yet the critical off-road clearances remained the same or better. The extra room in the new body was welcome and resulted in more passenger and cargo space. Longer, wider springs made for a better ride with International spending a lot more time on ride-tuning with the new Scout. The ride and drive were also enhanced by upgrades to the steering system and the availability of the Saginaw 700D power-steering system. Heavy-duty front and rear springs were optional. The standard Scout II had a 950-pound cargo capacity (including passengers) and a 4,600-pound GVW. Trailer wiring and a Class I or II hitch were optional.

International decided to include more accoutrements in the base Scout II than in previous models, but it was still pretty basic. There were two interior trim levels

▲ As with the 80 and 800, the Scout went into military service. International had a government sales office that managed to sell a batch of 1971s for military use. This Scout was used by the Military Police at Fort Story, Virginia. Reportedly it was a base model with a four-cylinder. Scout IIs of various years were seen in military livery and competed in the mid-'70s competition for an off-the-shelf 4x4 that could serve as a tactical vehicle in place of expensive, specialized units such as the M-151 jeep. *George Kirkham*

above the base, cleverly termed Deluxe and Custom. A 1/3–2/3 bench seat was included in all trim levels, but the upholstery differed. The base interior had little in the way of trim and a black seat. The Deluxe option included vinyl door panels, vinyl trim, and floor mats, all in a choice of four colors: Sage, Blue, Red, and Black (with or without nylon inserts in the seats). The Custom option provided carpet, tinted glass, and full undercoating. Bucket seats were also available in the same four color choices. And in an "it's about time" moment, there was a factory-air conditioning package. It was a pricey $402 option, but popular and effective.

The standard exterior featured painted bumpers and minimal trim or brightwork. The Deluxe Exterior Package included chrome bumpers and trim, dual chrome mirrors, bright side-trim moldings, and stainless steel wheel covers. A step rear bumper, with a spot for a ball hitch, was optional.

There weren't many changes to the 1971s while they were in production, but the Scout Product Committee Reports state the T-39 and T-49 BorgWarner automatics were replaced by the T-407 Chrysler TorqueFlite starting on August 23, 1971 at around VIN G464500. Similarly, on August 16, the AMC-built 258 began replacing the 6-232. Both items were officially slated for 1972 model-year production, but since the '71 production year ran longer than normal, some 258s and Torqueflites appeared in '71s. Relatively few 1971 Scout IIs were built and the exact number is difficult to verify because the available records do not distinguish between 1971 800B and 1971 Scout II. Based on counting Line Sequence Numbers, our estimate is that of the 19,952 Scouts listed as produced for the 1971 model year, approximately 14,500 were Scout IIs.

1972

The 1972 Scout II model year was marked by few major changes. Prices went up and down, but the base MSRP for a 4x4 Traveltop in March was $3,339, while a 4x4 Cab-Top was $3,184. Visually, the grille went from body color to a silver gray. The Deluxe exterior put some chrome trim on the grille, and Aegean Blue (6758) 1971 paint was replaced by Cosmic Blue (6766).

There was more to talk about mechanically, including replacement of the AMC-sourced 6-232 with its bigger brother, the 6-258. The burley and reliable T-407 automatic (Chrysler TF-727) transmission was fully integrated into the 1972 lineup and would become a staple for the Scout II through the remainder of production. It was stout enough to replace both the T-39 and T-49 BorgWarner and

▲ Visually, the 1972 models did not change much from the '71s, but one of the more noticeable differences was the Silver Grey grille treatment as seen on this Alpine White (9120) Cab-Top 4x4.

▶ Here is a 1971 "never happened." This Deluxe Traveltop is experimentally adorned with the chrome wheels and Polyglas tires used on the '71 Scout 800B Comanche. The backstory for this image is unknown, but International was developing Doll-Up packages for the Scout II at this time. *WHS 111547*

▲ International got into the striping business pretty early in the Scout II era. This '72 Custom was done for the Archery Trade Association Show that year. International was hitting many, if not most, of the big outdoor shows, knowing that many sales came from the outdoorsy segment. The three gold stripes and gold accents set it off nicely. The gold-accented hubcaps had not yet been fitted.

▲ This '73 Scout II was photographed while still in service with the Wentzville, Missouri, Volunteer Fire Department as a brush rig. The Scout II did see some conversion to fire apparatus of various types, but by this time, they were less common. Heavier-duty pickups held up better in this sort of work than more compact rigs due to the amount of weight they carried. *George Kirkham*

could be one reason why 1972 Scouts were advertised with up to a 3,500-pound tow rating (8,000-pound gross combined vehicle weight rating, or GCVWR) while the documentation was somewhat vague about the '71 models with the BorgWarner transmissions.

Optional for the 1972 models was the FA-44 heavy-duty front axle, which was none other than the Dana 44. Rated for a 3,200-pound gross axle rating (versus 2,500 for the standard FA-13 Dana 30 unit),

it was offered for snowplow operation and difficult off-roading situations. It cost $49 and came with heavy-duty front springs.

1973

The '73 Scout debuted at $3,421.73 for the base 4x4 Traveltop and $3,266.73 for the Cab-Top. Visually, there were a few things to talk about. The grille was totally redone, with vertical slots replacing the horizontal. A

▲ A good number of changes came for 1973, including a new grille. This '73 Custom has the bright exterior trim that included moldings for the grille. It may have come with the new Hy-Vo single-speed transfer case controlled via a knob on the dash. When combined with the auto-locking hub feature, it was ideally suited for those who needed simple inclement-weather traction. It wasn't quite as user-friendly as automatic full-time four-wheel drive, but it was less expensive, less complex, more reliable, and it offered better fuel economy. WHS 111546

▲ In the midst of all the glamour and recreation, don't forget the Scout was still a working rig. The City of Memphis had a fleet of 60 Scouts with two-yard dump beds for refuse collection. The unit pictured is a '72 Cab-Top with the rear body removed for the small dumper. It isn't clear if IH sent the Scout as shown or if a conversion company did the work. The idea was that five dump Scouts working with one larger trash truck (also an International product) were more economical than a larger fleet of medium trucks. WHS 99199

good many of the paint choices changed, with only two carryover colors: Alpine White and Flame Red. Appearing for the first time from IH were accessory striping kits in several styles and colors that could be installed by the customer or the dealer. A Panel-Top option appeared again after a long absence, but it lasted only through 1974 due to lack of customer interest.

On the inside, after about May 1 production, Sage vinyl replaced black on the base models to reduce interior temperatures on sunny days. A center console was introduced for bucket seat–equipped Scout IIs and as a parts department accessory for earlier Scout IIs. The $206 Custom Interior Package for 1973 Traveltops added an insulated cargo mat.

There were a goodly number of mechanical changes for 1973, most notably the absence of the four-cylinder engine. The 6-258 became the base engine, with the V-304A and V-345 as the upper options. This pretty much matched the rest of the industry—a six as the base, a smaller V-8 as the middle, and a "big" V-8 as the top dog. Dual exhaust became a $36.50 option for the V-8s.

SCOUT II, TRAVELER, AND TERRA: ANYTHING LESS IS JUST A CAR

▶ The Panel-Top reappeared in 1973. The panel sides still had the window openings and rubber, but a metal panel replaced the glass. Option code 16952, listed as "panel sides in lieu of glass," was a $40.50 option over a standard Traveltop. A bulkhead panel could be added to turn the Scout into a mini panel van. This was a seldom-chosen option offered only in 1973 and 1974.

A single-speed, chain-drive transfer case became standard, and the two-speed Dana 20 became a $40 option. The new TC-143 single-speed unit designed and built by IH and was a long time coming. A proprietary IH transfer case had been an on-and-off project for a while, but in September 1968, engineering approval had been given to continue the development. This option was chosen over a proposed chain-drive transfer case from BorgWarner that would later be used by Jeep. Called the Hy-Vo, the IH design became the base transfer case for the new Scout and the D-Line trucks. The advantage was a simple dash control to engage four-wheel drive. For those who used their Scout mostly for all-weather transport more than hardcore off-roading, it was a convenient setup when added to the automatic-locking hub option.

A brake improvement came in the early part of the 1973 model year with the introduction of a finned front brake drum. These were fitted starting with chassis number CGD11409 during November production. The finned drums ran cooler than the earlier ones and could be retrofitted to earlier Scout IIs.

1974

The 1974 model year brought significant changes, starting with yet another grille treatment. The almost obligatory annual change included a chrome overlay.

The big mechanical news was the addition of standard power front disc brakes—essentially the

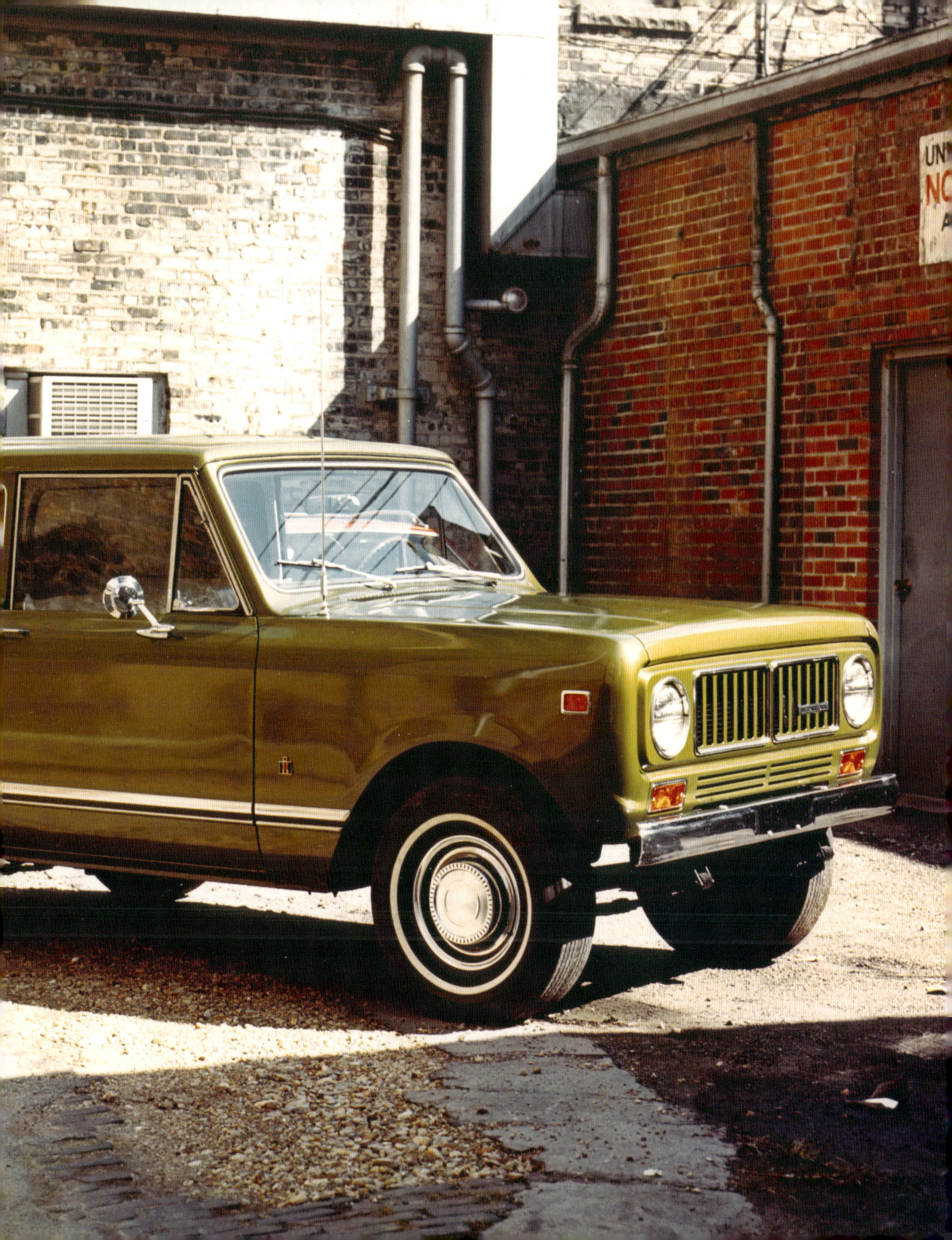

▲ Though this looks like a Deluxe exterior, it's missing the bright grille surround that was a part of that package. Deviations from the standard was not unusual in brochure images that used preproduction or prototype vehicles—which is why brochure images are not always the best standard for judging originality. WHS 111548

▲ The 1974 model year brought a new emphasis on body appliqués. With the option code 10842, this is one of the earliest that could be ordered factory-installed, and it could be purchased in a white or a woodgrain color. This striking Scout also has the two-tone option, which cost $29.50 in early 1974 if using standard colors. This appliqué cost $90, as did the woodgrain version. Note the new grille treatment for 1974. WHS 110807

same brakes used in the half-ton Light Line trucks. The standard GVW went up from 4,600 to 5,200 pounds, and the V-8s to 5,400 pounds. Tow rating increased to 5,000 pounds. The T-13/T-14 (Warner T-90) three-speed transmissions were replaced with fully synchronized three-speeds that IH called the T-332 and Warner Gear called the T-15D. This new transmission was capable of handling every engine in the lineup and became the base unit for all engines.

Underneath, the FA-43 3,200-pound front axle (a version of the Dana 44) became the heavy-duty option for six-cylinder Scouts only. The V-8 option automatically came with the FA-44 axle. Due to a shortage of FA-44 axles from Dana, many 1974 Scouts with V8s had FA-13 (Dana 30) axles substituted. Disc brakes became standard for 1974.

The base price for a six-cylinder Traveltop went up to $3,676 and the Cab-Top was at $3,521. New options included a $322 radial tire package with 15x7 chrome slotted rims and HR78-15 narrow whitewall radials. A V-345 four-barrel California emissions certified engine debuted. The larger carb was designed to bring back some of the power lost to emission-control devices and tuning. This year also brought the specter of separate California-certified engines into full swing.

▲ As the Scout line matured, International put an increasing emphasis on towing and load-carrying to appeal to the RV crowd of the mid-1970s. Here is the International lineup from 1974 that pretty much covers all the RV bases. *WHS 110693*

▶ A Sunburst Yellow (4403) '74 Traveltop with all the bells and whistles. Compare the woodgrain side appliqué and two-tone white-over-yellow to the red and white job pictured nearby. The '73 and '74 grilles are very similar, but the '74 had a much thicker bead of chrome around the grille, as well as the Silver Grey coloring. Also, a '74 will have a rectangular outside mirror versus a round one on the '73.

SCOUT II, TRAVELER, AND TERRA: ANYTHING LESS IS JUST A CAR

▲ Here's a wild one presented in 1975 as a possible future options package. Called the Sahara, there were two versions: one with the zebra appliqué over the Buckskin (4408), and another over Winter White (9219) with a top painted black. Both had Rallye wheels, Deluxe exterior trim, rear-step bumper, roof rack, and swing-away spare-tire carrier. Was it too wild for primetime?

For the first time, the vinyl side appliqué options appeared from the factory, this year in the form of white or woodgrain-style side panels or white stripes. More appliqués were available at the dealer parts counter and many dealers installed them on stock vehicles to enhance sales. See appendix for a guide to the factory installed appliqué. An air-assist front-spring package was optional for snowplow-equipped Scouts. Also, a front anti-sway device was made standard on 4x4s. One color change was noted when Sunrise Yellow was replaced by Sunburst Yellow. Other options and packages remained largely the same as 1973.

1975

IH made a lot of hay over the new 6,200-pound-standard GVW for 1975. The official spiel touted a wonderful increase in capacity, but in reality the move avoided a bunch of emission-control regulations for a couple of more years by bumping the Scout into a "truck" weight class.

▲ One of the more interesting toys offered for the Scout in the mid-1970s was the Mini-Shelter. It allowed just enough room with the tailgate down to use the back of the Scout for sleeping. In 1975, approximately when this image was taken, the Mini-Shelter (549360C91) cost $79.95.

Dubbed the XLC models, most 1975s have an "XLC" decal to advertise the increase in load capacity. Essentially what had been the optional heavy-duty front springs became standard, as did the FA-44 front axle regardless of engine, though a heavy duty FA-13 axle was used on four-cylinder applications but not listed in most documentation.

The AMC sixes were gone after 1974, replaced by the 4-196, brought out of mothballs ostensibly for improved fuel economy and to improve the company bottom line.. The 4-196 was not immediately available for California and it wasn't finally certified for sale there until early 1975.

There was a big jump in prices for 1975, with a base Traveltop going to $4,712 and the Cab-Top to $4,489. Seven new colors were unveiled, with only two carryovers: the ever-popular Glacier Blue and Sunburst Yellow. A new Custom Interior Package appeared too, the highlight of which was color-keyed, soft vinyl trim for the sides of the cargo area, and a new cargo mat and spare-tire cover. The door trim panels now included a wood grain insert

▲ In most years of the Scout II, the base and Deluxe exteriors had different grilles. This shot depicting the two versions of the 1975 grilles really shows the differences. On the left is the Deluxe, with its bright grille trim, chrome bumper, and bright moldings around the window and elsewhere. On the right is the base unit, with a painted Silver Grey bumper and grille without bright trim. While the '75 grille was similar to the '73 and '74, the black headlight bezels are the giveaway.

and the dash had a woodgrain panel. New high-back bucket seats were on the option list, and Tanbark replaced Sage as the standard seat color. And then there was the obligatory annual grille change, which was similar to the 1974 but with square headlight bezels.

Late in the year, the T-18C models, dubbed T-44 (4x2) and T-45 (4x4), were replaced by improved Warner Gear transmissions and a new wide-ratio option. The T-428 was a close-ratio four-speed Warner T-19A with similar gear ratios to the old T-44/45 but was synchronized in first. The wide ratio T-427 (T-19) had a deep, truck-style "granny" first gear. It delivered significant "gearing grunt" and was useful for low-speed off-roading.

The V-345 got the same light-duty treatment the V-304A had received a few years before. By deleting a few of the heavy-duty internal features, International could reduce costs on the new V-345A engine. Electronic ignition, introduced on some engines in 1973 and 1974, was now standard for all V-8s.

In order to promote better fuel economy, a 3.07:1 ratio debuted in 1975, replacing 3.31:1 as standard for the V-304A and V-345A V-8 engines with the

▲ In addition to the reintroduction of the 4-196 and the elimination of the 6-258 engine, big news for 1975 included the increase in standard gross vehicle weight to 6,200 pounds, putting the Scout well into the half-ton category and thus avoiding catalytic converters—for a few years at least. Of course advertising spun this as a bonus, but in reality it was simply a case of making the heavy-duty springs standard and recertifying the chassis for the new GVW. International saved a bunch of money and hassle by avoiding the converters for as long as possible. Along with the new GVW came a rather stylish XLC decal on the rear fender. These were seen into 1976 but slowly were phased out, even though XLC terminology was still used in advertising and sales literature. WHS 110523, WHS 110524

close-ratio four-speed or automatic. It made for good highway cruising but wasn't optimal for towing or off-roading. The versatile 3.31:1 ratio was replaced by the 3.54:1 but reappeared as an option in 1980.

1976

The bicentennial year saw a big shakeup for Scout with the introduction of the long-wheelbase Terra pickup and Traveler wagon. The low-volume 100-inch Cab-Top was discontinued as a result. Among the changes for Terra and Traveler early in production was the availability of two top colors. At first only a white top was available, but on September 24, 1975, a Midnight Brown top was made optional for both the Terra and Traveler, the white top being standard and the brown a no-cost option (code 10675). Speaking of tops, sliding rear quarter windows were not immediately available at the debut of the Traveler, but were announced on March 10. It would prove a popular option throughout the Traveler's run.

The 1976 model year also yielded some of the first Scout II specials. The Spirit of '76 and Patriot models followed on the heels of the specials done for the U.S. Olympic Ski Team. Their red, white, and blue striping came with a special list of options (see Chapter 5). These were the first of a growing number of specialty conversions available in the Scout line.

Prices jumped again to $5,438 for a base Traveltop. Base price for the 4x4 Terra gasser was $5,394 and the Traveler was $5,844. There were significantly

▲ A peak inside a spanking-new '75 reveals a Deluxe interior in the Tanbark color scheme, a full-width 1/3–2/3 bench seat, and air conditioning. The Deluxe included the woodgrain dash and door trim, vinyl/nylon seats, color-keyed vinyl floor mats, padded door trim, rear seat (optional on the base), and a padded cargo area floor mat. The Deluxe Interior was a $181 option in 1975; AC was $464.

▶ Here's a Custom-level 1975 interior, again in Tanbark, this time with the bucket seat option. It is without the $40 center console (16959). The Custom package added to the Deluxe by including carpeting, carpeted kick panels, and a snazzier horn pad. It cost $285, and the buckets were an additional $121. Unless the buyer ordered the Custom all-vinyl interior (only in the Saddle color), they got these rather nice nylon-insert seats with a herringbone pattern. The all-vinyl option was only an additional $10.50.

SCOUT II, TRAVELER, AND TERRA: ANYTHING LESS IS JUST A CAR

▲ The grandfather of the Traveler was this experimental 108-inch-wheelbase Scout 80 4x2 from 1961. There is certainly a family resemblance and it illustrates the length of time the idea gestated at International. This rig had an overall length of 162 inches versus 154 inches on the standard Scout 80. In retrospect, not many would disagree that this could have worked, even back in 1961.

less-expensive 4x2 versions at $4,543, $4,434, and $4,949. The 4x2 Traveltop, Terra, and Traveler were available through 1979, and these price differences were generally reflected throughout those years. The long-wheelbase 4x2s, particularly the Terra, were perhaps a little more popular than 4x2s in the short-wheelbase Traveltop line.

The diesel option and the 118-inch Scouts more or less shared a birthday in 1976. According to press releases, production of the diesel began on October 1, 1975, but in reality it was a bit earlier. Intense LST searches reveal the first diesel LST was dated September 26, 1975 (VIN FGD16314). It was among 16 prototypes built sporadically from September 26 to October 16 on whose typed LST was written "V345A GAS 2 BARREL" but scratched out and replaced with a handwritten "DIESEL." A Nissan diesel engine number replaced the gas engine number, the lowest

▲ A series of Styling Department images from April 1972 show a proposed Scout II pickup with an 18-inch stretch. The 18-inch section lines are barely visible on the body at either end of the little access door just behind the cab.

▲ Oh, the humanities! This is definitely a "What were they thinking?" moment. A close look at this June 1973 shot reveals it's not a 100-inch with an ugly tail extension but rather a 118-inch prototype with the rear wheel removed and blocks under the axle. A painted cardboard mockup covers the existing wheelwell and fakes a new rear wheel. The matching front wheel is also a fake. No doubt the bean counters thought they could do a bed extension without the expense of a wheelbase stretch. Someone without much imagination probably needed to see just how hideous that would look. After this very graphic example, it's unlikely the idea came up again.

diesel engine number found being 20600. These are all marked as pilot models and the service microfiches list them as gas models.

On December 22, two more diesel LSTs were marked as "PILOT" models: FGD25244 and FGD24245. While the diesel option was still handwritten, there was no scratched out gas-engine option and the diesel-only codes are typed in. Then, on January 19, 1976, regular production started, albeit slowly. Most early diesel production was in 118-inch Travelers or Terras, with a few 100-inch Traveltops tossed in. Diesel production continued sporadically until May, when it began to ramp up significantly.

The diesels were treated as a separate model and the cost of a Scout with the Nissan jumped considerably to $8,081 for a base model Traveltop, a substantial

▲ In December 1973, the Traveler was shaping up in this Styling Department rendition of what they called the "Greenhouse Traveler." Designing the top was fairly easy, but producing it was a bit more difficult. There were numerous updates and corrections before the hatchback Traveler went into production, but the look stayed pretty close to this Larry Nicklin rendering.

premium over a 4-196-powered unit or even a V-8. A diesel 4x4 Terra set you back $8,042 while the Traveler was $8,492.

The 198-ci (3.2-liter) naturally aspirated Nissan SD-33 diesel (which IH soon dubbed the 6-33) cranked out 81 net horsepower and International advertised an average economy of 22–28 miles per gallon with the T-428 manual. You could also get the diesel with the T-407 automatic. That reduced the already meager acceleration to a slug-like pace, but the fuel economy was still good at 21–25 miles per gallon (according to International). The diesel option included a few things that could be called "extras," but many of them were necessary for the diesel installation. The diesel package included an updated starting and charging system, tilt steering column, two-speed transfer case, 3.54:1 axle ratios, and standard T-428 close-ratio transmission. A few other changes in plumbing and wiring were necessary for the diesel, as were a few minor body changes. Most other Scout options were available, but some of the excluded ones were 3.07:1 axle ratio and right-hand drive. The T-407 automatic was $124 more.

Of course there was a new grille for '76 and this time is wasn't a revamp of the old one. It consisted of three divided sections with horizontal bars. Three new appliqué packages debuted too: the so-called two-tone in white (10842) or woodgrain (10843) for $108 (the $50 Deluxe Exterior Package was required), the $108 Feather Design (10846), and the Rallye Package (10845) for $611. The Rallye would prove the most popular Scout package of all time.

Solar Yellow (4410) emerged for 1976. According to Product News bulletin G-559 dated January 7, 1976, it was dropped early in 1976 production due to quality-control problems in the paint booth and was replaced by Sunburst Yellow (4403), which had been standard in 1975. An original Solar Yellow '76 Scout would be an interesting find today.

In February 1976, International debuted a line of Economy Scouts, a package that was available in all three models. They were as bare bones as the first Scout 80 base models of 1961. The list of standard features was very short and it knocked about $300 off the price: base for a 4x4 Economy Traveltop was $5,099, the Terra was $5,125, and the Traveler was $5,560. The package comprised a four-cylinder, three-speed, Hi-Vo-equipped 4x4 with a painted front bumper, a single outside mirror,

▲ Extended cabs were the new rage in the early and mid-1970s, and International wanted to latch onto that trend with their new pickup. Here is the Styling Department's proposal for a "Greenhouse Club." The totality of their thinking is unknown, but it appears this idea never made it past the concept/prototype stage. For one thing, it would have intruded into bed space, which was already only six feet. Second, the production result yielded a fair bit of space behind the seats and a good compromise between bed space and in-cab storage. This is another Larry Nicklin rendering.

▲ There are several things to look at in this December 1973 shot besides a comparison of bed sizes on the 100- and 118-inch pickups. This is the first iteration of the fiberglass top. It's got a smooth body color, but more than that, the square Traveltop window is evolving into the curved glass found on the production Terra and Traveltop. A filler piece on the glass simulates the desired look.

▶ At the time prototype Travelers were running around, the venerated Travelall was still in production. A bird's-eye view shows the size differential between the 100-inch Scout Traveltop, 118-inch Traveler, and 120-inch Travelall. Would a couple more inches of wheelbase have allowed International to add a second set of doors to the Traveler? Likely yes, but budget constraints did not allow for the project. *WHS 110699*

SCOUT II, TRAVELER, AND TERRA: ANYTHING LESS IS JUST A CAR

▲ By January 1974, the Traveler had come to life. At the time, it was called simply "LWB Traveltop," but the lines were pretty close to finalized. Detail choices on window trim, hatchback mechanisms, and top color and finish had a long way to go. Note that this concept shows matching top and body color—and it works, even though it's merely a painted mockup on cardboard or wood. Note also the square edge to the upper window frame, something that evolved into curved glass.

▲ As far as is known, this was the only time the Club Cab idea was ever brought anywhere close to reality. This is merely a painted cardboard mockup over a standard cab. The concept was being evaluated along with an early version of the Traveler in January 1974.

a single bucket seat for the driver, no lighter, locking hubs, undercoating, courtesy light, tinted glass, parking brake warning light, rear bumper, headliner, trim panels, wheel covers, or front floor mat.

International highlighted the towing capabilities of all the Scouts in 1976. A towing package (option code 10920, $172) and the right combination of options delivered a 10,000-pound GCVWR and a 3,500- to 5,000-pound trailer rating. The Towing Package included a Class II/III receiver hitch, trailer wiring, the big 61-amp alternator, and a larger 72–amp hour battery. A 5,000-pound trailer required the V-345A and 3.54:1 gears, with the T-428 close-ratio gearbox

▲ A poignant shot for International truck buffs if there ever was one. In this April 1974 image, one of the nearly finalized Terra prototypes is shown with a long-bed 1974 International 100 pickup. The 100 was International's lightest pickup outside the Scout II. It came in a short (115-inch) wheelbase with a six-foot bed, or an eight-foot bed with a 132-inch wheelbase. The GVW was only 5,200 pounds standard, but 6,000 pounds was optional. Either way, the 1975 and up XLC Scout II and Terra had it beat at 6,200 pounds. One has to wonder how much the people involved knew about the upcoming discontinuation of the light-truck line at this point. It was only a year away.

and heavy-duty clutch, or the T-407 automatic. The V-304A was only rated for up to a 3,500-pound trailer. A standard V-304A with the three-speed (heavy-duty clutch) and the step bumper with Class I hitch was rated for up to 2,000 pounds.

International really pushed the Traveler as the trailer-towing replacement for the legendary Travelall. To highlight these capabilities, two Travelers, a 4x2 and a 4x4, traveled toured dealers with RVs in tow. The 4x2 hauled a 27-foot Airstream at a GCVW of 9,840 pounds, and the 4x4 pulled a 24-foot Argosy at 8,410 combined weight.

In 1975, a tire and wheel package had been developed that became part of the Spirit Package in 1976 but was also available as an à la carte item on any Scout. The package featured 7x15 chrome slotted deep-dish wheels with 10.00-15 tires (usually mounting Goodyear Tracker A/T tires, with or without raised white letters). The wheels have since become known as the "Rallye" wheels because they came with the Rallye package, though not with the big off-road tires.

Early in the year, January 7 to be exact, IH announced in Product News Bulletin G-558 that the

SCOUT II, TRAVELER, AND TERRA: ANYTHING LESS IS JUST A CAR

▲ When the photo shoot for the board of directors meeting came in September 1974, the Traveler looked mighty nice, but the top was not anything like the production unit. The most obvious difference is the smooth finish and body color. Less noticeable is that the rear area is raised a little higher from the level of the cab area. This Glacier Blue (6772) Deluxe 4x4 wears a '75 grille and an extended-length version of the new-for-'74 appliqué in white. See appendix for a guide to the factory installed appliqué.

▶ A production Terra at work. This one has a Deluxe exterior with the heavy-duty step bumper (01643, $34) that included a 1 7/8-inch ball. Needless to say, that's an IH tractor and baler in a field of recently combined and windrowed wheat straw that will be baled. *WHS 110688*

▶ The Feather Appliqué (10846, $108) on a 1976 Winter White (2297) Traveler with a Midnight Brown top (10675, N/C) and luggage rack (16903, $86), but no sliding quarter window glass. It also has the radial-tire package (10938, $345) on Rallye wheels. Given the optional towing mirrors (16977, $29.50), it likely has the Class III tow package (10920, $172). *WHS 110462*

4-196 could be ordered with the T-407 automatic and that production on vehicles so-equipped would start May 3. The bulletin cautioned salespeople to suggest the option for mail delivery or low-stress applications that didn't involve trailer towing or severe off-roading. This was not a commonly ordered option but would be available through 1978.

It took a while, but by January 20, 1976, the Terra and Traveler became available in right-hand drive. This made them available for postal use and for export to countries that used the right-hand setup. Right-hand drive lacked a few popular options, however, including power steering, air conditioning, the TC-143 transfer case, AM-FM radio, electric clock, trailer-tow package, and the Rallye package. The option code was 10525. It cost $135 and came only with the standard interior trim.

International later claimed Scout sales jumped 45 percent for 1976, no doubt due to the demise of the light trucks and the introduction of the long-wheelbase models and the diesel.

1977

In 1977 a basic Traveltop 4x4 set you back $5,751, and a diesel 4x4 was $8,394. The base 4x4 Terra was $5,637 and a base Traveler was $6,122. The base diesel 4x4 Terra was $8,285 and the Traveler diesel 4x4 was $8,770. The long-wheelbase Scouts began cutting into Traveltop sales, and diesel sales increased as well.

▲ The 1976 model year brought many new elements to the Scout continuum. This 1976 shows the new appliqué and Rallye package. The package's various evolutions are the most generally popular and enduring of the Scout II packages. For 1976, this $611 package (10845) included a color-coded appliqué, 7x15 chrome-slotted wheels with chrome lug nuts and hub covers, HR-78 steel-belted radial tires, power steering, and larger-bore shock absorbers (1 3/16 versus 1 inch). The early renditions of the Rallye appliqué, however, are the least favored. *WHS 25823*

▲ Was a soft-top 1976 Rallye special being considered? Of course, any Scout could be ordered without a top. A top could also be bought and switched on, but new tops were coming out at this time. The next year, this Whitco top cost $257.95 from the dealer (244980R91, white), but could also be purchased on the aftermarket. It makes an appealing package, but ultimately the Spirit of '76 motif won out. *WHS 110687*

The grille change for 1977 consisted of five horizontal slots supported by three vertical supports. With the Deluxe Exterior Trim Package, the grille and headlight bezels were surrounded by bright trim. Sunburst Yellow and Pewter were dropped, and Siam Yellow (4414) and Elk (1608) were added.

On the inside, a new Parchment vinyl was added to the choices and Ivy Green was deleted. The nylon seat inserts were changed on the Deluxe and Custom option, but new all-vinyl interior options eliminated the inserts. The rear seat was removed from the Deluxe or Custom interior package and became a separate option. The front 2/3–1/3 bench now tilted on

▶ Another trial balloon? This June 1973 photo could depict another SR-2-like idea, perhaps something leading up to the Spirit of '76 or a shot to illustrate accessories.

150   INTERNATIONAL SCOUT ENCYCLOPEDIA

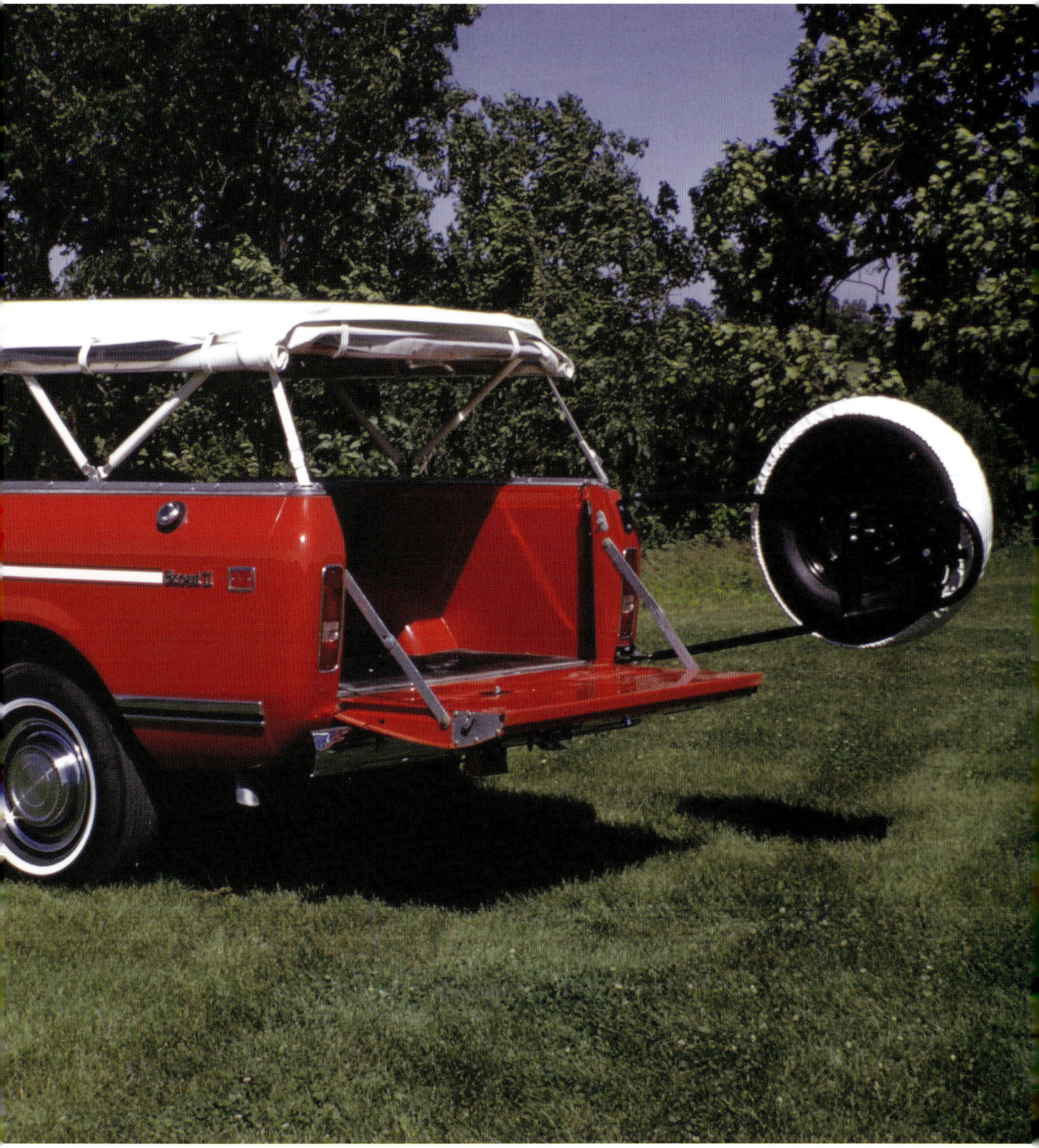
Scout II

▶ The introduction of the Nissan diesel for the 1976 model year heralded a new chapter in Scout II history. International wasn't quite the first SUV builder to try a diesel, but their efforts were successful in their slice of the market pie. *WHS 111001*

both sides, offering access from either door, not just the passenger side.

Cruise control was an option for 1977 but only with the V-8 and automatic transmission combination. What had once been an option, the 2,000-pound-pressure heavy-duty angle-link clutch (a particular design from Dana Spicer) was made standard, replacing an 1,800-pound unit, and a new 2,200 *super* heavy-duty clutch was optional above that. Product News Bulletin G-691

▲ The armored car concept was drawn up by the styling department in 1976 as part of a pitch to the government based on the ideas of Leo Windecker, who was just beginning to work with IH on composite body projects.

▲ At least one Armored Scout prototype was built, painted up with USAF markings. It appeared in October 1976 as a conversion on the long-wheelbase chassis and was designed to offer "ballistic resistance" (they were careful not to use the term "bulletproof") against up to 7.62mm NATO-spec M-80 ball ammunition. This was developed by Leo Windecker, who was also working on the fiberglass SSV models. The passenger compartment was protected, as were the engine compartment and underside. The stated purpose was "reliable and secure transportation to command personnel in hazardous situations or hostile environments." Curb weight was listed as 6,175 pounds with a gross vehicle weight of 7,500 pounds. It had six gun ports for return fire and a roof hatch for escape, observation, or return fire. The top/cab was built up much like any fiberglass piece, except it was thicker and infused with Kevlar sheets and featured ballistic-glass windows. Kevlar blankets were placed behind the existing Scout body, and the grille was shuttered to protect the radiator. The tires were listed as 7.00-15 run-flats, but it isn't clear if the prototype shown here is running them.
Betsy Blume/Ted Ornas Collection

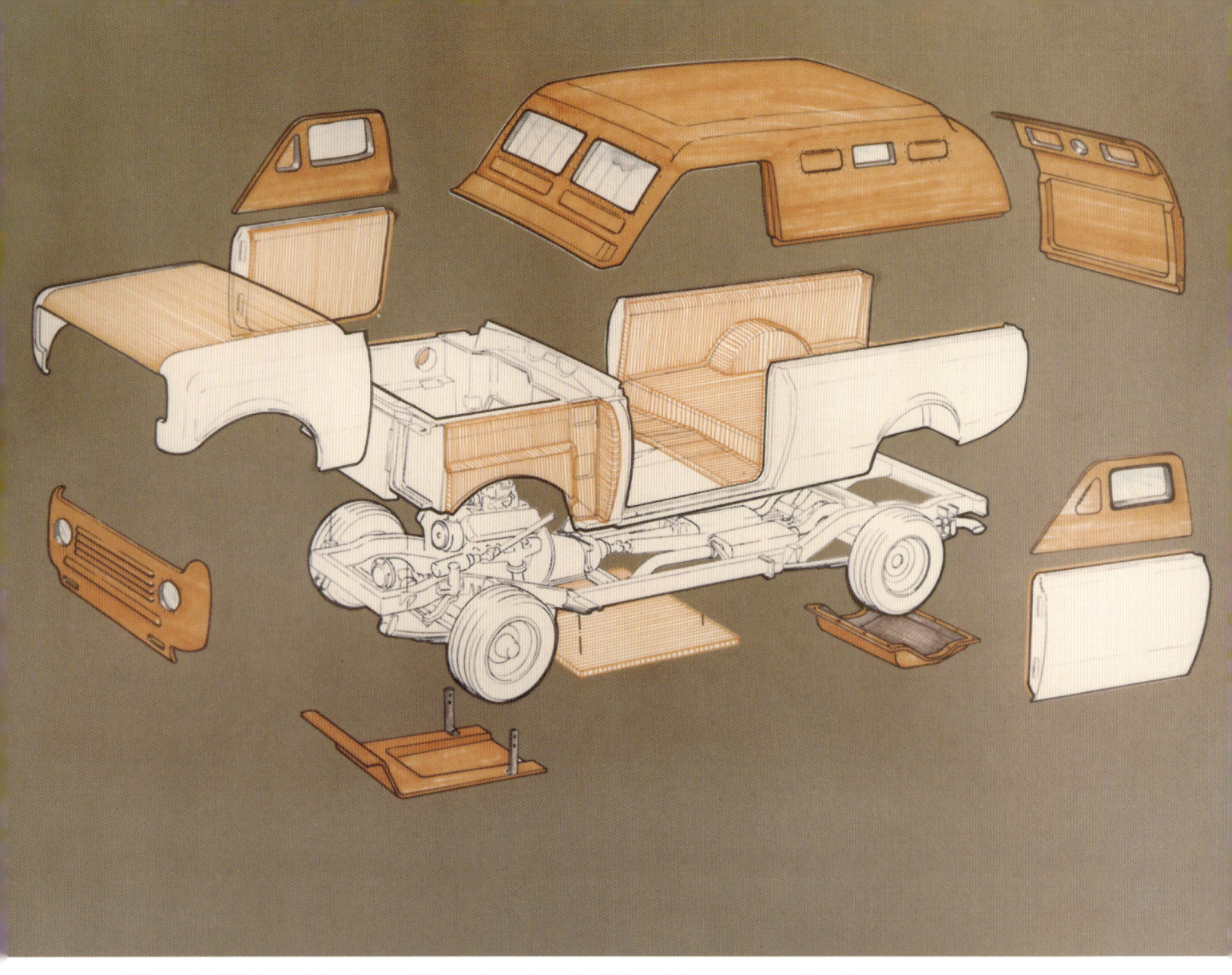

▲ This exploded view of the Armored Scout shows how the lightweight armor was placed. *George Kirkham*

announced that California emissions standards would eliminate the V-304A as an option there for much of the 1977 model year.

More specials appeared in 1977, including one that has since become iconic: the sporty SSII. The SSII gave IH some serious credibility in the four-wheeling realm because it was truly an off-roader's special. It was also suitable in an outdoor work realm (see Chapter 5). The SSII was part of a new marketing campaign that involved off-road racing (see Chapter 7).

A huge number of changes came to the long-wheelbase line for 1977, mainly involving special models. On October 18, 1976, the Midas Family Cruiser and Street Machines were announced. An ORV Package offered a few "off-roady" options to any Scout, including a brush bar with a mini winch, driving lights, a roll bar (Traveltop only), some unique side appliqués, and a CB radio. The Economy Scouts were mentioned in the advance ordering documents sent to dealers, but no other mention seems to have been made in 1977, at least not in the form outlined for 1976. It's possible they were a one-year wonder. There were base 1977 models called "XLC Economy" in the '77 price books, but they are not the same super-austere offering found in '76.

▲ Not exactly "high visibility" looking out of the Armored Scout prototype, but visibility was secondary to protecting the cargo and not getting shot. *George Kirkham*

▶ The Saddle Custom vinyl interior was new and the top-of-the-line starting in 1977. This gave the Scout a definite "Cadillac feel" on the inside. A similar option was Parchment, a lighter shade. Both could be chosen with the Deluxe (16928) and Custom (16834) interiors as a $221 option, plus another $13 for the Saddle material. *WHS 110694*

SCOUT II, TRAVELER, AND TERRA: ANYTHING LESS IS JUST A CAR

▶ Buyers who wanted the Deluxe interior for 1977 and liked the Tanbark vinyl and nylon paid $134. It included color-matched mats and woodgrain dash and shift quadrant plate. The bucket seat option was $79 and the console was another $48. The air conditioning was a $497 option and the electric clock added $23.50.
WHS 111380

Another whiz-bang option for 1977 was the Suntanner, a dealer-installed package that turned the Terra into a convertible pickup for a mere $246 (549968C91). According to International, a convertible pickup was an industry first for a civilian vehicle and would not be duplicated until the Dodge Dakota convertible pickup debuted in 1989. The Kayline kit was available from the aftermarket as well as from the dealer. IH literature lists them only in white, but Kayline materials list white and black, as well as red, blue, green, and tan at extra cost. Also new for the Terra this year were optional chrome bed tie-down rails (option 16556, $70), but they were not compatible with the Suntanner kit.

The aftermarket almost immediately produced fiberglass bed shells for the Terra, but on January 13, 1977, International announced an authorized unit from Ayr Way Industries, in Kendallville, Indiana, near Fort Wayne. It was listed as a $325 option, with tinted glass optional at $12.50 and a luggage rack for $28.

In February of 1977, IH announced the white spoke-wheel option. These wheels came with the upper-end SSII packages in combination with 10-15 Goodyear Tracker A/T tires. They became a popular option on all three versions of Scout. The Rallye wheel combination with the Tracker A/T was cancelled as a result. The spokers with Trackers was a $420 package in 1977 (option code 10964). The roll bar developed for the SSII was also available as an option for any short-wheelbase Scout.

1978

A new grille and a price hike were the two obligatory items to report for 1978. The grille was similar to 1977, with two horizontal slots, but the three vertical support bars were blacked out. The base price of a Traveltop gasser rose to $6,153 and the diesel to $8,790. The base gas Terra was listed at $6,108 for the early part of the model year and the base Traveler was $6,546. The diesels were $8,749 and $9,187, respectively. Painted front and rear bumpers were once again standard and chrome was optional, or included in the Deluxe Exterior Package.

Four new colors appeared for 1978, and Concord Blue (6787) replaced the very popular Glacier Blue. Other new hues were Woodbine Green (5902), Rallye Gold (4415), and Embassy Gray (8547). Electrostatic painting was introduced to the production line, resulting in much improved paint quality. Three new appliqué styles debuted, a two-tone treatment in three colors that more or less covered the entire side in white (10842, $169), black (10864, $169), or woodgrain (10843, $169) pattern, or offered a simple and decorative white accent stripe (10865, $109). The Rallye stripes changed to the type most favored by Scout fans today and were included in the usual Rallye package (10969, $745). A new Bright Finish Package (10898, $60) gave buyers a lower cost alternative to the Deluxe Exterior and included a bright grille and chrome bumpers front and rear. Inside the Scout, the high-back buckets used a new inertia lock on the tilt mechanism so there was no more groping for the release. The Custom Interior Package included carpeting on the cargo area panels and front kick panels.

On the mechanical front, you could order the V-304A with 3.73 or 4.09:1 axle ratios to enhance trail riding. The V-345A was available with the 3.73:1 cogs, and the diesel models went to 3.73:1 as standard to improve acceleration performance and hill climbing. Goodyear Tiempo All-Season radials were used in a radial tire package that could be ordered with the standard rim, the Rallye wheels, or as part of the Rallye package. A new 15-inch Sport steering wheel was made available for all models and became a popular option.

▲ For 1977, International announced an authorized fiberglass Terra bed shell by Ayr Way, of Kendallville, Indiana. It's shown here on a well-equipped Terra 4x4 that has the white two-tone stripes, Rallye wheels, Deluxe exterior option, and likely the trailer tow package, given the two-horse trailer hitched up behind.

▶ The Suntanner was neither a production model nor a special, but rather a roadster-like, dealer-installed package for the Terra. This is the concept drawing from 1975, done by either Dave Higley or Larry Nicklin in the design studio. It clearly depicts the package as it existed in reality.

1979

There were not many visual external changes to the 1979 models, though the SSII grille was made an option for all Scouts. Five of the nine colors offered for 1979 were new, including Persimmon (3199), Tahitian Red Metallic (2300), Lexington Blue (6788), Mint Green (5907), and Sunburst Yellow (4403).

The 4x2 Terra was dropped from the lineup, though a 4x2 Traveler and 100-inch were still offered. A base 4x4 Traveltop gasser set you back $6,899 and a diesel broke the $9,000 barrier at $9,241. The base Terra was $6,950 and the base Traveler was $7,563. The diesels were $9,446 and $9,696.

Among the new available features were two stereo packages, one an AM/FM head with three speakers and the other an AM/FM radio with an eight-track tape player. An auxiliary automatic transmission cooler was added to the options list. Inside, black replaced the Tanbark vinyl as the base upholstery. The entire interior color and trim was replaced by Highland Blue, Russet, Black, and Sierra Tan when you ordered the Deluxe or Custom interior option. The buckets could be ordered in a standard high-back or with a pillow effect. The optional door panels now included carpeting partway up with map pockets. The Deluxe and Custom interior dash finally lost the woodgrain and gained a burnished metal, engine-turned look. The lid for the center console was redesigned to include two cup holders and the Traveltops had a simulated vinyl headliner.

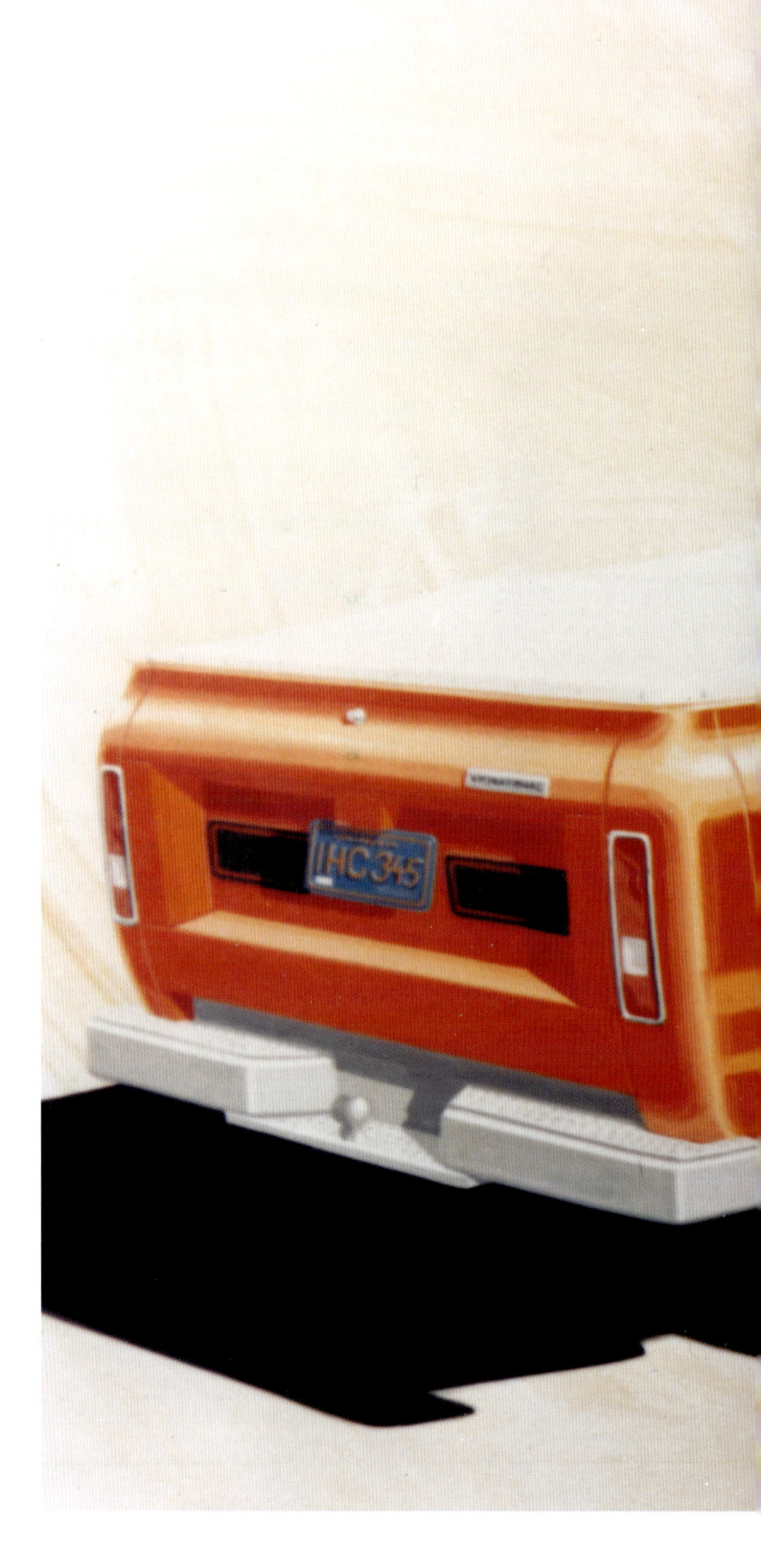

SCOUT II, TRAVELER, AND TERRA: ANYTHING LESS IS JUST A CAR   159

▲ One of the two Suntanner prototypes that made the rounds in 1976–1977, shown at the Phoenix Proving Grounds with its top up. The truck is a bit dolled up, with the hood stripes, brush bar, Marchal driving lights, and 11-15 Ridge Runner tires on white-spoked rims. The Suntanner was heavily promoted during the Mint 400 off-road race in 1976 and this image was taken somewhat prior. It isn't known how many of the packages were sold. The first announcement appears to be from February 1, 1977, in Parts Letter G-871. It was listed as part number 549968C91 listing for $264.15. The kit included the top, bows, tonneau, spare-tire cover, and top boot. The Marketing Department publicized the Suntanner very intensively in 1977 and the kit was available all the way to the end of production; publicity photos exist as late as 1979.

The 1979 model year also brought catalytic converters to the Scout line, mandating the use of unleaded fuel. The new EPA regulations bore down especially hard on 4-196-powered Scouts. A new 4,000-pound inertia weight class (a weight calculation used for EPA emissions dynamometer testing) meant that certain items had to be left off the options list to comply, including the heavy-duty rear-step bumper and the sliding quarter windows. The automatic-equipped 4-196 exited the stage for 1979 as well, reportedly for its inability to meet emissions regulations. The only V-345A offered was a four-barrel version, making up for some of the power losses due to the addition of the catalytic converter.

The T-407 automatic option was dropped from the diesel Scout lineup in the 1979 model year, though research did not determine exactly when this occurred. Strangely, intense VIN searches indicate virtually no

▲ Tom Thayer's '76 Terra with a Suntanner kit is one of the few intact, original survivors. Built in November 1975, it was ordered with a V-304A, T-407 automatic, TC-145 transfer case, 3.45:1 axle ratios, and the limited-slip rear axle. Resplendent in Fire Orange (3195), it has the Deluxe exterior (16835) and the Deluxe interior (16928) in Tanbark (16874). It shows only 29,000 miles. The Suntanner kit was installed by the original owner and Tom still has the original hardtop.

1979 diesel Scouts were produced after VIN JGD38694 on February 28 (a manual transmission Scout II shipped to Bisio Motors in Portland, Oregon). Evidently the numbers produced from July 1978 to February 1979 were enough to carry them through. Contributing factors could have included slow diesel sales and International's desire not to keep too much stock on hand with the new turbodiesel engine on its way.

On August 1, 1979, International announced the Scout Product Division with an effective date of November 1, 1979. The stated purpose for this reorganization within the Truck Division was to focus and coordinate efforts to recapture the larger SUV market share once held by the Scout. This was apparently the brainchild of Jim Bostic, who would become an IH vice president and head up the new division.

1980

With the new decade, IH broke the $10,000 barrier with the Scout Traveltop diesel offered for $10,244 at the start of the year. (The gas model was $7,748.) For 1980, the 4x2 models were no longer offered, though right-hand-drive Scouts were still listed. The Terra and Traveler base models were priced at $7,649 and

▶ The '78 Scout lineup had all the bases covered, from sport to utility and a lot else mixed in. The Terra, SSII, and Scout II Traveltop mount the Off-Road Tire and Rim package (10964, $446) that included 10-15 Goodyear Tracker A/T tires and white-spoke 8x15 wheels from Motor Wheel. This option required power steering if the tires were ordered à la carte (05281, $206), but power steering was included in the Rallye package or with the diesel engine. WHS 110943

▲ The immensely popular Rallye package changed its stripes for 1978, adding hash marks to the side appliqué and placing the letters on the side rather than on the hood. Note that the two-slot grille looks similar to a '77's, but the three support bars are blacked out. This color, Rallye Gold (4416), was new for 1978. WHS 110945

▶ Another new all-vinyl Custom interior debuted for 1979 with pillow-top bucket seat cushions. It's shown here in black, one of two optional colors (the other was Sierra Tan).

$8,783 with the diesels selling for $10,145 and $10,551, respectively.

The new SD-33T Nissan turbodiesel was added to the options list for the 1980 model year. It was announced in July 1979 that limited installation of turbodiesels into 1980 Scouts would start in September 1979 and full implementation would begin in October. The new engine offered much improved performance with same fuel economy as the previous naturally aspirated diesel. As of 1979, the automatic was no longer offered for the diesel Scouts.

An LST search shows the first turbodiesel Scout was a nicely optioned Winter White (9219) over Concord Blue (6787) 100-inch Traveltop (VIN KGD11388) with engine number 00039 (the lowest turbodiesel engine number listed was 00032) built on August 8, 1979. It was marked as the pilot model, and the next one, both by VIN and line sequence number, was marked "Master Line," with the next VIN and LSN being another pilot model. It would be a week before more turbodiesels rolled off the line and production gradually ramped up until the end of 1980, when only diesels were produced.

The turbodiesel engine numbers reset to zero. The last of the non-turbo numbers spotted in February 1979 were in the 26850 range. The last turbodiesel Scout engines were running in the 06300 range, with the highest found being 06334 mounted in KGD23061, built on October 20, 1980.

Three new Scout II specials emerged in 1980: the 434, 844, and the RS. On March 21, 1980, the 434 and 844 were announced as "fuel economy models." The luxury RS Traveler debuted at the same time.

Four new side appliqués also appeared for 1980: the Spear (Green/Blue 10843, or Orange/Orange 10844, $135), See-Thru Flare (Orange/White 10864, Orange/Black 10865, $191), the Wave Yellow/Orange/Orange 10841 and Yellow/Green/Blue 10842, both $191), and a new Rallye stripe design (White/Yellow/Orange 10969 and Black/Yellow/Orange 10969, both $152 but requiring either the Styled wheel package 10938 (Rallye wheels with Tiempo radials, $440) or the Off-Road wheel package 10964 (white spokers with 10-15 Trackers, $505). The wave appliqués were said to be limited to stock on hand and rarely seen. A variation of the old-style Rallye stripes remained available in Silver (10833) and Gold (10846) for $165, in addition to requiring one of the previously mentioned wheel-and-tire packages. As of February 1980, the Styled (Polycast) wheel package was available with Gold-accented wheels (10939, $440), again with the Goodyear Tiempo radial tires.

Many smaller changes appeared for 1980, including power steering becoming standard on all Scouts except right-hand-drive models. With the power steering came

SCOUT II, TRAVELER, AND TERRA: ANYTHING LESS IS JUST A CAR

▲ These white accent stripes (10865) were a new option in 1979, but they had appeared before in black on a special-edition SSII in 1977. It was a $116 option with the Deluxe exterior package.

a new 15-inch steering wheel. New chrome outside mirrors were also fitted, and a federally mandated 85-mile-per-hour speedometer replaced the previous 100-mile-per-hour style. A new air-conditioning system added 10 percent more cooling, quieter operation, and less interference with human shins.

More new standard colors appeared for 1980, including Black Canyon Black (0001), Copper (3201), Dark Brown (1032), Saffron Yellow (4417), Green Metallic (5901), and Concord Blue (6787), a return from '78. Stick-on side body moldings replaced the metal clipped-on pieces used previously. The standard grille and bumpers were black, but the optional Deluxe or Chrome exterior options included a silver grille with bright trim. Inside, the nylon seat inserts had a Tartan pattern in several colors.

Mechanically, besides the turbodiesel mentioned previously, the big news was the new transfer case. Dana had recently developed the Model 300, which became the IH TC-146. It was still in the family of the old Model 18s and 20s, but it was much quieter and had a lower 2.62:1 low range (versus 2.03:1). With the new transfer case came a new set of standard locking hubs in stylish black with a Scout logo on the knob. A problem occurred almost immediately. Spicer announced a shortage of Model 300 transfer cases so for a considerable period, 1980 Scouts continued to use Model 20s. The LST listed the parts substitution. It wasn't until mid September, 1979, production at around KGD13250, that Model 300 transfer cases began to appear in every Scout build.

The front end changed significantly for 1980. Not only was there an attractive new grille, but there were rectangular headlights. The rectangular headlight look was bemoaned on most other makes, but they did wonders for modernizing the look of the Scout, and, by golly, most Scout fans liked them. The Polycast wheels appeared for 1980 and they also greatly modernized the Scout's look. Polycast wheels had the appearance of an alloy, or "mag," wheel, but they were merely a steel wheels to which a decorative plastic insert was bonded. They were, and still are, a highly popular Scout accessory.

▲ At the time this picture was shot in 1979, the Wave appliqué (10841) was under consideration as an option for the 1980 model. It showed up as available in 1980 sales literature but is seldom seen. The stripes were designed by Dick Hatch (a great help with this book), and these are one of the few appliqués Dave Higley, his contemporary in the design department, didn't do.

As you will read in more detail farther on, by May of 1980 the Scout's fate was sealed by management. Either sell the Scout business or shut it down. As hope dimmed on the sale, steps were taken to minimize losses. One of those steps was to immediately increase the production of diesel Scouts. International had invested heavily in the new turbocharged version of the Nissan diesel. To avoid having to liquidate leftover engines, just about every Scout built late in May and on to the end was a diesel. By about May 25 and VIN KGD18700, it was a sea of diesel Scouts. Only a few gassers built to order rolled off the line. Even at that, International had to sell a lot of surplus diesel engines after the shutdown. Leading into the diesel run, there were a lot of gas 4-196 engines installed as well to clear the decks of those as well. More V-304s than V-345s too, as the 304 was mostly a Scout engine and the 345 still had some Medium Duty applications.

It's ironic that in the Scout's last year of production, International emptied both barrels in an attempt to end the rust issues that had plagued Scout from day one. Today the term "rust" and "Scout" are nearly synonymous, but rust in a 40- to 60-year-old vehicle is much easier to accept than it was for original owners back in the day who had problems inside a year. Certain high-vulnerability areas such as the rear inner quarter panels, quarter panel end caps, quarter panel splash shields, and lower grille were galvanized. Other parts such as the outer doors, liftgate, windshield frame, and outer rear quarter panels were coated with Zincrometal, a sprayed-on coating of zinc powder and plastic. Zinc-rich primer was used in certain vulnerable areas such as the hood seams, tailgate seams, drip-molding joints, and front fender seams. Hot wax was sprayed into many of the void areas, including the doors and tailgate. Add to that the optional undercoating and the 1980 Scout had the best rust protection in its history.

SCOUT II, TRAVELER, AND TERRA: ANYTHING LESS IS JUST A CAR   165

END OF THE ROAD: THE CORPORATE STORY

The fate of Scout had been hanging in the balance for a long while—a lot longer than many who built them even knew. The Scout's funeral dirge could be faintly heard upon opening the internal October 1979 report, *Scout Business Plan Overview*, in which the Scout's competitive strength was rated low and the recommendation was made to minimize resource investment in it. Further, despite some very strong pitches by proponents of investing in the brand, the business plan recommended exploring the sale of the division. International Harvester was spread too thin across too many product lines. There had been a time when the "everything to everyone" business model could work. By the late 1960s, and maybe earlier, that time had passed and manufacturing industries were paring down to specialize on core products.

▲ The final year of Scout may have been its best looking. Most people think the square headlights worked with the broad-shouldered look. This '80 Rallye in Green Metallic (5901) has the older-style Rallye stripes in gold (they were also available in silver), but they were slightly different than the previous versions. The stripe no long went over the fenders and onto the hood. The Rallye was no longer a package but simply an appliqué (10845, $165). It required the styled Rallye wheel package (10938, $440) or the shown Off-Road Tire and Wheel Package (10964, $505). This rig also has the standard exterior, which had gone to black bumpers for 1980, making for a sharp package.

Sometimes those transitions went smoothly; sometimes they did not.

The end of the light-truck line in 1975 was the first nail in the Scout's coffin. Many on the IH board wanted to end Scout at the same time, and from a strictly business point of view, they were probably right. The same forces that killed the light-truck line in 1975 killed the Scout in 1980. First and foremost, the profit margin per unit was too low. Because the light trucks and the Scout were not the main business at IH, they couldn't spread the costs of designing, building, and upgrading them very far. GM, for example, could spread these costs over hundreds of thousands, nearly millions, of similar vehicles each year. International didn't build much more than a hundred thousand Light Line rigs a year, even a good year, so the costs per unit were higher. As a result, they had to charge more for a product similar to those available elsewhere for less.

▲ The Spear appliqué remains one of the more popular legacies of the final year of the Scout. It's shown here on coauthor John Glancy's Black Canyon Black '80 Traveltop with the Deluxe exterior and bumper guards. This Black Canyon Black (0001) Scout has a mere 300 miles on the odometer and still has the plastic on the seats.

To cut costs, International either had to operate with less profit (or even at a loss), or cut quality. International tried a little of all three and the results were never pretty. It was the business version of the proverbial rock and the hard place. Another integral part of the IH business model pitted the various divisions against each other to compete for resources. As a result, factions in the centralized governing body were fighting to keep *their* special interest at the top of the heap. This resulted in an inertia that kept the IH ship headed directly for the iceberg while only delaying the tough decisions.

Many other factors led to the demise of Scout. Some were unavoidable . . . business-world "acts of God." Among them were impending federal emissions, crash test, and fuel economy standards. The entire light-duty automotive industry was bumping into these things, and again, the key was

◀ The Rallye stripes changed again for 1980 (shown on 1979 Scout), but the older style remained available in silver or gold. There were two versions of this style: those like this with a white border (10845) and those with a black one (10969). The version applied depended on the body color. Both were a $152 option.

to spread resulting costs as thinly across the product lines as possible to keep price parity with the competition. But the fewer the products, the thicker the costs had to be spread.

To make matters worse, the late 1970s brought an oil crisis and a recession that hit everyone. High interest rates curtailed investments and the general business outlook was grim. The farming industry was especially hard hit, which struck deeply at International Harvester's core agriculture business. Ultimately, this was the motivator necessary to cut through the egos, boardroom turf wars, and inertia. It went beyond just one division in a very large company and became a survival situation for the *entire* company, pure and simple.

There has long been a trend to try to pin the death of Scout on individual executives in the IH organization. While there are plenty of "heroes and zeros" to highlight and no shortage of ineptitude to report, ultimately it doesn't matter. The Light Line no longer fit the IH business model, and even with stellar management, it would have been a drain on the company. The Scout was probably doomed unless some radical steps were taken (see Chapter 6).

The United Auto Workers strike of late 1979 and early 1980 also is often blamed for the demise of Scout and it certainly was a major contributing factor. The UAW and the IH boardroom played a serious game of hardball, neither fully realizing, or choosing to ignore, the potential pitfalls. Again, the problems were bigger than this one event, but the strike may have been a balance tipper. Without it, the Scout may have survived on inertia until the SUV market picked up just a few years later.

Barbara Marsh's *A Corporate Tragedy: The Agony of International Harvester* provides an excellent account of the corporate details. Marsh doesn't write much about Scout, but she gives a good picture of what happened at the highest levels at IH and how that trickled down to the various divisions.

▲ One of the last new appliqués was this woodgrain pattern shown on a decked-out 1980 Traveler (10840, $199). This Traveler is seen in Saffron Yellow (4417) with the Deluxe exterior trim (16835, $213), Styled Wheel Radial Tire Package (Polycast wheels, 10938, $440), luggage rack (16903, $116), sliding quarter windows (16893, $104), and heavy-duty rear-step bumper (01643, $44).

END OF THE ROAD: THE SCOUT STORY

After the crippling UAW strike, which started on November 1, 1979, and ended April 21, 1980, the factions within IH that wanted to divest the company of the Scout line finally got their way after a lot of mixed signals to the division from upper management. Shortly after the strike ended, the word went out that the Scout division was up for sale. The official public announcement came in a press release dated May 15, 1980. An in-house announcement appeared earlier the same day in a Truck Division internal newsletter, marking the first time many within the company heard.

While there was optimism, probably somewhat misplaced, that the division could somehow be sold, the backup plan was to end production at the end of the 1980 model year. It isn't clear how well this backup plan was known to those outside the executive boardroom, but former line workers report the doom-and-gloom end-of-the-line rumors started when the strike did.

First in line to buy Scout was none other than Ed Russell, the man behind the Custom Vehicles International (CVI) Scouts. Flamboyant in the way only a Midwestern transplant to Texas can be, Russell showed up in a cowboy hat and boots to take the reins. Negotiations were announced in the latter part of May to employees and dealers. An article by Robert Ratliff in the June 1980 issue of *International Intercom* stated negotiations were underway and explained the reason for the "Scout Decision" and how it would ultimately benefit all concerned. That was followed up in the August/September issue by an article quoting Jim Bostic, the vice president and general manager of the Scout Division, on the details of the sale, which sounded advanced and quoted Russell as saying 1981 production would begin in January 1981.

For $8.5 million ($6 million cash), Russell's newly formed Scout Corporation would have bought the Scout tooling, trademarks, engineering designs (for past, present, and future models), and use of all patents related to the Scout. For at least three years, Scout Corporation would sell the units through the existing IH dealer network, though it was hinted dealer participation was not mandatory—important since some IH dealers considered Scout sales a vexing burden.

The new Scouts would be built in the old factory, which would be bought by a consortium of Fort Wayne investors and leased back to the Scout Corporation. IH would retain the Scout parts business. Service would stay at the dealers, though warranties would be administered by Scout Corporation. It was clear Russell intended to cultivate a new and much more broadly located dealer network, as International had been planning. The picture painted was extremely upbeat and rosy.

According to Dick Nesbitt, a designer who worked at CVI, Russell didn't have the money himself, but the backing came from Steyr-Daimler-Puch, the Austrian company behind the Haflinger and Pinzgauer 4x4s. It isn't clear what part of the deal fell through, but fall through it did, likely at Steyr-Daimler-Puch.

THE LAST DAY

It must have been rough going to work at the Scout plant that Tuesday, October 21, 1980. More than 40 years down the road, Scout enthusiasts can look at this day in the remote and painless context of history. The 464 people who lost their jobs, according to a newspaper article from the period, had a more painful and immediate view. IH kept as many people as they could by moving them to other jobs within the company, both locally and elsewhere, but a substantial number went hunting. The job losses hurt the Fort Wayne community too.

The average production day for one shift on the Scout line was around 110 units in the late 1970s. The line had a maximum capacity of 200 units per shift. A review of the LSTs for every Scout built that last half-shift shows a mere 35 Scouts were built. The first Scout built that day was KGD23073, with a line sequence number of D5181. The highest known Scout VIN is KGD23106 with a line sequence number of D5214, but it wasn't the last Scout.

The powers-that-were moved a Tahitian Red (2300) 100-inch, four-speed diesel Scout, VIN KGD23023, to the end of the line. Its original line sequence number was D5140, which would have been a Scout built the day before, but a handwritten "F5214" on the LST replaced it.

The line sequence number of the highest VIN was D5214. The Scout assembly line at Fort Wayne was called the D-Line, hence the D-prefix. According to sources that worked the line, an F indicated a fill-in vehicle whose construction was delayed for some reason. Often it was for a lack of parts, but it could be done easily and at the whim of the people in charge. The LST would be pulled from the normal sequence. Once it was decided where to insert it, the F would be followed by the LSN number after which the Scout would be reinserted into the sequence. Since D5214 was the last regular Scout in line, an "F5214" told line workers to insert it after that number.

Voila! The Last Scout.

▲ The Last Scout! The official last one, anyway. There aren't many smiling faces in this picture. International picked a Tahitian Red Traveltop to be the last, possibly because it was going to Mary Garst, a member of the IH board of directors, as a loaner. The Scout was sent to the Melrose Park, Illinois, TSPC, where a number of small modifications were made, including a winch, some custom pinstriping, and a rear-window defroster. From there it went to Garst in Iowa. The Last Scout was the 532,674th to roll off the line in Fort Wayne. *Mike Bolton Collection*

The big question was why that particular Scout was made to be The Last Scout. Well, there are some clues. It was ordered on September 18, 1980, for Mary Garst, a member of the IH board of directors. That position rated a loaner vehicle, and this wasn't Mary's first. As a highly respected board member, perhaps rearranging things and giving her the last official Scout was a measure of respect. Or, as rumor has had it, it could have been an attempt to mollify a vocal board member unhappy with the bonus given to a certain CEO in the midst of a financial crisis.

If the name Garst is familiar, the Garst & Thomas Seed Company started in 1930, and the later Garst Seed Company offshoot is now owned by Syngenta. Over many years, the various Garst agricultural properties were among International Harvester's unofficial agricultural product test areas.

The funny part of this story is that apparently Mary Garst was less-than-enamored of this particular Tahitian Red diesel Scout after it left her stranded a couple of times at the Des Moines airport by refusing to start in cold weather. Mary's husband was Steve Garst, son of Roswell, the seed company patriarch and a mover and shaker in agriculture and business. Steve soon adopted the Scout for his use, and when it came time to return it, he opted to buy it instead. He used it as a hunting, fishing, and farm vehicle until not long before his death in 2004. Scout collector and historian Mike Bolton bought the Scout from the Garst family in 2003.

Of the Last Day Scouts, 23 were 100-inch Traveltops, five were Travelers, and four were Terras. Two are unknown due to lost LSTs. Every one of the Last Day Scouts were diesels with T-428 four-speeds. The last Midas Scout was in this batch (KGD23090), a Dark Brown (1032) Traveltop with the Tan Family Cruiser package and destined for Dallas, Texas. Of the final 10 Scouts built, one (KGD23100) was the last Traveler and it was a well-equipped unit in Black Canyon Black (0001) with a Highland Blue Custom interior, Polycast wheels, and radial tires. It was shipped to Arizona. Also among that last 10 was the last Terra built (KGD23097). It was painted a special CT-399 (custom color catalog) color, Montauk Blue (6796), and went to a dealer in Fairmont, Minnesota.

So ended the Scout. All that was left was to sweep the floors, stop off for a last beer with workmates, and head home. While there was some follow-up work to do, International started breaking down the line almost immediately. The tooling that was useful to other divisions of International was moved where it was needed and the factory building became a vehicle-storage facility. In 1981 and later, stockpiles of Scout parts, test and prototype vehicles, and various other material was auctioned off. Eventually the building was sold and International Harvester, soon to become Navistar, and the Fort Wayne community moved past the Scout.

▲ The Last Scout—a darker view. Enthusiasts lament the last of a line of great vehicles, but for 464 of the people who worked the Scout line, it meant the end of a job or even a career. *Mike Bolton Collection*

SCOUT II, TRAVELER, AND TERRA DATA

MODEL YEARS 1971–1980

INTRODUCTIONS

1971: April 16, 1971
1972: October 15, 1971
1973: September 15, 1972
1974: October 1, 1973
1975: ~November 1, 1974
1976: ~November 1, 1975
1977: November 12, 1976
1978: ~November 1, 1977
1979: ~November 1, 1978
1980: November 19, 1979

PRODUCTION DATES

Dates in italics are estimates based on standard procedure, based on a line retooling shutdown in early July and a vacation break that sometimes coincided with that time and sometimes didn't.

1971: April 19, 1971–*October 29, 1971*
1972: September 7, 1971–July 13, 1972
1973: July 31, 1972–July 13, 1973
1974: August 6, 1973–August 1, 1974
1975: August 2, 1974–*July 12, 1975*
1976: July 25, 1975–*July 7, 1976*
1977: July 19, 1976–*July 1, 1977*
1978: *July 15, 1977–July 18, 1978*
1979: *August 1, 1978–July 6, 1979*
1980: July 23, 1979–October 21, 1980

VIN INFORMATION

Ranges include canceled VIN.

1971: G425128–G474896 *
1972: G484897–G519977
1973: G519978–522965 †
1973: CGD10001–43359 ‡
1974: DGD10001–41865
1975: EGD10001–41866
1976: FGD10001–47244
1977: GGD10001–49284
1978: HGD10001–49171
1979: JGD10001–54434
1980: KGD10001–23106

* Approximate range. Includes 800B in range; first Scout II pilot was G442553. First production Scout II was G447268.
† Approximate range of old G-numbers issued early in the 1973 model year production from July through September 1972 It is unknown how many Scouts were in this batch of approximately 3,000 vehicles built at Fort Wayne.
‡ New style numbers issued from September 30, 1972.

MODEL CODES: 1971–SEPTEMBER 30, 1972

VINs for the first year-plus of Scout II production used the same 13-digit VIN and model convention used since the 800 debut in 1966. The model code prefix was followed by the sequential numbers that began with a G to indicate Fort Wayne Assembly. The prefix helped identify the type and configuration of the Scout. The sequential numbers had started in 1965 at 165000. The sequential number ran consecutively year by year through 1972 and a little into the 1973 model production (1972 calendar year production) before the system was changed again.

Digit	1	2	3	4	5	6	Description
	1	1	3	8			Scout II Traveltop 4x2
	1	8	3	8			Scout II Traveltop 4x4
	1	1	2	8			Scout II Cab-Top 4x2
	1	8	2	8			Scout II Cab-Top 4x4
	A						Seen instead of "1" in 1972 §
					0		4-cyl. engine
					4		Possibly 4-cyl.
					6		6-cyl. engine
					8		V-8
						0	All

§ Yet another example of the mysterious doings at IH. See the 1974–1980 charts for context.

MODEL CODES: OCTOBER 1, 1972– MARCH 31, 1974

For '73, IH changed the system again, still using a 13-digit VIN, but in a way that made more sense than the previous. The prefix numbers changed completely and the sixth digit allowed one to determine the year from the VIN. The sequential number started fresh every year at 10001. According to period literature, the first digit indicated when the model went into production. This seems like a redundant element, given the model-year indicator is also the sixth digit. At the end of March 1974, IH changed some details (see next chart).

Digit	1	2	3	4	5	6	7	8	Description
	3								1973 design
	4								1974 design
		S							Scout series
			8						Scout
				H					4x2 Traveltop
				S					4x4 Traveltop
				B					4x2 Cab-Top

Digit	1	2	3	4	5	6	7	8	Description
				L					4x4 Cab-Top
						0			Gasoline Engine
						6			6-cyl. engine
						8			8-cyl. engine
						C			1973 model
						D			1974 model
							G		Fort Wayne plant
								D	Scout Line

MODEL CODES: APRIL 1, 1974–MARCH 31, 1977

During this period, the 13-digit system changed yet again. The second through fourth digits indicated the model and body style. The prefix numbers changed slightly with a letter and the sixth digit allowed one to determine the year from the VIN.

Digit	1	2	3	4	5	6	7	8	Description
	D								1974 design
	E								1975 design
	F								1976 design
	G								1977 design
		0	0	1					Scout II Half Scout 4x2
		0	0	2					Scout II Cab-Top 4x2
		0	0	3					Scout II Traveltop 4x2
		0	0	4					Scout II Traveltop 4x4
		0	0	5					Scout II Cab-Top 4x4
		0	0	6					Scout II Traveltop 4x2
		0	0	7					Terra 4x2
		0	0	8					Traveler 4x2
		0	0	9					Terra 4x4
		0	1	0					Traveler 4x4
		0	2	3					Scout II Traveltop 4x2

Digit	1	2	3	4	5	6	7	8	Description
		0	2	6					Scout II
									Traveltop 4x4
		0	2	7					Terra 4x2
		0	2	8					Traveler 4x2
		0	2	9					Terra 4x4
		0	3	0					Traveler
					0				Gas engine
					1				Gas engine
					2				Gas engine
					3				Nissan diesel
						C			1973
						D			1974
						E			1975
						F			1976
						G			1977
							G		Fort Wayne plant
							D		Scout assembly

MODEL CODES: APRIL 1, 1977–MARCH 31, 1980

Another slight change was made to the 13-digit system to reflect changes in the model lineup. It stayed the same for the rest of the Scout run.

Digit	1	2	3	4	5	6	7	8	Description
	G								1977 design
	H								1978 design
	J								1979 design
	K								1980 design
		0	0						Custom Scout
		0	1						Custom Scout
		0	2						Austere Scout
		0	3						Austere Scout
				3					Traveltop 4x2
				5					SSII 4x4
				6					Traveltop 4x4
				7					Terra 4x2
				8					Traveler 4x2
				9					Terra 4x4
				0					Traveler 4x4
					2				4-cyl. or V-8 gas
					3				Nissan diesel
						H			1978
						J			1979
						K			1980
							G		Fort Wayne plant
							D		Scout assembly

TOTAL SCOUT II 100-INCH MODEL PRODUCTION

These numbers are best regarded as approximate. The best sources available were used, but those sources were disparate. Calculating from VIN ranges produces a higher number, but there was always a certain number of cancelled VINs.

1971: ~14152 **

1972: 30,820

1973: 33,752

1974: 29,926

1975: 21,366

1976: ~25,840 ††

1977: ~27,072 ††

1978: ~26,369 ††

1979: ~35,286 ††

1980: ~8,519 ††, ‡‡

** In 1971, total 800B and Scout II production was 19,952 units. There is evidence of approximately 5,800 800Bs, which was deducted from the total to arrive at this figure for Scout II.

†† Total Scout production included both Scout II and Terra/Traveler production. The long-wheelbase Scout production figure was subtracted to get this number.

‡‡ Reflects the UAW strike (December 1979 to March 1980).

SCOUT II 118-INCH MODEL ANNUAL SALES, 4X2 AND 4X4 §§

1976: Terra–2,513
Traveler–5,445
1977: Terra–2,376
Traveler–6,298
1978: Terra–2,314
Traveler–7,137
1979: Terra–1,623
Traveler–3,857
1980: Terra–616 ***
Traveler–4,450 ***

§§ These are annual sales of particular models, not production- or model-year sales.
*** Includes 1980 and the following year sales to 1982 as remaining stock was sold.

SCOUT DIESEL MODEL PRODUCTION (100- AND 118-INCH)

1976: 1,602
1977: 1,038
1978: 1,045
1979: 537
1980: 5,389

AVERAGED SCOUT II 100-INCH PRODUCTION, 4X2 VS. 4X4 MODELS

1971–1979: 94.1 percent 4x4 vs. 5.9 percent 4x2
1980: Scout II 4x2 not produced

AVERAGED SCOUT II 118-INCH PRODUCTION, 4X2 VS. 4X4 MODELS

1976–1979: 73.6 percent Terra 4x4 vs. 26.4 percent 4x2
1976–1979: 89.5 percent Traveler 4x4 vs. 10.5 percent 4x2

AVERAGED SCOUT II CAB-TOP PRODUCTION VS. TRAVELTOP

1971–1975: 96.7 percent Traveltop vs. 3.3 percent Cab-Top
1976-80: Cab-Top Scout 100-inch not produced

SCOUT II SPECIFICATIONS (100- AND 118-INCH)

Axles, Front

4x2: FA-3, dead, tubular steel, 2.5-in. dia., 0.375-in. wall (1971–1979)
4x4: FA-13, live, Spicer 30, open knuckle (1971–1974 standard)
4x4: FA-13, live, Spicer 30, heavy duty, open knuckle (1975–1977 in some four-cylinder applications)
Optional 4x4: FA-43/FA-44, live, Spicer 44, open knuckle (1972–1974)
Standard 4x4: FA-43/FA-44, live, Spicer 44, open knuckle (1975–1980)
Front Axle 4x4 V-8: FA-44, live, Spicer 44, open knuckle ('72-74 standard)

Axles, Rear

Standard 4x2 and 4x4: RA-18, Spicer 44, semi-float (1971–1980)
Optional 4x2 and 4x4: RA-28, Spicer 44, semi-float, w/ Trac-Lok (1971–1980)

Axles, Ratios

1971: 4-196 – 4.27:1 standard (3.31:1 3.73:1 optional)
6-232 – 3.73:1 standard (3.31:1 4.27:1 optional)
V-304 – 3.73:1 standard (3.31:1, 4.27:1 optional)
V-345 – 3.73:1 standard (3.31:1 optional)
1972: 4-196 – 4.27:1 standard (3.31:1 3.73:1 optional)
6-258 – 3.73:1 standard (3.31:1 4.27:1 optional)
V-304 – 3.73:1 standard (3.31:1, 4.27:1 optional)
V-345 – 3.73:1 standard (3.31:1 optional)
1973: 6-258 – 3.73:1 standard (3.31:1, 4.27:1 optional)
V-304 – 3.73:1 standard (3.31:1, 4.27:1 optional)
V-345 – 3.73:1 standard (3.31:1 optional)

1974: 6-258 – 3.73:1 standard (3.31:1, 4.27:1 optional)
V-304 – 3.73:1 standard (3.31:1)
V-345 – 3.73:1 standard (3.31:1 optional)

1975: 4-196 – 4.09:1 standard (3-speed manual, no option)
4-196 – 3.54:1 standard (4-speed manual, 4.09:1 optional)
V-304/345 – 3.54:1 standard (T-332 3-speed and T-427 4-speed, no option)
V-304/345 – 3.07:1 standard (other 4-speeds/ automatics, 3.54:1 optional)

1976: 4-196 – 4.09:1 standard (3-speed manual, no option)
4-196 – 3.54:1 standard (4-speed manual, 4.09:1 optional)
6-33 Diesel – 3.54: (4.09:1 optional)
V-304/345 – 3.54:1 standard (T-332 3-speed and T-427 4-speed, no option)
V-304/345 – 3.07:1 standard (other 4-speeds and automatic, 3.54:1 optional)

1977: 4-196- 4.09:1 standard (3-speed manual, no option)
4-196 – 3.54:1 standard (4-speed manual, automatic 4.09:1 optional)
6-33 Diesel – 3.54: (4.09:1 optional)
V-304/345 – 3.54:1 standard (T-332 3-speed and T-427 4-speed, no option)
V-304/345 – 3.07:1 standard (other 4-speeds, automatic, 3.54:1 optional)

1978: 4-196 – 4.09:1 standard (T-332 3-speed, no option)
4-196 – 4.09:1 standard (T-427 4-speed, 3.54:1, 3.73:1 optional)
4-196 – 3.54:1 standard (T-428 automatic, 3.73:1, 4.09:1 optional)
6-33 Diesel – 4.09:1 standard (T-427 4-speed, 3.73:1 optional)
6-33 Diesel – 3.73:1 standard (T-428 4-speed, automatic, 4.09:1 optional)
V-304/345 – 3.54:1 standard (T-332 3-speed and T-427 4-speed, 3.73 optional)
V-304/345 – 3.07:1 standard (other 4-speeds and automatic, 3.54:1 optional)

1979: 4-196 – 4.09:1 standard (T-332 3-speed, no option)
4-196 – 4.09:1 standard (T-427 4-speed, 3.54:1, 3.73:1 optional)
4-196 – 3.54:1 standard (T-428 automatic, 3.73:1, 4.09:1 optional)
6-33 Diesel – 4.09:1 standard (T-427 4-speed, 3.73:1 optional)
6-33 Diesel – 3.73:1 standard (T-428 4-speed, 4.09:1 optional)
V-304/345 – 3.54:1 standard (T-332 3-speed and T-427 4-speed, 3.73 optional)
V-304/345 – 3.07:1 standard (other 4-speeds and automatic, 3.54:1 optional)

1980: 4-196 – 3.73:1 standard (T-332 3-speed, no option)
4-196 – 3.73:1 standard (T-427 4-speed, 3.54:1 optional)
4-196 – 3.54:1 standard (T-428, 3.73:1 optional)
6-33T Diesel – 3.73:1 standard (3.54:1 optional)
V-304/345 – 3.31:1 standard (T-332 3-speed and 4-speeds, 3.54:1 optional)
V-304/345 – 2.72:1 standard (automatic, 3.31:1, 3.54:1 optional)
V-345 (w/ Tow Option) – 3.73:1 (4-speeds, automatic)

Brakes

Front 4x2/4x4:11x2 in. drum (standard 1971–1973)
Front 4x2/4x4: 11.75x1.18 in. disc (standard 1974–1980)
Rear 4x2/4x4: 11x1.75 in. drum (standard 1971–1973)
Rear 4x2/4x4: 11.03x2.25 in. (standard 1974–1980)
Front Lining Area, Drum Brakes 4x2/4x4: 93.7 sq. in. (1971–1973)
Front Swept Area, Disc Brakes 4x2/4x4: 226 sq. in. (1974–1980)
Rear Lining Area 4x2/4x4: 82 sq. in. (1971–1973)
Rear Lining Area 4x2/4x4: 101 sq. in. (1974–1980)

Capacities

GVW 4x2/4x4: 4,600 lbs. (Cab-Top 4-196 or 6-cyl. standard, 1971–1973, FA-3/13 axle)
GVW 4x2/4x4: 5,200 lbs. (Cab-Top optional, Traveltop standard, V-8, standard 1971–1973)
GVW 4x2/4x4: 5,400 lbs. (V-8 models, Traveltop, standard 1974)
GVW 4x2/4x4: 6,200 lbs. (all 1975–1980)
FA-3 Front-Axle Capacity 4x2: 2,500 lbs. (1971–1979)
FA-13 Front-Axle Capacity 4x4: 2,500 lbs. (standard 1971–1973)
FA-44 Front-Axle Capacity 4x4: 3,200 lbs. (optional 1972–1974, standard 1975–1980)
RA-18/28 Rear-Axle Capacity 4x2/4x4: 3,500 lbs.
Main Fuel Tank: 19 gal.

Clutch

Standard 4-196: 10 in., 6-spring, 91.8 sq. in. (1971–1972)
Standard 6-232: 10.5 in., 6-spring, 103.2 sq. in. (1971)
Standard 6-258: 10.5 in., 6-spring, 103.2 sq. in. (1971–1974)
Standard V-304 V-8: 11 in., 6-spring, 123.8 sq. in. (1971–1974)
Standard V-345 V-8: 11 in. heavy-duty, 6-spring, 123.8 sq. in. (1971–1974)
Standard All Engines: 11 in., 6-spring, 123.8 sq. in. (1975–1980)
Optional All Engines: 11 in. heavy-duty, 6-spring, 123.8 sq. in. (1975–1980)

Typical Dimensions Traveltop and Cab-Top

Wheelbase: 100 in.
Overall Length: 166.2 in. (w/ bumpers)
Overall Width: 70 in.
Overall Height, Traveltop 4x2/4x4: 66.4 in.
Overall Height, Cab-Top 4x2/4x4: 66 in.
Approach Angle 4x2/4x4: 48 deg. (standard tires)
Departure Angle 4x2/4x4: 23 deg. (standard tires)
Ramp-Breakover Angle 4x4: 24 deg. (standard tires)
Front-Axle Ground Clearance 4x4: 10.25 in. (FA-13 axle)
Front-Axle Ground Clearance 4x4: 7.6 in. (FA-44 axle)
Rear-Axle Ground Clearance 4x2/4x4: 6.9 in. (standard tires)

Typical Dimensions Terra and Traveler

Wheelbase: 118 in.
Overall Length: 184.2 in. (w/ bumpers)
Overall Width: 70 in.
Overall Height, Terra and Traveler 4x2/4x4: 66 in.
Bed Dimensions, Terra: 72x54 in.
Approach Angle 4x2/4x4: 49 deg. (standard tires)
Departure Angle 4x2/4x4: 22 deg. (standard tires)
Ramp-Breakover Angle 4x4: 20 deg. (standard tires)
Front-Axle Ground Clearance 4x4: 7.6 in. (FA-44 axle, standard tires)
Rear-Axle Ground Clearance 4x2/4x4: 7.0 in. (standard tires)

Electrical

System Voltage: 12V

Standard Alternator Capacity: 37A (1971–1980)

Standard Alternator Capacity Diesel: 40A (1976–1980)

Optional Alternator Capacity: 61–63A (1971–1980)

ENGINE 4-196E, 1971–1972

Type: 4-cyl. slant-four OHV, IH 4-196

Displacement: 196.44 ci

Bore and Stroke: 3.88x3.656 in.

Compression Ratio: 8.10:1

Block and Head Material: cast iron

Number of Main Bearings: 5

Weight: 545 lbs. complete/dry

Gross HP Rating: 110.8 hp @ 4,000 rpm

Net HP Rating: 102.7 hp @ 4,000 rpm

Gross Torque Rating: 180.2 lb-ft @ 2,000 rpm

Net Torque Rating: 176.1 lb-ft @ 2,000 rpm

ENGINE 4-196E, 1975–1978

Type: 4-cyl. slant-four OHV, IH 4-196

Displacement: 195.45 ci

Bore and Stroke: 4.135x3.656 in.

Compression Ratio: 8.02:1

Block and Head Material: cast iron

Number of Main Bearings: 5

Weight: 545 lbs. complete/dry

Net HP Rating: 86 hp @ 3,800 rpm

Net Torque Rating: 157 lb-ft @ 2,200 rpm

ENGINE 4-196E, 1979–1980

Type: 4-cyl. slant-four OHV, IH 4-196

Displacement: 195.45 ci

Bore and Stroke: 4.135x3.656 in.

Compression Ratio: 8.02:1

Block and Head Material: cast iron

Number of Main Bearings: 5

Weight: 545 lbs. complete/dry

Net HP Rating: 76.5 hp @ 3,600 rpm

Net Torque Rating: 153 lb-ft @ 2,000 rpm

ENGINE 6-232, 1971

Type: 6-cyl. OHV, IH 6-232

Displacement: 232 ci

Bore and Stroke: 3.75x3.50 in.

Compression Ratio: 8.50:1

Block and Head Material: cast iron

Number of Main Bearings: 7

Weight: 478 lbs. complete/dry

Gross HP Rating: 145 hp @ 4,300 rpm

Net HP Rating: 120 hp @ 4,000 rpm

Gross Torque Rating: 215 lb-ft @ 1,600 rpm

Net Torque Rating: 202.5 lb-ft @ 1,000 rpm

ENGINE 6-258, 1972–1973

Type: 6-cyl. OHV, IH 6-258

Displacement: 258.1

Bore and Stroke: 3.75x3.895 in.

Compression Ratio: 8.00:1

Block and Head Material: cast iron

Number of Main Bearings: 7

Weight: 484 lbs. complete/dry

Gross HP Rating: 140 hp @ 3,800 rpm

Net HP Rating: 113 hp @ 4,000 rpm

Gross Torque Rating: 235 lb-ft @ 1,200 rpm

Net Torque Rating: 191 lb-ft @ 2,000 rpm

ENGINE 6-258, 1974

(Emissions engine)

Type: 6-cyl. OHV, IH 6-258

Displacement: 258.1

Bore and Stroke: 3.75x3.895 in.

Compression Ratio: 8.00:1

Block and Head Material: cast iron

Number of Main Bearings: 7

Weight: 489 lbs. complete/dry

Net HP Rating: 97 hp @ 3,600 rpm
Net Torque Rating: 190 lb-ft @ 1,200 rpm

ENGINE V-304A, 1971–1972
Type: V-8 OHV, IH V-304A
Displacement: 303.68 ci
Bore and Stroke: 3.875x3.218 in.
Compression Ratio: 8.19:1
Block and Head Material: cast iron
Weight: 700 lbs. complete/dry
Gross HP Rating: 193.1 hp @ 4,400 rpm
Net HP Rating: 180 hp @ 4,400 rpm
Gross Torque Rating: 272.5 lb-ft @ 2,400 rpm
Net Torque Rating: 262 lb-ft @ 2,800 rpm

ENGINE V-304A, 1973–1974
Type: V-8 OHV, IH V-304
Displacement: 303.68 ci
Bore and Stroke: 3.875x3.218 in.
Compression Ratio: 8.19:1
Block and Head Material: cast iron
Weight: 700 lbs. complete/dry
Net HP Rating: 137 hp @ 3,600 rpm (single exhaust)
Net HP Rating: 139 hp @ 3,600 rpm (dual exhaust)
Net Torque Rating: 233 lb-ft @ 2,400 rpm (single exhaust)
Net Torque Rating: 239 lb-ft @ 2,400 rpm (dual exhaust)

ENGINE V-304A, 1975–'78
(Dual exhaust standard, 1976–1978)
Type: V-8 OHV, IH V-304
Displacement: 303.68 ci
Bore and Stroke: 3.875x3.218 in.
Compression Ratio: 8.19:1
Block and Head Material: cast iron
Weight: 700 lbs. complete/dry
Net HP Rating: 140 hp @ 3,800 rpm (single exhaust)
Net HP Rating: 144 hp @ 3,600 rpm (dual exhaust)
Net Torque Rating: 243 lb-ft @ 2,400 rpm (single exhaust)
Net Torque Rating: 247 lb-ft @ 2,400 rpm (dual exhaust)

ENGINE V-304A, 1979–1980
(50 states engine)
Type: V-8 OHV, IH V-304
Displacement: 303.68 ci
Bore and Stroke: 3.875x3.218 in.
Compression Ratio: 8.19:1
Block and Head Material: cast iron
Weight: 700 lbs. complete/dry
Net HP Rating: 122 hp @ 3,400 rpm
Net Torque Rating: 226 lb-ft @ 2,000 rpm

ENGINE V-345, 1971–1972
Type: V-8 OHV, IH V-345
Displacement: 344.96 ci
Bore and Stroke: 3.875x3.656 in.
Compression Ratio: 8.05:1
Block and Head Material: cast iron
Weight: 725 lbs. complete/dry
Gross HP Rating: 196.7 hp @ 4,000 rpm
Net HP Rating: 143.5 hp @ 3,600 rpm
Gross Torque Rating: 309 lb-ft @ 2,200 rpm
Net Torque Rating: 262 lb-ft @ 2,000 rpm

ENGINE V-345, 1973–1974
Type: V-8 OHV, IH V-345
Displacement: 344.96 ci
Bore and Stroke: 3.875x3.656 in.
Compression Ratio: 8.05:1
Block and Head Material: cast iron
Weight: 725 lbs. complete/dry
Net HP Rating: 144 hp @ 3,600 rpm (149 hp in 1974)
Net HP Rating: 157 hp @ 3,600 rpm (dual exhaust)

Net Torque Rating: 263 lb-ft @ 2,000 rpm
   (285 hp in 1974)
Net Torque Rating: 287 lb-ft @ 2,000 rpm
   (dual exhaust)

ENGINE V-345 4-BBL, 1974
(California only)
Type: V-8 OHV, IH V-345
Displacement: 344.96 ci
Bore and Stroke: 3.875x3.656 in.
Compression Ratio: 8.05:1
Block and Head Material: cast iron
Weight: 725 lbs. complete/dry
Net HP Rating: 167 hp @ 3,600 rpm
Net HP Rating: 179 hp @ 3,600 rpm (dual exhaust)
Net Torque Rating: 273 lb-ft @ 2,000 rpm
Net Torque Rating: 288 lb-ft @ 2,000 rpm
   (dual exhaust)

ENGINE V-345A, 1975–1978
(California; dual exhaust only, 1976–1978)
Type: V-8 OHV, IH V-345
Displacement: 344.96 ci
Bore and Stroke: 3.875 x 3.656 in.
Compression Ratio: 8.05:1
Block and Head Material: cast iron
Weight: 725 lbs. complete/dry
Net HP Rating: 158 hp @ 3,600 rpm
Net HP Rating: 163 hp @ 3,600 rpm (dual exhaust)
Net HP Rating: 146 hp @ 3,600 (California, dual exhaust)
Net Torque Rating: 288 lb-ft @ 2,000 rpm
Net Torque Rating: 292 lb-ft @ 2,000 rpm
   (dual exhaust)
Net Torque Rating: 275 lb-ft @ 2,000 (California, dual exhaust)

ENGINE V-345A 4-BBL, 1979–1980
(50 states engine)
Type: V-8 OHV, IH V-345
Displacement: 344.96 ci
Bore and Stroke: 3.875x3.656 in.
Compression Ratio: 8.05:1
Block and Head Material: cast iron
Weight: 725 lbs. complete/dry
Net HP Rating: 150 hp @ 3,500 rpm
Net Torque Rating: 263 lb-ft @ 2,000 rpm

ENGINE 6-33 DIESEL, 1976–1979
Type: Nissan inline six OHV, naturally aspirated
Displacement: 198 ci
Bore and Stroke: 3.27x3.94
Compression Ratio: 22:1
Block and Head Material: cast iron
Weight: 622 lbs. complete/dry
Net HP Rating: 81 hp @ 3,800 rpm
Net Torque Rating: 138 lb-ft @ 1,200–1,600 rpm

ENGINE 6-33T DIESEL, 1980
Type: Nissan Inline six OHV, turbocharged,
Displacement: 198 ci
Bore and Stroke: 3.27x3.94
Compression Ratio: 20.8:1
Block and Head Material: cast iron
Weight: 673 lbs. complete/dry
Net HP Rating: 101 hp @ 3,800 rpm
Net Torque Rating: 175 lb-ft @ 2,200 rpm

PERFORMANCE
Turning Radius Traveltop 4x2/4x4: 33.1 ft.
Turning Radius Terra/Traveler 4x2/4x4: 38. 8 ft.
0–60 1972 Scout II: 18.4 sec. (6-258, 4-speed, 3.73:1)
0–60 1975 Scout II: 15.3 sec. (304 V-8, automatic, 3.54:1)

0–55 1975 Scout II: 20 sec. (4-196, 4-speed)
0–60 1976 Traveler: 14 sec. (V-345, automatic, 3.07:1)
0–60 1977 Terra: 33.1 sec. (6-33 diesel, 4-speed, 4.09:1)
0–60 1978 SSII: 11.7 sec. (V-345, automatic, 3.73:1)
0–60 1978 Scout II: 12 sec. (V-345, automatic, 3.54:1)
0–60 1980 Terra: 14 sec. (4-196, 4-speed, 3.73:1)

SUSPENSION AND STEERING

Front Springs Length/Width 4x2/4x4: 43x2.00 in.
Front Springs No. Leaves 4x2/4x4: 6
Optional Front Springs No. Leaves 4x4: 6
Standard Front Springs Rate 4x2/4x4 w/ FA-3 or 13 axle: 260 lbs.-in. (1971–1974)
Standard Front Springs Rate 4x2 w/ FA-3 axle: 335 lbs.-in. (1975–1980)
Optional Front Springs Rate 4x2/4x4: 300 lbs.-in. (FA-3/13 axle) (1971–1974)
Standard Front Springs Rate 4x4 w/ FA-44 axle: 378 lbs.-in. (1971–1974)
Standard Front Springs Rate 4x4 w/ FA-44 axle: 300 lbs.-in. (1975–1980)

Optional Front Springs Rate 4x4 w/ FA-44 axle: 378 lbs.-in. (1975–1980)
Standard Front Springs Capacity 4x2/4x4 w/ FA-3 or 13 axle: 1,100 lbs. (1971–1974)
Optional Front Springs Capacity 4x2/4x4 w/ FA-3 or 13 axle: 1,200 lbs. (1971–1974)
Standard Front Springs Capacity 4x4 w/ FA-44 axle: 1,550 lbs. (1971–1974)
Optional Front Springs Capacity 4x4 w/ FA-44 axle: 1,600 lbs. (1975–1980)
Manual Steering Type 4x2/4x4: S-8 (Gemmer 24J)
Manual Steering Ratio: 24:1
Power Steering Type 4x2/4x4: S-281 (Saginaw 700D or D-6)
Power Steering Ratio: 17.5:1
Steering wheel Diameter 4x2/4x4: 17 in. (standard 1971–1979)
Steering wheel Diameter 4x4: 15 in. (optional 1975–1979, standard 1980)

Rear Springs Length/Width 4x2/4x4: 56x2 in.
Standard Rear Springs No. Leaves 4x2/4x4: 4
Optional Rear Springs No. Leaves 4x2/4x4: 4
Standard Rear Springs Rate 4x2/4x4: 184 lbs-in (1971–1974)
Standard Rear Springs Rate 4x2/4x4 w/ FA-44 axle: 270–500 lbs.-in. (1971–1974)
Standard Rear Springs Rate 4x2/4x4 Traveltop: 258 lbs-in. (1975–1980)
Standard Rear Springs Rate 4x2/4x4 Cab-Top: 270–500 lbs.-in. (1975–1980) (optional on Traveltop)
Optional Rear Springs Rate 4x2/4x4: 258 lbs-in. (1971–1974)
Standard Rear Springs Capacity 4x2/4x4: 1,000 lbs. (1971–1974)
Standard Rear Springs Capacity 4x2/4x4: 1,550 lbs. (1975–1980)
Standard Rear Springs Capacity 4x4 w/ FA-44 front axle: 1,375 lbs. (1971–1974)
Optional Rear Springs Capacity 4x2/4x4: 2,080 lbs. (1971–1974)
Optional Rear Springs Capacity 4x2/4x4: 1,550 lbs. (1975–1980, standard on Terra 197–1980)

TIRES

Tire manufacturers varied. Common brands were General, Firestone, Goodyear, and Goodrich. Common sizes shown per data book; others available.

1971 4x2/4x4: E78-15, standard
(4- and 6-cyl., highway tread)
F78-15, standard (V-8,
highway tread)
G78-15, optional (highway
or off-road tread)
H78-15, optional (highway or
off-road tread)
Special Tread Tires:
Firestone Town and Country
Sup-r-Belt
General Gripper 780
Goodrich Silvertown
Trail Maker
Goodyear Suburbanite Polyglas
1972 4x2/4x4: E78-15, standard (4- and 6-cyl.,
highway tread)
F78-15, standard (V-8, highway tread)
G78-15, optional (highway or
off-road tread)
H78-15, optional (highway or
off-road tread)
Special Tread Tires:
Firestone Town and
Country Sup-r-Belt
General Gripper 780
Goodrich Silvertown
Trail Maker
Goodyear Suburbanite Polyglas
1973 4x2/4x4: E78-15, standard (4- and 6-cyl.,
highway tread)
F78-15, standard (V-8,
highway tread)
G78-15, optional (highway or
off-road tread)
H78-15, optional (highway or
off-road tread)
Special Tread Tires:
Firestone Town and Country
Sup-r-Belt
General Gripper 780
Goodrich Silvertown
Trail Maker
Goodyear Suburbanite Polyglas
1974 4x2/4x4: F78-15, standard (highway tread)
G78-15, optional (highway or
off-road tread)
H78-15, optional (highway or
off-road tread)
HR78-15, optional
(radial highway tread)
Special Tread Tires:
Firestone Town and Country
Sup-r-Belt
General Gripper 780
Goodrich Silvertown
Trail Maker
Goodyear Suburbanite Polyglas
1975 4x2/4x4: H78-15, standard (highway tread)
HR78-15, optional (radial, highway,
or dual-purpose tread)
Special Tread Tires:
Firestone Town and Country
Sup-r-Belt
General Gripper 780
Goodrich Silvertown
Trail Maker
Goodyear Suburbanite Polyglas
1976 4x2/4x4: H78-15, standard (highway tread)
HR78-15, optional (radial, highway,
or dual-purpose tread)
Special Tread Tires:
Firestone Town and Country
Sup-r-Belt
General Gripper 780

		Goodrich Silvertown
		Trail Maker
1977	4x2/4x4:	H78-15, standard (highway tread)
		HR78-15, optional (radial, highway, or dual purpose tread)
		10-15, optional (off-road)
		7.00-15, optional (Load Range C, highway tread, or mud and snow)
		Special Tread Tires:
		Firestone Town and Country Sup-r-Belt
		General Gripper 780
		Goodrich Silvertown Trail Maker
		Goodyear Tracker A/T
1978	4x2/4x4:	H78-15, standard (highway tread)
		HR78-15, optional (radial, highway, or dual purpose tread)
		10-15, optional (off-road)
		7.00-15, optional (Load Range C, highway tread, or mud and snow)
		Special Tread Tires:
		Firestone Town and Country Sup-r-Belt
		Goodyear Tracker A/T
1979	4x2/4x4:	H78-15, standard (highway tread)
		H78-15, optional (mud and snow)
		P225/75R-15, optional (radial, Goodyear Tiempo)
		10-15, optional (off-road)
		7.00-15, optional (Load Range C, highway tread, or mud and snow)
		Special Tread Tires:
		Goodyear Tracker A/T
1980	4x2/4x4:	H78-15, standard (highway tread)
		H78-15, optional (mud and snow)
		P225/75R-15, optional (radial, Goodyear Tiempo)

215HR-15, optional (radial, Michelin XVS
Special Tread Tires:
Goodyear Tracker A/T

TRANSMISSION AND TRANSFER CASE

Standard 4x2: 3-speed manual, T-13 (Warner T-90) (1971–1973)

Standard 4x4: 3-speed manual, T-14 (Warner T-90) (1971–1973)

Standard 4x2/4x4 V-8: 3-speed manual, T-14 (Warner T-90) (1971–1973)

Standard 4x2/4x4: 2-speed manual, T-332 (Warner T-15D) (1974–1980)

Optional 4x2: 4-speed manual, T-44 (Warner T-18C) (1971–1975)

Optional 4x4: 4-speed manual, T-45 (Warner T-18C) (1971–1975)

Optional 4x4: 4-speed manual, T-427 (Warner T-19) (1976–1980)

Optional 4x2/4x4: 4-speed manual, T-428 (Warner T-19A) (1976–1980)

Optional 4x4 V-8: 3-speed automatic, T-39 (BorgWarner M-11) (1971)

Optional 4x4 V-8: 3-speed automatic, T-49 (BorgWarner M-11, upgraded) (1971)

Optional 4x4 V-8, 6-cyl.: 3-speed automatic, T-407 (Chrysler TF-727) (1972–1980)

Transmission Shift Type: floor (all transmissions)

Ratios 4x2/4x4 4- or 6-cyl., T-13/T-14: 1st - 3.39:1, 2nd - 1.85:1, 3rd - 1.00:1, R - 4.531:1

Ratios 4x2/4x4 V-8, T-14: 1st - 2.80:1, 2nd - 1.85:1, 3rd - 1.00:1, R - 4.53:1

Ratios 4x2/4x4, T-44/T-45: 1st - 4.02:1, 2nd -1.97:1, 3rd - 1.41:1, 4th - 1.00:1 R - 4.42:1

Ratios 4x2/4x4, T-332: 1st - 2.99:1, 2nd - 1.55:1, 3rd - 1.00:1, R - 2.99:1

Ratios 4x2/4x4, T-428: 1st - 4.02:1, 2nd -2.41:1, 3rd - 1.41:1, 4th - 1.00:1 R - 4.73:1

Ratios 4x4, T-427: 1st - 6.32:1, 2nd - 3.09:1, 3rd - 1.68:1, 4th - 1.00:1 R - 6.96:1

Ratios 4x2/4x4, T-39: 1st - 2.40:1, 2nd - 1.47:1, 3rd - 1.00:1, R - 2.00:1

Ratios 4x2-4x4, T-407: 1st - 2.45:1, 2nd - 1.45:1, 3rd - 1.00:1, R - 2.20:1

Transfer Case: 1-speed, chain, TC-143 (standard 1973–1979)

Transfer Case: 2-speed, TC-145 (Spicer 20; standard 1971–1972, optional 1973–1979)

Transfer Case: 2-speed, TC-146 (Spicer 300; standard 1980)

Transfer Case Ratios TC-143: 1.00:1

Transfer Case Ratios TC-145: Low - 2.03:1, High - 1.00:1

Transfer Case Ratios TC-146: Low - 2.62:1, High - 1.00:1

SCOUT II STANDARD COLORS

Color	Code	'71	'72	'73	'74	'75	'76	'77	'78	'79	'80
Frost Green Metallic	5865	X	X								
Aegean Blue Metallic	6758	X									
Alpine White	9120	X	X	X	X						
Sundance Yellow	4394	X	X								
Gold Metallic	4357	X	X								
Yuma Yellow	4385	X	X								
Flame Red	2289	X	X	X	X						
Cosmic Blue	6766		X								
Ceylon Green	5882			X	X						
Burnished Gold	4393			X	X						
Mayan Mist	4400			X	X						
Glacier Blue	6772			X	X	X	X	X			
Sunrise Yellow	4401			X							
Sunburst Yellow	4403				X	X				X	
Pewter Metallic	8539					X	X				
Grenoble Green	5888					X	X	X			
Dark Brown Metallic	1606					X	X	X	X	X	
Buckskin	4408					X	X	X			
Fire Orange	3195					X	X	X			
Terra Cotta	2297					X	X	X	X		
Winter White	9219					X	X	X	X	X	X
Solar Yellow	4410					X					
Elk	1608							X	X		
Siam Yellow	4414							X	X		
Woodbine Green	5902								X		

Color	Code	'71	'72	'73	'74	'75	'76	'77	'78	'79	'80
Concord Blue	6787								X		X
Embassy Grey	8547								X	X	X
Mint Green	5907									X	
Lexington Blue	6788									X	
Rallye Gold	4415								X	X	
Persimmon	3199									X	
Tahitian Red Metallic	2300									X	X
Saffron Yellow	4417										X
Copper	3201										X
Black Canyon Black	0001										X
Dark Brown	1032										X
Green Metallic	5901										X

FINAL ENGINE EVOLUTION

At the start of the Scout II era, International's engine line was showing its age a bit. The V-8s were going on 12 years old, which is an eternity in the auto world, but they were still meeting the ever higher standards for exhaust emissions and fuel economy. Emission controls added a sense of urgency to many upgrades. It became harder and harder for older designs to meet the latest emissions standards. This scenario spelled the end for many of Detroit's longest-lived and best-loved engines, which is one reason why the International engines were being phased out in Scout planning after 1980.

SLANT-FOUR: OLD RELIABLE

The International slant-four was well on its way to legendary status by 1971. According to figures from the Indy Engine Plant, some 235,705 of the engines (the 4-152, 4-152T, and 4-196) had been built up to that time (266,858 total from 1960 to 1984). When the Scout II debuted in 1971, the 4-196E was the base engine for the line. That lasted until the 1973 model year—it was eliminated for 1973 and 1974 in favor of the AMC-sourced inline sixes and the reasons why are interesting.

◀ When the 4-196 reemerged for the 1975 model year, it had many new features, not the least of which was standard electronic ignition. Others included the same cooling mods the V-392 got starting with July 1972 production. This involved internal water-flow changes; it's easy to tell an improved-cooling engine from a non-improved-cooling engine by the position of the bypass hose on the cylinder-head side of the engine. If it runs from the water pump to the cylinder head, the engine has non-improved cooling. If it goes from the water pump to the block, just below the head surface, it's an improved cooling engine. The new cooling system also included a closed system with a recovery bottle and a 17-inch fan. A dry-element air cleaner with a thermostatic device to aid warmups was made standard. The rubber engine mounts were improved to dampen some of the big four-cylinder's naturally bad harmonics.

According to Terry Hankins, a highly placed source at the Indy Engine Plant back in those days, the V-8s, for both the light- and medium-duty lines were in high demand. Eliminating the four-cylinder, done on the same tooling as the V-8, freed up IH to produce that many more V-8s. Some of the justification was that the fuel economy differences between the IH four and the AMC six were minimal, plus a six and two V-8s in the lineup offered a nice spread that matched the competition.

The reasons behind the 4-196's revival are equally interesting. International's financial problems, the

▲ The 4-196 was a critical part of the Scout II story, even though it was on vacation for a couple of years in 1973–1974. It fit into the engine compartment just like a V-8, so swapping between the four or the eight during production wasn't a big deal. A 4-196 is shown here in 1977. The number of four-cylinder Scout IIs was relatively small and grew smaller after the introduction of the 6-33 Nissan diesel. For the last year of Scout production, only 912 4-196 engines were produced.

end of the light-truck line that resulted, along with a surplus of engine capacity, ultimately dictated a return of the 4-196. It made more sense to keep the engine line at Indy running hard than to run it at less than full speed while making cash payments to AMC. The 4-196 returned to the line for 1975 with a few modifications and stayed with the Scout until its demise in 1980. It was altogether fitting that the engine which essentially made the Scout died with it. Only a handful of slant-fours were built after the Scout died, the last in 1984. They had been offered as stationary engines all along, but that business wasn't big enough to justify the expense of keeping them in production, and the V-8s were about gone at this time too.

The only serious evolution the 4-196 made in the Scout II era was related to its parent, the V-392. That

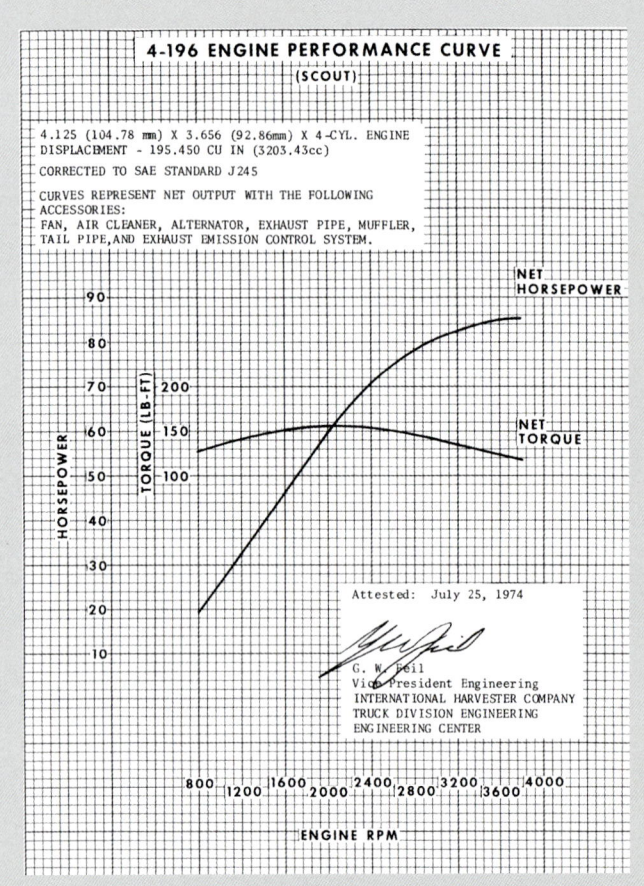

▲ The 4-196 net-power and torque graph for the 1975 model year. It was rated for 86 horsepower and 157 lb-ft, down from 102.5 horsepower and 176 lb-ft due to changes in tuning and mandated changes in test methodology (even though both were net-power tests). Gross power of the 4-196 was 111 horsepower. It was claimed this engine could bring a Scout II from 0 to 55 mph in 20 seconds with a close-ratio T-428 four-speed.

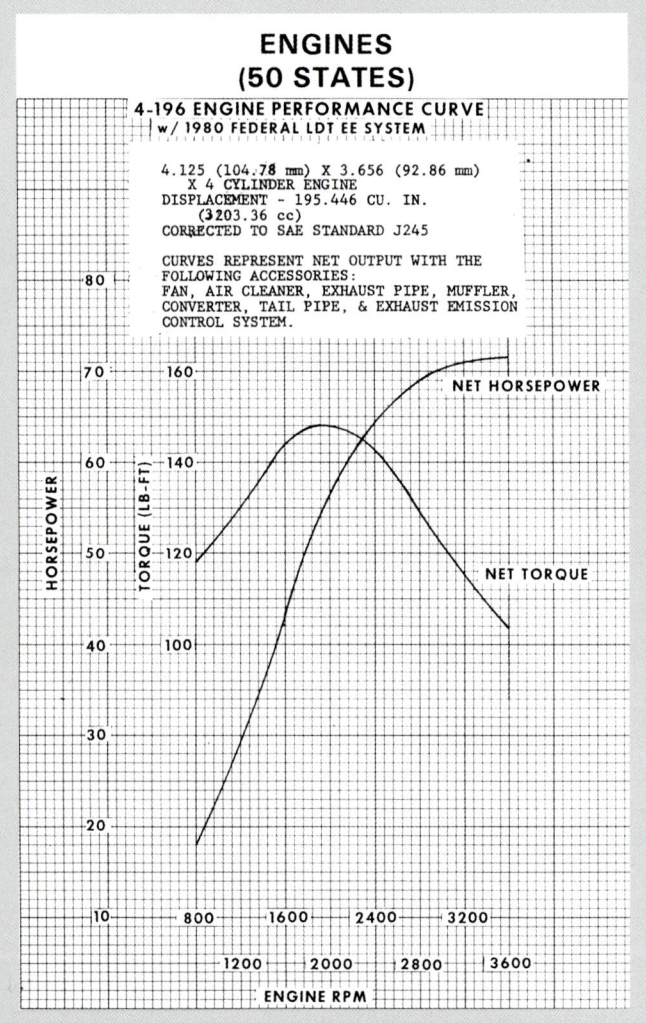

▲ Here is the net power and torque with a full complement of emissions controls, including catalytic converter, from 1980. Power was down to 76.5 horsepower. Peak torque didn't drop off all that much, but the curve changed radically from long and flat to more peaky with a quicker drop-off.

medium-duty V-8 engine got cooling-water flow changes in 1972 and the 4-196 followed right along when it went back into production. Emission controls could also be considered an "upgrade" and, increasingly, changes were made to the tuning and external devices so that power levels gradually dropped. The 4-196E began life in the Scout II era rated at 86 net horsepower, reentered the scene in 1975 at the same power level, and left in 1980 at 76.5 horsepower with a catalytic converter and exhaust gas recirculation (EGR) valve.

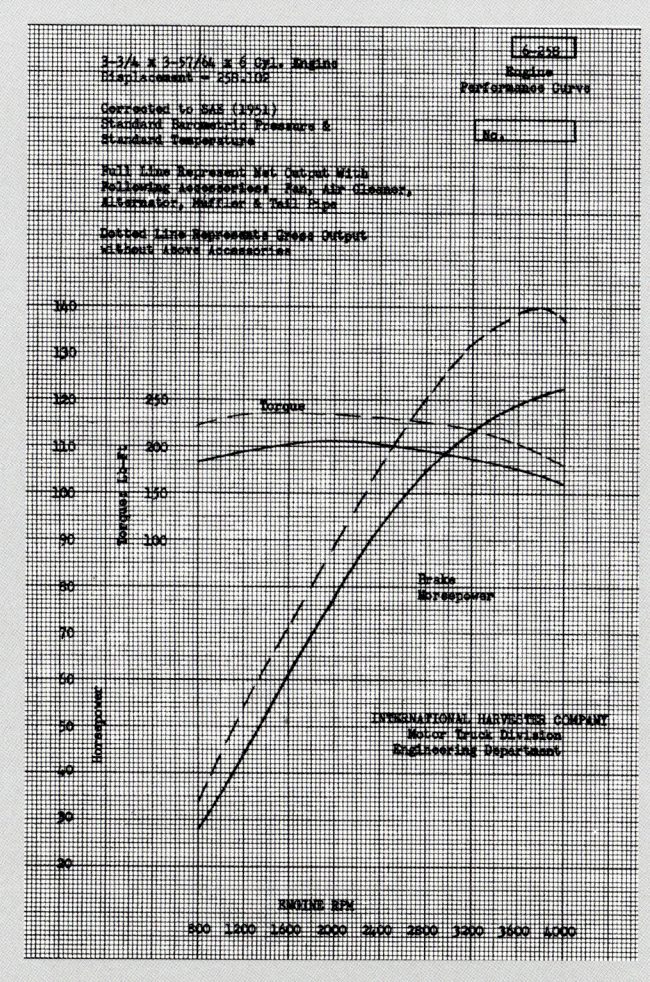

▲ The 258 had a different set of curves than the 232. Both gross and net power are shown for this 1972 chart. Net power was 113 horsepower and torque was 191 lb-ft. Generally speaking, the torque comes on stronger in a lower rpm range on the 258 due to the longer stroke.

AMC TO GO

After the brief 1971 Scout II model year, the AMC 258 (designated IH 6-258) replaced the 6-232 as the middle engine option. The 258 was a longer-stroke version (3.875 inches versus 3.5) moving torque a littler lower in the rpm range. Scout buyers didn't see any power or torque increases, despite the increased displacement on paper. This was due partly to more emissions tuning on the 258 compared to the 232, and partly to a change in SAE net power testing methodology from 1971 to 1972. Most owners reported a seat-o-the-pants difference, however.

The 258 left the Scout stage after 1974 and was missed by many. It wasn't an IH engine, but it was a good fit for the Scout. The biggest downer to IH purists was that it was also used by International's archenemy in the SUV market: Jeep.

THE V-8 LEGACY

The V-8 line remained very much the same through the Scout era. The V-304 evolved into the V-304A before the Scout II debuted. The 345 V-8 followed the same light-duty route for 1975. International contemplated adding the V-392 to the Scout line at the beginning, but the gas crunch eventually put the last nail in that coffin.

Like the fours and sixes, the V-8 endured the performance-robbing outrages of emissions controls. This started with tuning and air injection, transitioned to EGR, and culminated in 1979 with catalytic converters. Along the way, the V-345 acquired a four-barrel Carter ThermoQuad carburetor in most markets, either optional or as part of the California emissions package.

Speaking of the Golden State, as with other IH engines, there was often a California V-8 and a "49 states" engine. From 1976 through 1978, all V-8s in the line featured standard dual exhaust, which

▲ A publicity shot of a preproduction Scout II chassis shows how the V-345 began the Scout II era.

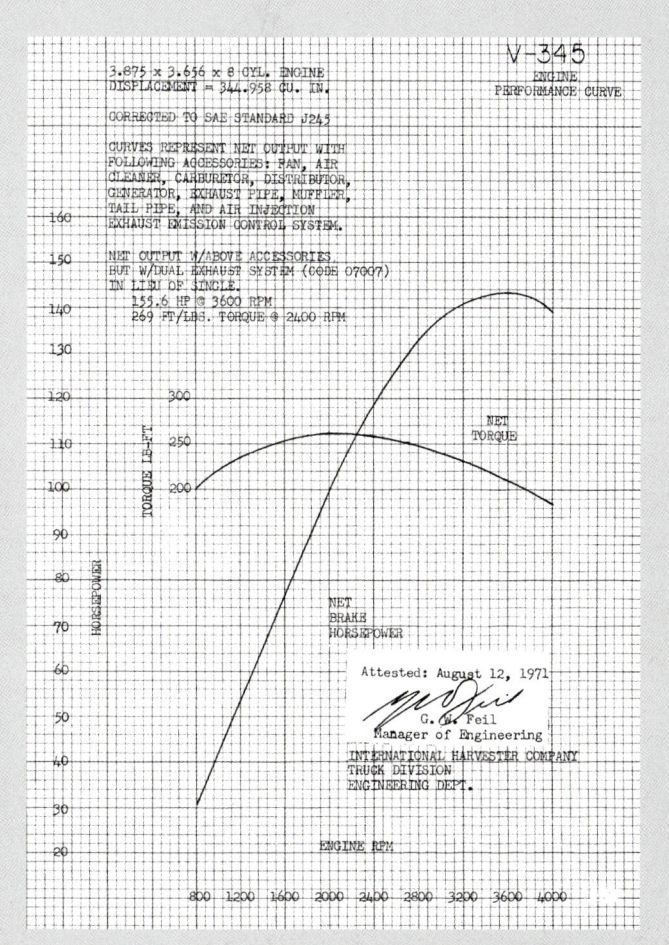

▲ The power and torque graph for a 1972 V-345 shows an engine with the emissions controls and a single-exhaust power rating of 143.5 horsepower and 262 lb-ft. Though not graphed, the dual-exhaust peak ratings are listed on the sheet as 155 horsepower and 269 lb-ft. Remember that in 1972, net-horsepower ratings under SAE ratings standards J-245 and J-1995 were required rather than the previous gross ratings or other methods of rating net power. Power output for the two-barrel engines remained roughly the same through 1978. After that, catalytic converters took a bigger bite of the output.

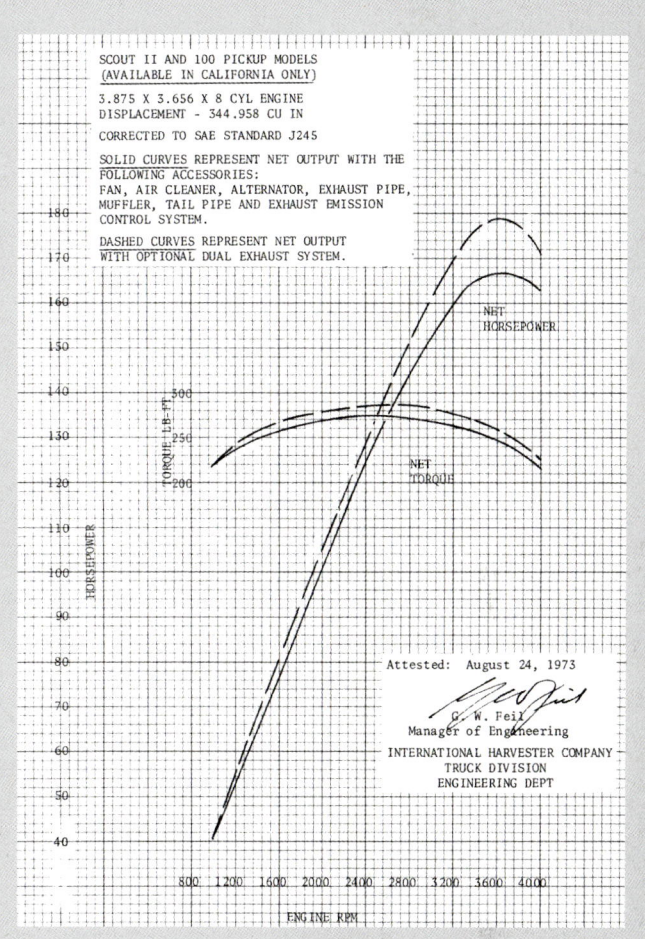

▲ This graph shows the differences for the four-barrel engine with and without dual exhaust. In 1974, the four-barrel engine was optional in "49 states" Scouts, but mandatory in California, where it was the only IH engine certified. With 179 horsepower and 292 lb-ft, the four-barrel, dual-exhaust 345 was the most powerful factory engine used in the Scout II. The California emissions thing was a thorn in everyone's side. Many engines met the standard, but California required individual certification in their labs for each engine offered, requiring several hundred thousand dollars. This tended to reduce the line of engines available in that state.

helped counter the power losses from emissions tuning. Previously, optional duals had been available from 1973 on, but they were discontinued with the advent of the catalysts.

The International gas V-8s didn't last much longer than the Scout. The 345 and 392 hung on a little past International Harvester itself. The last V-345s and V-392s were used in S-Series medium-duty trucks for 1986 after the company had transitioned to the Navistar corporate identity, with the last of the small V-8 engines built in 1988, according to engine plant records. By then, Navistar had gone all diesel and that was that. A nearly 30-year run had ended.

THE BLACK SMOKE OPTION

For the 1976 model year, IH put a small diesel on the Scout options list where it remained until the end. International began seriously evaluating diesel

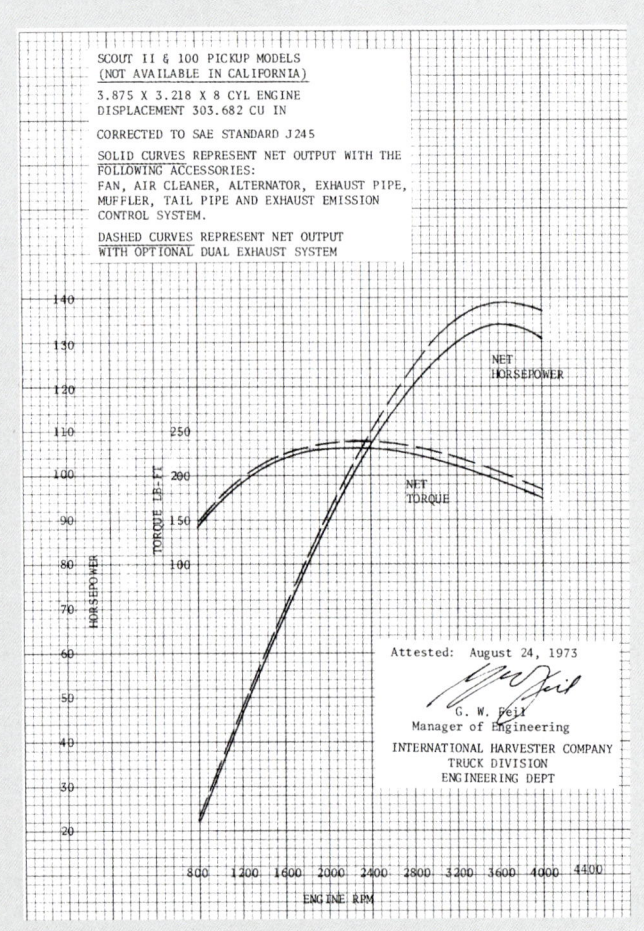

▲ The V-304A was rated at 140 net horsepower using the 1972 J-245 standards. Like the V-345, it remained more or less at this rating until the cats appeared in 1979. The chart shows the difference between the standard two-barrel single exhaust and optional two-barrel dual-exhaust engines. The V-345 two-barrel didn't make much more power than the V-304, but it did generate substantially more torque.

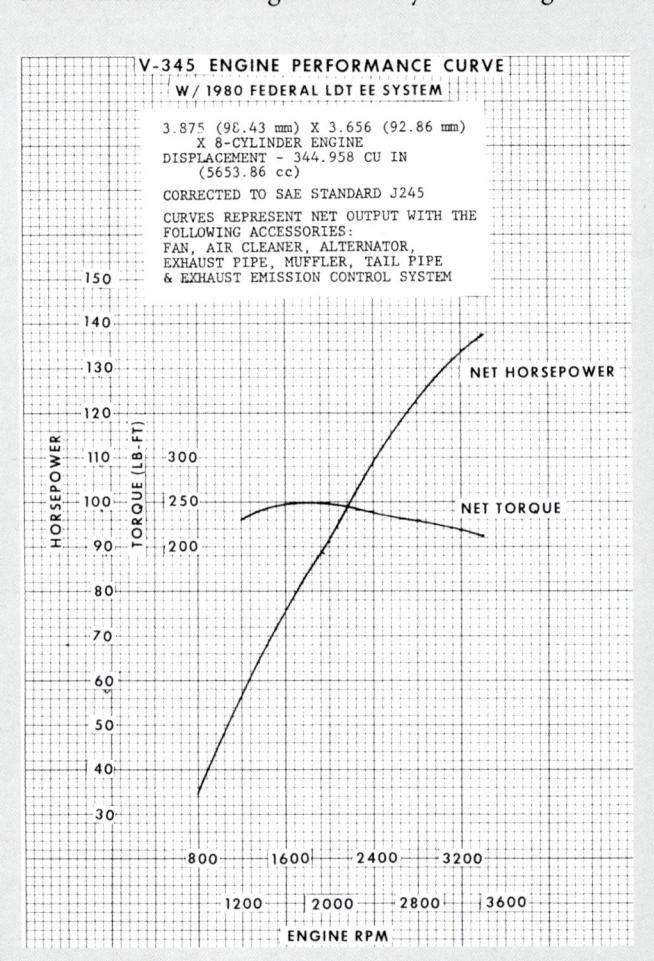

▲ The V-345 ended its run in Scout rated at 150 horsepower and 263 lb-ft—not too shabby considering it had a catalytic converter and EGR.

▲ The V-8 engines were relatively accessible in the Scout II with the forward-tilting hood. It can be tough to tell a V-304 from a V-345 visually. The V-304 block is a little lower due to the shorter stroke, but it isn't clearly noticeable sitting in the Scout. A factory four-barrel carb always indicates a V-345. A four-barrel was never offered for the V-304.

▲ Yeah, somebody forgot to turn off the rust on this '73 Scout but it's an important element of Scout history. As far as we can tell, and there is a significant amount of evidence to support the theory, this is the first Nissan diesel test mule. The Scout was ordered by Engineering with the intention of making it a diesel test mule. It was built in June of 1973 as a bare-bones Cabtop with a 258 six. It was quickly sent to the Melrose Park prototype lab for the installation of a Chrysler-Nissan CN6-33 diesel. That engine was built on May 15, 1973, so was still fresh when installed. In the engine shot, take note that it's wearing the air filter from the 258 gas six that originally powered it. This Scout is a time capsule of period documentation and still has some of the temporary adaptations and test instruments. It still wears Engineering Department decals and other test mule accoutrements. Along the way, a Traveltop was added as well as an updated interior. Updates like these were common for test mules for the long hours drivers put into running them on the test track.

◀ Though the power increase with the turbocharged engine was modest, the cooling and lubrication systems were significantly changed. Differences with the turbo engine included an AiResearch TO-3 turbo, oil-cooled pistons with a drop in the compression ratio, a stronger crankshaft with improved bearings, a higher-capacity oil pump, a larger oil cooler, and the necessary tuning to the Bosch Type A injection pump and Bosch injectors. These engines generally produced 5- to 7- psi boost.

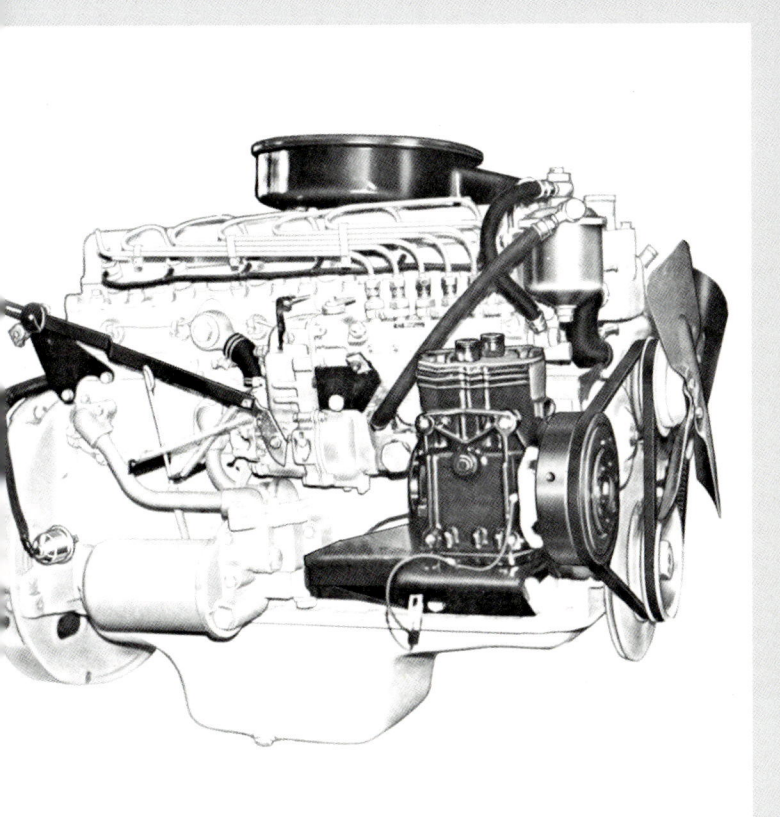

engines for the Scout as early as 1971. Initially, the research was intended to provide an export-friendly Scout for overseas markets, many of which preferred diesels. The gas crunch of 1973 put the domestic market on the table because a diesel was the fuel-economy choice.

Reportedly, engineers test-fitted at least four six-cylinder diesels in the Scout: the Peugeot XDP 6.90, Perkins 6.247, IH D301, and Chrysler-Nissan CN6-33. The Peugeot was rejected for packaging problems. A tight fit, high weight (almost 1,000 pounds), and poor fuel economy also nixed IH's own Neuss-built D301. The Perkins 105-horsepower fit well, but IH rejected it for perceived supply issues and relatively poor fuel economy. That left the Chrysler-Nissan CN6-33, several more of which were installed for more extensive testing that also proved successful. Some of the diesel test rigs survive intact, including the first CN6-33 powered Scout II and the Perkins powered one. The remains of the Peugeot-powered Scout have been found as well.

The Nissan SD series diesel engine family had debuted in 1964 with a 2.0-liter four. Larger

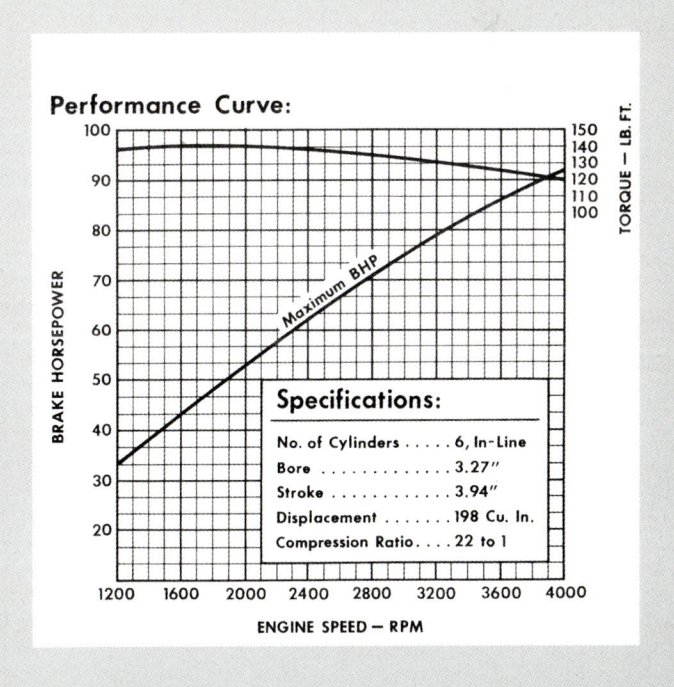

▲ The naturally aspirated 198-ci (3.2-liter) Nissan diesel gave International a unique position in the market. A couple of small diesels were offered in previous American-market light 4x4s, but in 1976 the Scout was ahead of the pack. The original intent was for an export option and, as a result, the engine was a bit lower in power and torque than if it had been chosen for the American market. Overseas markets of that era tolerated low power in the name of fuel economy, but not so much the North American market. Still, International was more successful with the Scout diesel than Jeep was with the Perkins-powered CJs
in the 1960s.

▶ The naturally aspirated 6-33 diesel was similar in torque and power output to the four-cylinder International gassers. The Nissan had a longer and broader torque curve and at least a 30 percent advantage in fuel economy (on fuel that was 20 percent cheaper).

SCOUT II, TRAVELER, AND TERRA: ANYTHING LESS IS JUST A CAR   197

▲ The Nissan inline diesel was a similar fit as the previous year AMC inline gassers. By the time this engine was installed in 1980, engine color had gone from the first Chrysler-Nissan Yellow to Nissan Blue and the valve cover was marked "Nissan Diesel" along with its Nissan SD-33 model designation.

displacement fours appeared later, as well as the 3.3-liter inline six. All of these engines saw extensive use in commercial, stationary, and marine applications around the world. In 1969, Nissan and Chrysler created a joint venture to distribute the versatile SD series in North America.

The naturally aspirated CN6-33 ("CN" for Chrysler-Nissan) was a four–main bearing IDI (indirect injection) diesel with dry sleeves. It produced a modest 81 net horsepower and 138 net lb-ft. Fuel was pushed by a license-built Bosch A Series pump and injectors. In a typical Scout, that 81 horsepower produced 60 miles per hour from zero in a very leisurely 25 seconds, a little slower than the gas four, but the diesels could easily break the 20 highway miles per gallon barrier. Magazine testers generally obtained 18-plus miles per gallon in mixed city/highway driving.

The early NA CN6-33/SD-33, which soon became known simply as the 6-33, came in yellow, some with "Chrysler-Nissan" on the valve cover. Later versions, made after the Chrysler-Nissan deal expired, had only "Nissan." The blue engines came in 1978, but as old stocks were used up, yellow NA engines were occasionally installed as late as 1979.

A turbocharged version debuted for 1980: the SD-33T. Power went up from 81 to 101 horsepower and torque jumped from 138 lb-ft to 175 lb-ft. The new power and torque outputs were a considerable improvement over the emissions-choked gas four of the period. Not only did the diesel provide more power, it got 20-plus miles per gallon. Most testers reported a slight uptick in fuel economy with the turbodiesel and a three- to four-second drop in 0–60 times versus the NA engine. Generally speaking, the new TD was faster than the gas four.

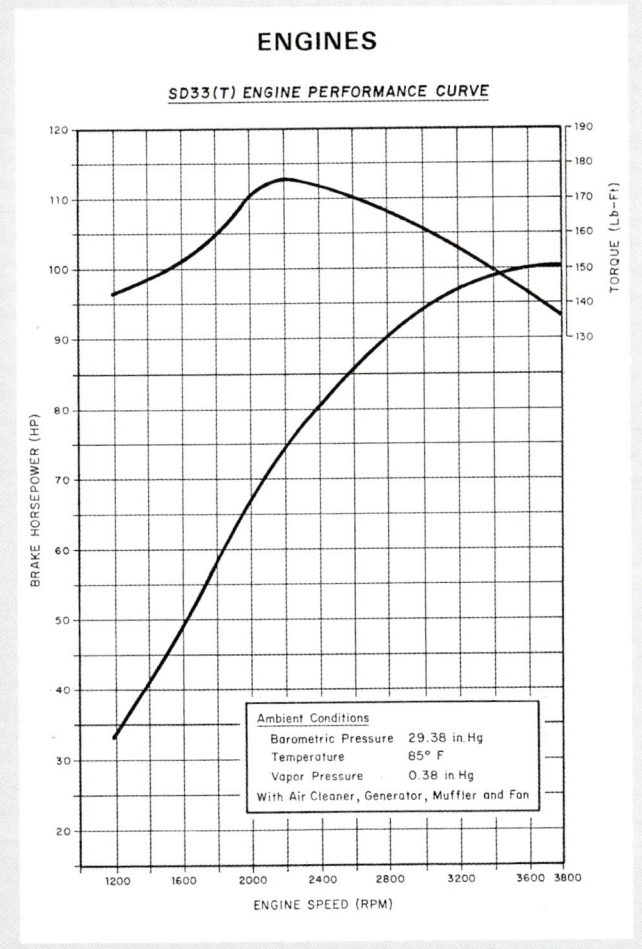

▲ The power and torque graph for the 6-33T is significantly different than the naturally aspirated version. It's a little more "peaky" on the torque side but definitely produces higher power and torque peaks.

THE LAST SCOUT

▶ Mike Bolton made a small-scale reproduction of the same sign, which he uses when displaying the historic Scout today. *Mike Bolton Collection*

▼ The Last Scout as it rolled off the line on October 21, 1980, reportedly near midday. One wonders the thoughts behind the faces. Some are smiling, some not.

Jim Hirschberg (jimhirschberg.com)

▲ The unrestored Last Scout in 2015 on what was once its home ground. It spent a lifetime on what is now the Whiterock Conservancy, once the Garst family estate near the Middle Raccoon River before the property was donated to the state. The Warn 8274 8,000-pound winch on a Warn Classic bumper and the fog lights were installed by the TSPC in Melrose Park, Illinois, as was the custom pinstriping. The Last Scout now shows 59,646 miles. *Jim Hirschberg (jimhirschberg.com)*

▲ The restored Last Scout in the spring of 2020, just days after emerging from a painstaking four and a half year restoration. You will notice some differences from the unrestored image nearby, the grille for one. The original silver grille had been broken and replaced with a black one, so it went back to the original silver. All the correct trim was added to replace what was lost or damaged. While it wasn't particularly rusty as Scouts go, as a farm and hunting vehicle, it had been bumped and banged around a lot and there was extensive body work required. Some significant, but poorly repaired, frame damage was also discovered and made right. *Jim Hirschberg (jimhirschberg.com)*

▲ Mike Bolton agonized over whether to restore the Last Scout or not but in the end, he decided it would survive longer if it was restored, maintaining or reproducing all the original TSPC additions. Because his had been a decades long quest to acquire this Scout and research all it's aspects, Bolton felt that he was the best person to oversee a correct restoration and make whatever compromise decisions had to be made. Few that know Mike Bolton would disagree. The restoration started in the fall of 2015 and was finished just a week before the second edition of this book was finished in the Spring of 2020. Joe Stiltz and Wyatt Deist of Resurrection Rides of Coon Rapids, Iowa, where the Last Scout lived it's whole life, did the masterful frame off restoration and you can look forward to seeing this Scout in all it's glory at Scout and IH shows. *Jim Hirschberg (jimhirschberg.com)*

▶ Reproducing the custom pinstriping done at the Melrose Park TSPC in 1980 was tricky. Forrest "Frosty" Harrington, well known in Iowa for such things, was up to the task. The original pinstriper is unknown but likely worked in and around the Chicago area. The style looks very similar to a 1978 show Scout that was prepped in facilities around Chicago. Some of the line worker's signatures on various parts of the vehicle had almost faded away, so Frosty also reproduced them as only a fine artist can. *Jim Hirschberg (jimhirschberg.com)*

▲ Garst's Scout was ordered with the Deluxe interior with a pillowtop bench seat in Sierra Tan. Also ordered was the newly improved air conditioning system. Before it was delivered, the Melrose Park TSPC added a digital clock, an item that could be found in a number of IH products including big trucks and the Midas Scouts. They also added a Craig AM/FM stereo system. For the restoration, Bolton made some changes. The first includes carpeting, which didn't come with the Deluxe interior. The repro floor mats don't fit any better than the originals and carpets just looked better. The Craig stereo was dead and couldn't be replaced, so long ago Bolton decided to replace it with a radio delete plate that the late Ted Ornas had signed in 2007. By the time the resto was finished, Ted's signature had faded, so he is still contemplating exactly what to do about that. For the moment, a facsimile is in place. *Jim Hirschberg (jimhirschberg.com)*

▲ This poignant Garst family picture was taken on Steve Garst's last hunting get-together with close friends in 2002 before his recently diagnosed Alzheimer's had fully taken hold. Steve is in the tan jacket and standing next to the Scout. Other notables include U.S. Senator Tom Harkin (back row, third from Garst), who served five terms from Iowa. This is the last known image of Steve Garst and his beloved Scout together. *Liz Garst*

▲ Some sad line worker wrote "10-20-80 The End" on the bellhousing of the Last Scout when the chassis was assembled on October 20, 1980. It appears this Scout was pulled off the line at some point that day to be reinserted on October 21 and thus become the Last Scout. *Jim Hirschberg (jimhirschberg.com)*

SCOUT II, TRAVELER, AND TERRA: ANYTHING LESS IS JUST A CAR

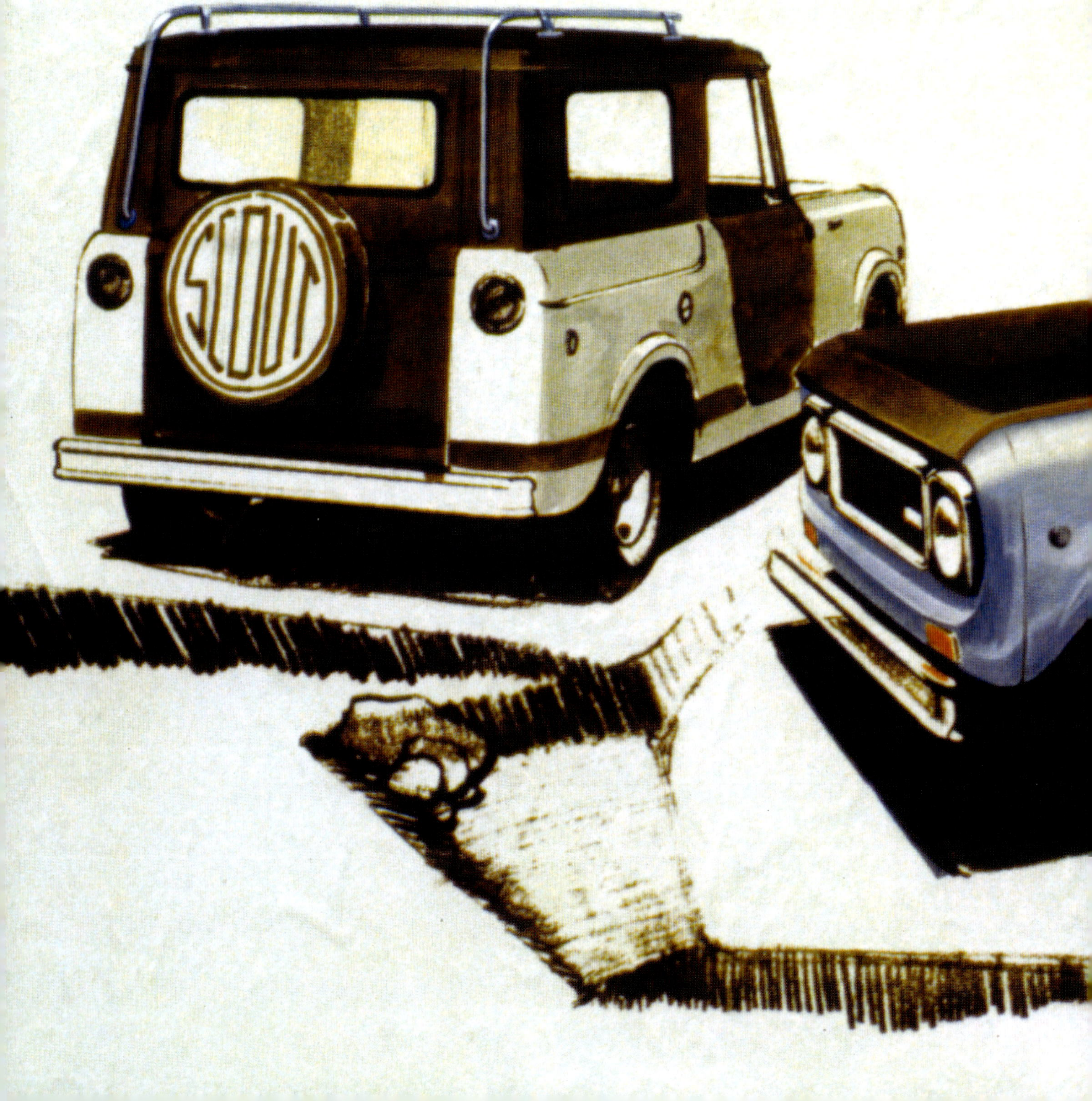
SCOUT

Chapter 5
SCOUT SPECIALS ENCYCLOPEDIA

▲ One of the earliest renderings by Kikuyo Hayashi in the development of what would become the Aristocrat. This rendering shows an early Trailblazer idea marked as the "Rawhide."

SCOUT CAMPER

Model Year: 1963
Production Dates: ~November 1962–June 1963
Introduction: October 25, 1962
Numbers Produced: ~88–100*
Known Serial Number Range:
    Within FC60686–FC61376†

Identification: Visually and via LST (see below)

*Estimated. See text.

†This was the range of the first 60 built, which is to say most of those built. More were done after, but a range is not available.

The Scout Camper was an innovative attempt to turn the Scout 80 chassis into a mini-motorhome. The Camper was first discussed early in 1962 and approved in November of that year after a sample was viewed by Product Planning. The fiberglass parts were designed, built, and installed by well-known fiberglass boat pioneer Winner Boats of Dickson, Tennessee (a.k.a. Dickson Boats) via a spun-off division called Fibertron. Letters to district managers indicated Camper construction was to be transferred to a new facility in Georgia, but by January 1963, that had not happened and it isn't clear if it ever did before the Campers were discontinued.

▲ At the time this book was being written, Chris Brooks held the honor of being the only person to have restored a Scout 80 Camper. It's a '63 Deluxe that was ordered in Harvester Red (201). A little LST research revealed that his rolled off the line on November 29, 1962, and was the only red one—every other one was white. His was also one of only two in that group that didn't have a dealer shipping address; given the timing, it seems likely this rig went to an RV show in December 1962, something that's hinted at in sales documents. *Chris Brooks*

The basis of the Scout Camper was a 4x4 Scout with a walk-through bulkhead, bucket seats, roll-down windows, front and rear bumpers, heavy-duty RA-9 rear axle with heavy-duty springs/shocks, and larger rear brakes. The rear-body end panel, to which the tailgate and rear lights were mounted, was left off with the base price of the Scout adjusted accordingly. The 4x4 Scouts were to be shipped to Dickson from Fort Wayne once the order was made, converted to Campers there, and then transshipped to the ordering dealer.

Three models were offered, each with à la carte options. The Standard Camper added $649 to the price of the Scout (as of January 16, 1963) and the Deluxe added $1,220. Documents indicate a Commercial version was also to be offered as a high-capacity delivery body, which was basically just the shell. It isn't clear how many of these were built, if any.

The fiberglass shell was permanently attached and extended the tail by three feet. The new rear panel contained the taillights. Initially, there were some issues with the legality of the lighting, so IH

▲ The Deluxe Camper had windows in the side panels, though they could be ordered à la carte on a Standard. The optional hitch setup is shown here, but with all the extra weight and four-cylinder power, too much trailer was not advisable. When photographed in 2015, this Deluxe Camper was owned by Scout Connection of Iowa and was is very good original condition. It's in the standard Whitecap White (the case for all but one Camper). The serial number indicates it was the 11th Camper built, roll-tested on December 12, 1962. Originally destined for Memphis, it ended up in Washington State.

▲ With the side panels down, the bunks and tents were deployed for sleeping. Optional privacy curtains closed off the view through the front windows. *Chris Brooks*

and Fibertron redesigned the arrangement before sales began. The large rear door contained a spare tire mount and was embossed with "International" on the outside.

Both the Standard and Deluxe Campers utilized the same shell and both had foldout side panels that became bed platforms, each with a frame and tent. The Standard Camper had no windows in the foldout panels and the interior was open and ready for gear storage or owner modifications. Padded side cushions for the wheelhouses were optional for the Standard Camper ($14), as was a folding dinette table ($23).

The Deluxe Camper included windows in the foldout panels and contained a modular fiberglass kitchen unit that included a fiberglass sink, electric stove, three-burner propane stove, foldaway toilet and holding tank, full-pressure water system and tank, dining area, and storage box.

Optional for both were a supplementary 12-volt electrical system and second battery ($96.50), 15,000 BTU heater ($376), a hammock-like third bunk for between the foldouts ($22.10), shower attachment ($25), rear tent attachment ($165), rear awning ($49.50), luggage rack ($36.50), privacy curtains

▲ Coffee anyone? Note the location of the spare tire in the door. This proved to be a problem, as the door was not up to the task of carrying the weight of the wheel and tire. WHS 11088b

($22.50), water purification system ($63), and insect screens for the front doors ($9.50). A trailer hitch was available for light trailers ($12).

The Camper option was mistakenly announced as being available for both 4x2 and 4x4 Scouts, but the four-lug hubs, light-duty Dana 27 (RA-4) rear axle, light springs, and small brakes under the 4x2 were not up the task of carrying the extra weight. By the time the mistake was realized early in 1963, only two 4x2 campers had been ordered and both were changed to 4x4 chassis and the 4x2 option removed from any new literature.

The first retail deliveries occurred late in January 1963, but orders were fewer than International had hoped for by a long shot. Once deliveries occurred, cracking issues with the fiberglass were noted as well as a large number of fit and finish problems. Fibertron developed fixes and an updated unit was produced

▲ Sure, this scene is a tired old stereotype, but it shows the layout of the Camper. The sink and icebox were to the woman's left and the table for two is to the left of the picture. The stove was a Dometic three-burner propane unit with the five-gallon tank mounted under the Scout—and rather shoddily, says Chris Brooks. WHS 111181

▲ How convenient! You could cook and use the toilet at the same time! Even more convenient, the icebox was opposite the stool! This is the Scout Connection's Deluxe Camper on the inside and it's very well preserved.

for testing, but by then the point was moot. In May 1963, International pulled the plug on the Camper. The few remaining orders were fulfilled and that was that. IH offered to allow Fibertron to continue producing campers on their own, with the restriction that "International" be removed from the rear door. That offer, apparently, was not accepted.

Records of exactly how many Scout Campers were built are unavailable. Documentation from March 1963 indicates there had been 88 orders to that time. The May 31, 1963, cancelation mentioned "in the neighborhood of 80 units" had been built with a few orders still left to fill. Based on this information, it's unlikely that more than 100 Campers were built. Sixty can be found in the LST files well into May production, but they do not comprise the entire count.

Recently, five extant Campers were discovered in various conditions: four Deluxe and one Standard. No doubt there are others (or pieces of others) out there, but they surely are very few. It's likely some were repurposed into regular Scouts after the camper parts failed.

Camper LSTs reveal some interesting things, including "omit RR tailgate & SP TR MT SE 816025," which likely refers to omitting the tailgate and entire rear sheet metal. Also look for "omit cab 16704" and either the RA-9 or RA-23 heavy-duty axle.

RED CARPET SERIES

Model Year: 1964
Production Dates: July 22, 1964–November 2, 1964
Introduction: November 1964
Number Produced: 3,450
Known Serial Number Range: FC101111–FC106798
Identification: Model code 781805 from first line of LST plus special features

The Red Carpet Series (RCS) was a preemptive strike to counter new models from Ford and Jeep expected to appear late in 1965. The Scout 80 was being replaced by the 800 for 1966, but International wanted to increase showroom traffic and media buzz late in 1965 to clear the 80 models, and production of the 100,000th Scout seemed a good enough reason. The Red Carpet marked the first of many dressed up Scouts that would become known as "Doll-Ups" internally.

The project was conditionally approved in April 1964 after the MTC viewed a dolled-up Scout built by a combined team from Sales and Styling. That first model was more like the Champagne Series, but during the meeting, Sales formally requested another Doll-Up in white and red to celebrate the 100,000th Scout, which tentatively would roll off the line July 13, 1964. The MTC approved both ideas subject to further planning. Final approval was given on June 1, 1964, even though the project was already well on its way. Red Carpet production was to start in August 1964, with 64 to be built per day from August into October for an expected run of 3,000 units. Production was announced to start on August 3 but actually began on July 22, 1964.

The earliest RCS serial number found in LST records was FC101111, the LST marked "Pilot Model," and the roll-test date was July 24. The earliest build date found is FC101151, with a July 22 roll-test date. The last RCS found by serial number (FC106798) was also dated the last day records show an RCS was built: November 2, 1964. While an actual count of the number built was not made for the purposes of this book, records show Red Carpet Series rolling down the line in batches,

100,000 RED CARPET SERIES Scout SPECIFICATIONS and PRICES

CODE	DESCRIPTION	LIST	INVOICE NET	SURCHARGE
7818052	Scout 80 (4x4) #902 Solid White	$2,021.00	$1,637.01	$163.70
01577	Outside Mounted Tire Carrier	11.00	8.80	.88
01611	Rear Bumper—Chrome	20.25	16.20	1.62
16641	Heater & Defroster	60.00	48.00	4.80
16676	Roll-Down Windows	35.00	28.00	2.80
16709	Bucket Seats *less* Bulkhead	66.00	52.80	5.28
16734	Full Length Travel-Top	129.00	103.20	10.32
16823	Full Width Rear Seat	49.00	39.20	3.92
185259	7.60 x 15 4-PR W/S/W Front Tires	24.50	19.60	1.96
225259	7.60 x 15 4-PR W/S/W Rear Tires	36.75	29.40	2.94
614559	7.60 x 15 4-PR W/S/W Nylon Mud & Snow Tires	N/C	—	N/C
01600	Front Bumper—Chrome	10.50	8.40	.84
10534	Five Chrome Wheel Discs	20.50	16.40	1.64
16665	Interior Custom Trim	79.00	63.20	6.32
16666	Interior Custom Trim	19.00	15.20	1.52
	Total Chassis and Attachments	$2,581.50	$2,085.41	$208.54
	Surcharge	208.54	208.54	
	*F.O.B. Factory—Fort Wayne, Ind.	$2,790.04	$2,293.95	

*Plus Freight, Handling and Local Taxes

CODE		LIST PRICE
16665	**Interior Custom Trim** (A): (Travel-Top with Front Bucket Seats and No Rear Seat)	$79.00
	1. New headliner *less* vibradamp	
	2. Sun Visor—left side	
	3. Door trim panels with new trimmed arm rests	
	4. Front bucket seats with new trim	
	5. Hard fiberglas instrument panel trim pad	
	6. New carpeting in front compartment	
	7. Chrome mirror—left side—mounted on outside of door	
	8. Sound deadening tape and spray on sound deadener	
	9. Rubber sound deadener on floor of front compartment	
	10. Special paint on instrument panel, steering wheel, steering column, horn button and steering column seal.	
	11. Chrome plated control knobs	
16666	**Interior Custom Trim** (B): (Rear compartment with full width rear seat)	19.00
	1. Rear seat with new trim	
	2. Rear arm rests with new trim	
	3. Rear compartment carpeting with two chrome trim strips	
	4. Rubber and Latex floor sound deadener in rear compartment	
16667	**Interior Custom Trim** (C): (Cab top, or no top, with bucket front seats and No Rear Seat)	68.00
	1. Same as Code 16665, except omit headliner.	

June 1, 1964

▲ Plain white wrapper. The Red Carpet was elegant in its monotone simplicity. Gene Robinson's restored Red Carpet highlights its special features, which include the wheel covers, exclusive to the Scout, and used for many years. Lots of chrome and a "Custom" emblem on the doors, plus the "100,000" decal (missing on Robinson's), set it apart. It was the first Scout Doll-Up but not the last. Originally, the Red Carpet mounted "big" 7.60-15 nylon mud and snow tires. During the 1965 model year, the Tire and Rim Association changed the way tires were sized and the same tire became an 8.45-15. The tire size may appear listed both ways in 1965, but they are the same tire with different markings.

sometimes with sequential numbers, sometimes not (but close), and sometimes with sequential Line Sequence Numbers. There are also sometimes large gaps until another batch comes through. The production of the Champagne Series Scouts immediately followed the RCS on November 16, 1964, with a starting serial number of FC109102.

A Red Carpet brochure dated June 1, 1964, was sent to dealers in special red velvet packaging as an inducement to early orders. In theory, each dealer was to get at least one Red Carpet Scout as a showroom model. The brochure listed the special model code, option codes, and prices for the items in the package.

▲ There wasn't much difference from a standard Scout from the rear, still very understated, but that was just the styling target International was trying to hit. List price for the Red Carpet was $2,790.04.

▶ Most of Red Carpet's whiz-bang was on the inside. Red interiors were a popular '60s phenomenon and International pulled out all the stops by painting the dash red and using red levers, knobs, and steering wheel contrasted by chrome accents. International went further than they had yet gone to make the Scout a civilized beast. Extra sound-deadening, full headliner, carpet, and upholstered door panels made it the quietest Scout to date.

▲ The full-tilt passenger bucket seat allowed rear passenger access. International was an early proponent of seatbelts, and unlike some manufacturers that kicked and screamed against the federal mandate for their fitment, IH offered them long before the mandate.

The Red Carpets were distinctive in their Whitecap White (902) monotone exterior and red interior, which included a red dash pad, red dash, red steering column, red carpeting, and a red and white headliner. A "Custom" badge affixed to the door was similar to those used for high-end Travelalls of the period. A special "100,000" decal was placed on the door below the "Custom" emblem. It's also sometimes seen on the dash, as well as a "Made Especially For" decal with the owners name.

Both the red dash pad and two-tone headliner are subject to deterioration (especially the headliner), so may not be present in surviving RCS Scouts. Neither are reproduced today and should be considered "unobtainium." The headliner can be reproduced by a good upholsterer if the trim pieces are still there, but

▶ An original Red Carpet headliner is a rare find. Noted Scout collector and historian Mike Schmudlach's unrestored Red Carpet has survived with only a few screws added, like as not within a few years of when it was built. The typical problem was that the headliner sagged very soon after it was installed. The cure was to add some support screws before it sagged to the point where it was damaged.

the dash pad is solid gold. The Champagne and the Red Carpet used the same dash pad, albeit in different colors, so they can be swapped back and forth and painted as necessary. The special wheel covers were created for these two Scouts, but they were an option on the 800 lines all the way into the Scout II era and are relatively easy to find.

The Red Carpet models may not be easy to identify once some or all of the special features deteriorate or are removed by previous owners. In addition, some special RCS parts may be found swapped into non-RCS Scouts. The serial number tag is not descriptive at all because it does not contain the model prefix code. The answer lies in the LST. The Scout 80 RCS has the special model code 781805 (see the Scout 80 Data in Chapter 2) on the first line and the LST stamped "DOLL-UP-SCOUT" at the top. In addition, the accessory codes list the top-line interior options. The package option codes were the same as the Champagne Series and so were the basic pieces, albeit in different colors.

It's long been assumed that only 3,000 RCS units were built, but according to MTC documents dated September 29, 1964, the number was actually 3,450.

▶ International capitalized on the 100,000th Scout with some high-quality decals. The Custom badge was taken from the high-end Travelall. The Scout badge was silver, however, while the Travelall's was gold. *Betsy Blume*

SCOUT SPECIALS ENCYCLOPEDIA 215

CHAMPAGNE SERIES

Model Year: 1965
Production Dates: ~November 16, 1964– June 14, 1965
Introduction: November 1964
Numbers Produced: ~3,000
Known Serial Number Range: FC109102–FC122755
Identification: Via LST, "Doll-Up" notation and features

The Champagne Series (CS) Scouts were the second Doll-Up used to bolster sales in the months leading up to the Scout 800. The CS Scout 80 followed the Red Carpet pattern in most ways but had a variety of available exterior colors. The concept was approved in April 1964 in an MTC meeting. A customized workup prior to that was approved by the MTC. Final approval was given on June 1, 1964.

The first concept vehicle was essentially a CS, but the MTC wanted to do the Red Carpet first. Many of the special parts were the same or similar, notably the chrome bumpers and wheel covers. The treatment was similar, but the colors were not. The CS used what IH described as a "soft pastel champagne" color upholstery and offered a choice of exterior colors, including the signature Champagne Mist (1501). The majority of CS Scouts were Champagne Mist, but LSTs can be found indicating most available colors.

An LST search yielded the first CS with a serial number of FC109102 and a build date of November 16, 1964. As with the RCS, they are found in batches, sometimes sequentially serial-numbered, but often in groups on the same day with close or sequential LSNs. The last CS serial number a search yielded was FC122755 with a roll-test date of June 14, 1965. That also happens to be the date the very first Scout 800 pilot models rolled down the line.

Among the distinct features are the foam dash pad and headliner. Any dickering on a purchase price or collector value should reflect the presence and condition of these parts. As with the RCS, a headliner can be fabricated by a good upholsterer if the trim pieces are still there, but not the dash pad. The CS and RCS used the same dash pad, albeit in different colors, so they can be swapped back and forth and painted as necessary. Same goes for the wheel covers, which were also used on Custom models through the 800, 800A, and 800B eras and even into the early Scout II era.

The Scout 80 CS models may not be easy to identify once some or all of the special features deteriorate. Refer to the LST, some of which have "Cham" or "Champagne" handwritten on the ship-to box. Unlike the RCS Scouts, the Scout 80 CS doesn't have a special model code, but the LST will be stamped "DOLL-UP-SCOUT" at the top. In addition, the option codes are top-line interior options. The important ones are 16665 Interior Custom Trim A ("Doll-Up Traveltop 16665" on

CHAMPAGNE SERIES SCOUT SPECIFICATIONS

CODE		LIST PRICE
16665	**Interior Custom Trim (A):** (Travel-Top with Front Bucket Seats and No Rear Seat)	$84.00
	1. New headliner *less* vibradamp	
	2. Sun Visor—left side	
	3. Door trim panels with new trimmed arm rests	
	4. Front bucket seats with new trim	
	5. Hard fiberglas instrument panel trim pad	
	6. New carpeting in front compartment	
	7. Chrome mirror—left side—mounted on outside of door	
	8. Sound deadening tape and spray on sound deadener	
	9. Rubber sound deadener on floor of front compartment	
	10. Special paint on instrument panel, steering wheel, steering column, horn button and steering column seal.	
	11. Chrome plated control knobs	
16666	**Interior Custom Trim (B):** (Rear compartment with full width rear seat)	29.00
	1. Rear seat with new trim	
	2. Rear arm rests with new trim	
	3. Rear compartment carpeting with two chrome trim strips	
	4. Rubber and Latex floor sound deadener in rear compartment	
16667	**Interior Custom Trim (C):** (Cab top, or no top, with bucket front seats and No Rear Seat)	71.00
	1. Same as Code 16665, except omit headliner.	

▲ There appear to be very few surviving Champagne images outside of those in the brochures, this being one of the few original images left over from the brochure shoot. While the Champagne Metallic shown here is the color the CS is most famous for, the package was available on Scouts in three other colors. To confuse matters, by the time the Champagnes came out, any Scout was available in the four CS colors and the chrome bumpers and wheel covers could be ordered à la carte. From the outside, this would make them hard to identify. The interior is the giveaway, with the Champagne-colored dash, dash pad, and seats. *WHS 54307*

the LST) for the front and included the Champagne-colored upholstery, color-keyed dash, dash pad, carpets, sound deadening, door trim, headliner, and other small accoutrements. The 16666 Interior Custom Trim B gave you the included matching rear-compartment pieces, including the color-keyed seat, sound deadening, and carpets. The 16667 Interior Trim C was for a Cab-Top or and Roadster and was the same as 16665-A, but omitted the headliner, so, yes, you could have a Champagne Roadster or Cab-Top. Research uncovered several Cab-Tops with these options throughout the build period, but no Roadsters.

The paint codes on the LST are another giveaway and include Champagne Mist (a.k.a. Champagne Metallic, 1501), Aspen Green (5683), Apache Gold (4298), and Moonstone Blue (6606) for the CS line. The CS would lead directly to the Custom 800 line, which were Doll-Ups available to order and not a limited edition. Many thousands of very similar 800 Doll-Ups were built into 1966 but not sold or advertised as Champagne Editions, but rather under the Custom moniker. The CS was a transitional Scout. When it was introduced, it was marketed as a "Special Model." As Sales and Marketing moved towards the Custom trim level Scout for the next generation, the CS became less a "Special" and more a higher trim level option. In doing that, they moved towards making Doll-Ups an everyday part of production. Between the RCS and the CS, International learned the profit margin could be higher on Doll-Ups *and* it satisfied public demand for more well appointed Scouts.

An interesting research sideline is that of the CS LSTs found, a high proportion had the 4-152T turbo engine. These may have been dealer-stock builds, with sales and marketing people deciding such a high-end Scout needed a high-end engine.

SPORTOP

Model Years: 1966–1968
Production Dates: November 1965– ~December 1967
Introduction: February 2, 1966
Serial Number Range: ~FC157586–FC160573
~G165294–~G289377
Numbers Produced:

	'66	'67	'68	Total (to Nov.)
Fiberglass Top	2,242	336	1	2,579
Convertible Soft Top	669	214	85	968
Annual Totals Both	2,911	550	86	3,547

Identification: Via LST. Sixth digit in model code: 1 for Convertible Sportop, 2 for Hardtop Sportop.

The Sportop Scout 800 was a big part of the push to civilize the Scout and provide a bridge from the humble Scout 80 to the upcoming X-Scout. As early as August 1963, product planners had discussed an upmarket convertible Scout with a "sports car motif." Sales advised a market existed for such model, so a prototype was developed and displayed late in 1963 with general approval from dealer personnel and Scout owners. Official project approval was given in September 1964 with the working name "Sport Scout."

The Styling folks got to work because Sales wanted units in time for the 1965 model year intro in late 1964. That was not to be. The factory was running out of room and the many special features envisioned for the Sport Scout required more assembly line space than was available. Later, cost analysis showed the project as envisioned was highly impractical and a lower investment proposal was adopted. The final delay came when some outside special parts producers experienced production delays. Along the way the Sportop name was adopted.

Volume production began in November 1965; the earliest records found had a November 12 roll test. Serial numbers on these early Sportops were in the low FC1578XX range. A large batch was built in November and had consecutive, or nearly consecutive, serial numbers. In December, the "FC" serial numbers transitioned to the 13 digit VIN as mandated by Federal law. Relatively few FC-numbered Sportops were built.

The Sportops were officially announced on February 2, 1966, but some were shipped to dealers before that and production ramped up very rapidly. Delays with convertible-top parts meant, initially, a high percentage of hardtops. Eventually, they outnumbered convertibles by more than three to one that year.

▲ The Sportops were the most stylish Scouts built to date. The redhead is enjoying what is probably a prototype hardtop in late summer 1965. The fiberglass top was surprisingly difficult to get right. The Sportop emblem was not available at the time this prototype was photographed, so it was not attached to the rear side of the roof. A convertible Sportop prototype is in the background. *WHS 111659*

▲ The cheesecake view of Kevin Coulter's 1967 V-8 Convertible Sportop shows not only the top, but the unique rear body adaptations required to fit it. The convertibles were the more expensive of the two units to produce, carried the highest price (about $300 more than a Travel-Top), and were the slowest sellers.

The Sportops were a sales flop. Many thought they were overpriced for the market and Sales pressed the planning committees for ways to bring the unit prices down. MTC letters from 1966 and 1967 are positively awash with moaning over the costs of building Sportops. While the basic vehicle ran down the same line as other Scouts, many of the special parts (particularly for the convertible) had to be installed on a separate line in another building. Those costs could be amortized only with volume sales—something that did not materialize.

In April 1967, the makers of the fiberglass tops, Smith Plastics Inc., cited increased manufacturing costs, requiring a substantial price increase to IH from $143 to $272 each. Looking to redesign the top while reducing the per-unit cost, IH got quotes from Goodyear Aerospace, General, and Molded Fiberglass for a new top with a slightly different design. The new-look top would have eliminated the ledge at the rear as well as a few small parts. A metal top built in-house was also considered, but tooling costs for all options gave great pause. It was noted that the proposed hardtop changes would make it necessary to completely redesign the convertible or delete the option.

By October 1967, Sportop parts were running low and the newly formed Scout Product Committee contemplated discontinuing production but decided to build a few more from the remaining parts stock even though they had to order a few bits and pieces from suppliers, resulting in a small number of Sportop

▲ The top-down view shows the top covered by a bonnet, which matched the interior color, in this case Champagne.

▶ The Sportop was available as a 4x4 or a 4x2, as shown here. Note the Sportop emblem on the rear roof side. The 4x2 Sportops were low-volume back in the day and are now almost extinct. *WHS 111067*

Scouts built in the 1968 model year, all apparently built in November and December of 1967. There were still a lot of parts in inventory and some thought was given to build out as many as 500 convertibles and 260 hardtops but the decision was a resounding "NO! The SPC officially killed the Sportop in an October 1, 1968 meeting. Nationwide dealer inventory was listed as 32 units, which were no doubt sold through the remainder of 1968 and into 1969. Most of the leftover parts were ordered to be scrapped.

The Sportop was a unique product. (You can decide if "unique" means ugly as sin or pearls before swine.) The practical view is the numbers weren't there to support the array of special parts needed.

▲ The lap of luxury—or about as close as the Scout came to it at the time. The interior was an equivalent to the Custom Doll-Ups but with some unique features necessary for the model, among them the curved panels in the rear that cover the wheel boxes.

ARISTOCRAT

Model Year: 1969
Production Dates: ~March 1, 1969–August 15, 1969
Introduction: April 1969
Number Produced: ~2,500
Known VIN Range: ~G353467–G371551
Identification: Model codes 784807 (4-cyl.), 784817 (6-cyl.), 784827 (V-8). Doll-Up notation on LST. Visually (see below). Usually, some abbreviated form of "Aristocrat" handwritten in ship-to box.

In May 1968, the Styling Department produced proposals for a new Doll-Up. The idea was to offer it in the 1969 and possibly 1970 800A model years. They were given interesting project names such as Cabana, Sun Scout, Coachman, Royal Coachman, Swinger, Trailblazer, Comanche, and Prairie Dog. The Coachman, Trailblazer, and Prairie Dog concepts were given go-aheads for further development.

The Coachman concept had black simulated-vinyl paint on its Travel-Top, hood, and liftgate, with black striping around the wheel openings and along the rocker panel. A new silver-gray metallic paint would be developed for the main body and for highlights on the interior. An attractive cover would be developed for the rear-mounted spare tire. It would have a roof rack with handles that extended down in the back. The interior would be similar to the 800A but all in black—seats, carpet, and panels.

▲ Another Hayashi rendering, probably done at the same time as the Rawhide. This was named Prairie Dog and one can see the SSII and even the Hurst Shawnee in it.

▲ The Trailblazer prototype, sometimes called the "Brown Door," was very similar to what the Aristocrat would eventually become, save the earth tones inside and out. *WHS 111240*

▶ If it weren't for the colors, one would be excused for thinking this was an Aristocrat. The familiar accoutrements are there. Judging by the small size of the tailpipe, its position, and lack of a V emblem on the fender, this must be a four-cylinder rig. *WHS 111233*

SCOUT SPECIALS ENCYCLOPEDIA

The Trailblazer was envisioned much the same as the Coachman but using different colors. This rig would have had brown metallic tones rather than black, and the doors would be brown.

The Prairie Dog was an early hint at the later Scout II–era SSII or Shawnee. It was to be doorless and topless with a styled roll bar. The coloring was planned as black and white, and fiberglass panels would be developed to highlight the radical styling. It was intended as a two-seater, with red bucket seats and no backseat. Special bumpers were to be developed.

In July, the Styling Department gave the SPC estimated release dates for the three ideas. The Coachman came up first with a public announcement date of March 1969. The most elaborate of the three, the Prairie Dog, was the last, with a May 1970 announcement. Further engineering research showed the simulated-vinyl paint would require a costly setup. The SPC gave the go-ahead to continue development without the simulated-vinyl paint and to develop chrome reverse seven-inch-wide rims for a tire/wheel package using the H70-15 red-stripe tires in vogue at the time.

In August 1968, the SPC viewed Coachman and Trailblazer prototypes. The Coachman had changed in some respects, namely coloring. The exterior was

▲ The Coachman prototype had a black interior with silver-gray accents that matched the exterior color. Again, other than the colors, this is an Aristocrat. *WHS 111189*

▲ The final evolution of the Coachman, August 1968. By this time it was called the Aristocrat after Coachman was discovered to have trademark entanglements. Notice this prototype does not have the chrome axle-nut cover. Note also the difference in the rocker striping from the previous Coachman images. WHS 111230

now two-tone, with the Travel-Top and hood painted a deep blue metallic and the rest of the body a silver metallic with a blue rocker stripe. The interior colors also changed from black to the same blue vinyl used in the upcoming D-Line trucks. It also had a special headliner. The metal interior pieces were painted a mix of blue and silver, and sliding rear windows with bright trim had been designed especially for the unit.

The SPC approved the project so engineering could develop the Coachman in time for a production date of March 1, 1969, and a sales announcement in April. A production run of 2,500 units was affirmed for the 1969 model year. IH wanted the Coachman available with all the engine and transmission combinations, including the upcoming PT-6 six-cylinder and new T-39 automatic. Marketing was asked to advise whether the Coachman name was available for use. On February 2, as the project neared completion, the SPC found the names "Coachman" and "Royal Coachman" were not available as trademarks. A little quick thinking came up with "Aristocrat."

The Aristocrat was given a public introduction at the Playboy Club Hotel in Lake Geneva, Wisconsin, in April 1969. The handful of vehicles there included the original prototype and some of the first production rigs. A deluge of generally good press followed the intro, though that had as much to do with the new six-cylinder and Scout's first automatic as anything. International used the Aristocrat launch to introduce the new PT-6 engine before it was available in the rest of the line. The official announcement for the engine stated general availability would be in June 1969.

Research of LST records did not reveal the first Aristocrat, but it did uncover the last, VIN G371551, built August 15, 1969. Based on dates, one would expect the first Aristocrats to be in the G34xxxx range. The same research documented about a thousand Aristocrat LSTs from mid-April 1969 to the end of production in August.

So what happened to the Trailblazer and Prairie Dog? The Trailblazer prototype made the rounds of the auto shows and was reportedly well received. It was evident the Coachman/Aristocrat design was better liked by execs, so the Trailblazer idea languished until the project was officially discontinued in January 1969. It doesn't appear a prototype Prairie Dog was ever made, and given the large number of special parts, and a very narrow market segment, it isn't difficult to conclude why.

An interesting sideline to the Aristocrat was the development of the chrome deep-dish wheels and wide-oval Polyglas tire package. Originally intended for the Aristocrat only, it eventually was offered as an option on all Scouts starting in 1969. The project was completed well in advance of the Aristocrat but was delayed so as not to blunt the effect of the new Aristocrat.

Relating to the new wheels was Dana's advice to de-rate the gross axle-weight rating (GAWR) by 1,000 pounds when the deep-dish, increased-offset wheels were used because the increased track width (1.44 additional inches) put added stress on the wheel bearings. Also, the normal two-piece axle required a chrome axle-nut cap for a good appearance. Both problems were negated by the advent of the one-piece flanged axle in early 1969. While the prototypes were seen with a two-piece axle with chrome caps, it's unclear whether any production rigs were so equipped.

▲ The familiar Aristocrat interior borrowed a blue vinyl that would be introduced in the 1969 Light Truck line. A manual transmission can be seen in this prototype shot in August 1968. Because the Aristocrat was used to launch the T-39 automatic transmission, many were so-equipped.

The original Aristocrat run was to be 2,500 units, 250 of them to mount four-cylinder engines and the rest featuring the newly introduced PT-6 and 304 V-8s. After 154 of the four-cylinder Aristocrats were built, Marketing asked for non–special order production to stop, citing an inability to sell four-banger Doll-Ups. This makes the four-cylinder Aristocrat a very rare bird (though perhaps not a much sought-after one).

As with most other Doll-Ups, several numbers were listed on the LST to indicate the engine and transmission combination. Not all LSTs have this, and the first one a search of the records uncovered is from early June 1969 production around G361246. Prototype number one, it seems, was the PT-6 and T-39 automatic and number three was the PT-6 with the T-14 three-speed. Oddly, research turned up only four LSTs for the PT-6 with the T-45 four-speed transmission and a model code of 784817, but none had prototype numbers on them. Prototype number four was a V-304 with the T-14, number five was the V-304 with the T-45, and number six was the V-304 with the T-39. These three would have had model code 784827. Research found only one four-cylinder LST with a model prefix of 784807, and given it was late production, it had prototype number two. This vehicle still exists in the hands of a collector.

▲ Resplendent in its original Dark Blue Metallic (6748) and Silver Iridescent (8528), Tom Thayer's immaculately restored V-8 Aristocrat is a showstopper, right down to the reproduction red-stripe tires. Even when equipped with a V-8, as this one is, Aristocrats do not have the "V" emblem.

SCOUT 800A SR-2

Model Year: 1970
Production Dates: April 20, 1970–July 1970
Introduction: March 25, 1970
Number Produced: 1,975
SR-2 w/ 4-196: 1
SR-2 w/ 6-232: 567
SR-2 w/ V-304: 1,407
Known Serial Number Range: G409191–G419659

Identification: First line of LST reads "Scout 800A SR2." Known model codes from first line of LST: 782887 (V-304) and 782867 (6-232).

The SR-2 was another promotional special intended for the 1970 model year. The earliest reports come from November 9, 1969, when a concept vehicle was displayed as a collaborative idea between Marketing and Styling. There was a connection to the simultaneous introduction of a new Whitco soft-top.

▲ The SR-2 was originally intended to be a convertible using the new Whitco top, which is how it's presented here in a Styling Department image from December 1969. The notice to dealers in March 1970 did not mention the roadster option specifically and stated the hardtop was standard. The February 26, 1970, announcement from the Chicago Auto Show contradicts that somewhat, highlighting the availability of both versions and announcing they would be available "in the spring." Besides this prototype, it's unlikely any roadster SR-2s actually rolled off the line in Fort Wayne, as that option was intended to be handled at the dealer level. This is very likely the same preproduction vehicle shown at the Chicago Auto Show, February 26–March 2, 1970.

The initial idea was for a V-304 V-8 roadster with the T-39 automatic and the newly debuted Whitco top in white. A new body color was chosen, Flame Red (2289, an orange-red hue) with white accent stripes on the side, rear, and hood. It would have chrome bumpers with guards, bucket seats, radio, and rear seat, Warn front hubs, and door trim panels. It would use the same chrome reverse wheels and H70-15 Goodyear Polyglas tire package as the Aristocrat. The first estimate was for a run of 1,000 units, but the specifications were to be released so more could be ordered as desired.

The first changes came in early December, when the bumper guards were deleted from the package along with a big "SR-2" side appliqué on the rear quarter panel. On February 5, 1970, concern was expressed there might not be enough market interest to absorb 1,000 soft-top roadsters, so a Travel-Top in white was made standard and the Whitco soft-top became a dealer option. In addition, another color was made available, Burnished Gold (4393), and the engine choice was opened up to the 6-232 or the V-304, and the transmission choices were made to include any of the available gearboxes. Production records show a single 4-196 SR-2 was produced, but no other information is available.

As Scout Doll-Ups go, the SR-2 was rather simple to produce. According to SPC documents, the SR-2 was to have four model codes and several prototype codes. LST research shows the model code for a V-8 hardtop was 782887 and a 6-232 was 782867. Prototype code 1 is a 6-232/T-14, code 2 is a 6-232/T-39, code 3 is a 6-232/T-45, code 10 is a V-8/T-14, code 11 is a V-8/T-39, and code 12 is a V-8/T-45. The LST also has the "DOLL-UP" stamp at the top.

The final equipment list for the SR-2 included an outside spare-tire carrier, chrome front and rear bumpers, front-locking hubs, radio, undercoating,

▲ The standard Flame Red (2289) SR-2 Travel-Top. Among the survivors, this is much more prevalent than Burnished Gold (4393), of which no factory images seem to exist. It's also difficult to determine final production numbers for red versus gold. *WHS 110869*

chrome license light cover, racing stripes, dual fuel tanks, rear vinyl floor mats, custom trim for the Travel-Top, backseat with no bulkhead, and chrome deep-dish wheels with H70-15 Goodyear Polyglas white-letter tires.

A number of vehicles have turned up with SR-2-like stripes on vehicles known to not be "official" SR-2 spec chassis. Many wild conclusions have been reached regarding these Scouts. The answer is found in SM-30 from March 23, 1970. "It is intended," the document reads, "that this attachment package (Code 10904) is available as an extra-cost option after completion of the initial 1,000 SR-2 units." The attachment package code 10904 is found midway on the LST and is listed as "Special Promotional Vehicle." The stripe kit, apparently the same as the SR-2's, was available as a dealer option or over-the-counter sales item.

SCOUT SPECIALS ENCYCLOPEDIA   229

▲ The interior of the SR-2 was right off the standard IH parts shelves, but it came only in the Custom trim level and in a black that coordinated well with the red or gold exterior. A rear seat and floor mat was standard, as was the AM radio. This is probably a prototype. *WHS 110858*

▲ The chrome rally wheels with Goodyear Polyglas tires, white Travel-Top, striping, sliding rear quarter windows, and deluxe exterior features were all part of the SR-2 exterior package. From a mechanical standpoint, a second fuel tank and Powr-Lok rear axle were also included, as was undercoating. *WHS 110868*

The handful we have seen appear to be the same but without the SR-2 labeling.

Suggested factory list prices were $3,735 for the six-cylinder SR-2 and $3,857 for the V-8, with a choice of color for each. A T-39 transmission added another $290 to the price and the T-45 four-speed was $91.50. With an order, the axle ratio could be chosen in 3.31:1, 3.73:1, or 4.27:1, but on the LSTs researched, all were 3.73:1. California orders were required to have order code 15927 specified—the California emissions package.

The commonly seen final numbers for the SR-2 are a total of 2,500–2,000 Flame Red and 500 Burnished Gold. Research has not substantiated those numbers, but it did uncover a total of 1,975 in production records. The April 20 production start date likely is correct. It is also very likely they turned out a bunch in that time frame, possibly the initial 1,000 as originally planned. End of production is less clear, but the last ones found in LST search were built in mid-July.

▲ Our searches did not yield an exact starting date for the production of the Burnished Gold (4393) SR-2 but the earliest LST record we found was for G411852, built May 8, 1970. Various documents mention the production of the SR-2 "Goldies" started later than the red ones. The original intent was for 500 Burnished Gold SR-2s but we have not found the exact number. We surmise less than 500. This Burnished Gold SR-2 is a mostly original survivor owned by noted Scout collector Mike Grieble. It's an SR-2-12, which translates to a 304 powered unit with the T45 4-speed. *Brendan Grieble*

230 INTERNATIONAL SCOUT ENCYCLOPEDIA

SCOUT 800B COMANCHE

Model Year: 1971

Production Dates: November 17, 1970–March 26, 1971

Introduction: October 23, 1970

Number Produced: ~1,500

VIN Range: G431470–G445187

Identification: Via features and model code 883867 or 883887 from LST

The Comanche was the last 800-line Doll-Up. The original package concept was outlined by the Scout Product Committee in their June 11, 1970, meeting as having a American Indian/Southwest theme with three variations for different sales zones. Each featured the same basic layout on a Winter White Traveltop body but with different accent colors to be released at different times. The three original accent colors listed were Ever-Green (5880), Gold Rush (4398), and Blue-Lite (6768). Each was to use a different American Indian tribal name. The wheels were to be painted the accent color and have stainless steel beauty rings.

▲ This rendering shows one of the three American Indian–themed promotional vehicles done by stylist Dave Higley in May 1970 in Ever-Green.

▲ The Ever-Green concept made it to prototype stage, and it's a pretty nice-looking concept. One of the big production hitches was painting just the top portion of the roof and the wheels that special color. The 5880 special color is listed in the MT-90 parts book as a custom color but is not in the CT-399.

▲ Here, in its final form, is the "Gold Rush" concept that made the cut and was called the Comanche in production. Bob Young's stunning, all-original, unrestored vehicle gives us a superb look at how the Comanche was configured. It's a Prototype 5B, which is a common variant, featuring the V-304 and T-39 automatic with 3.73:1 axle ratios. What's incredibly rare and special is that the first owner saved the original General Scrambler tires. They are very fragile at this point and Young uses them only for shows.

Original plans called for 2,000 total units but was later changed to 1,500 (500 of each color). The SPC approved the plan and wanted production to commence on or about November 15, 1970. By August, the plan was changed and limited to one type, the Prairie Gold (Prairie Gold and Gold Rush are the same color) variant, with three possible names: Comanche, Cherokee, or Cheyenne. Instead of color-keyed painted wheels, the wide chrome wheels from the SR-2 were used, mounting less-expensive H70-15 General Scrambler tires instead of the Goodyear Polyglas.

In October 1970, a brochure sent to dealers announced the Comanche and gave ordering details. The Comanche was not built necessarily to customer order but in six variations for the dealer to choose.

The ordering time frame was rather narrow, so only a really clued-in customer with a motivated salesman's help got to place a specific order. In many cases, the dealers just ordered a variety of types for their lot.

There were six numbered combinations, with three subvariants (lettered A through C). The numbered variations were the engine and transmission combination (listed below), three 6-232 model codes (883867), and three V-304 model codes (883887). The letter variations indicated the axle ratios: A for 3.31:1, B for 3.73:1, and C for 4.27:1. The LST also has the "DOLL-UP" stamp at the top.

Code	Engine	Transmission
1	6-232	3-speed T-14
2	6-232	3-speed automatic T-39
3	6-232	4-speed T-45
4	V-304	3-speed T-14
5	V-304	3-speed automatic T-39
6	V-304	4-speed T-45

▲ This rear view is highlighted by the seldom-seen 1971-era General Scrambler tire. The sliding quarter windows, normally an option, were a part of the package.

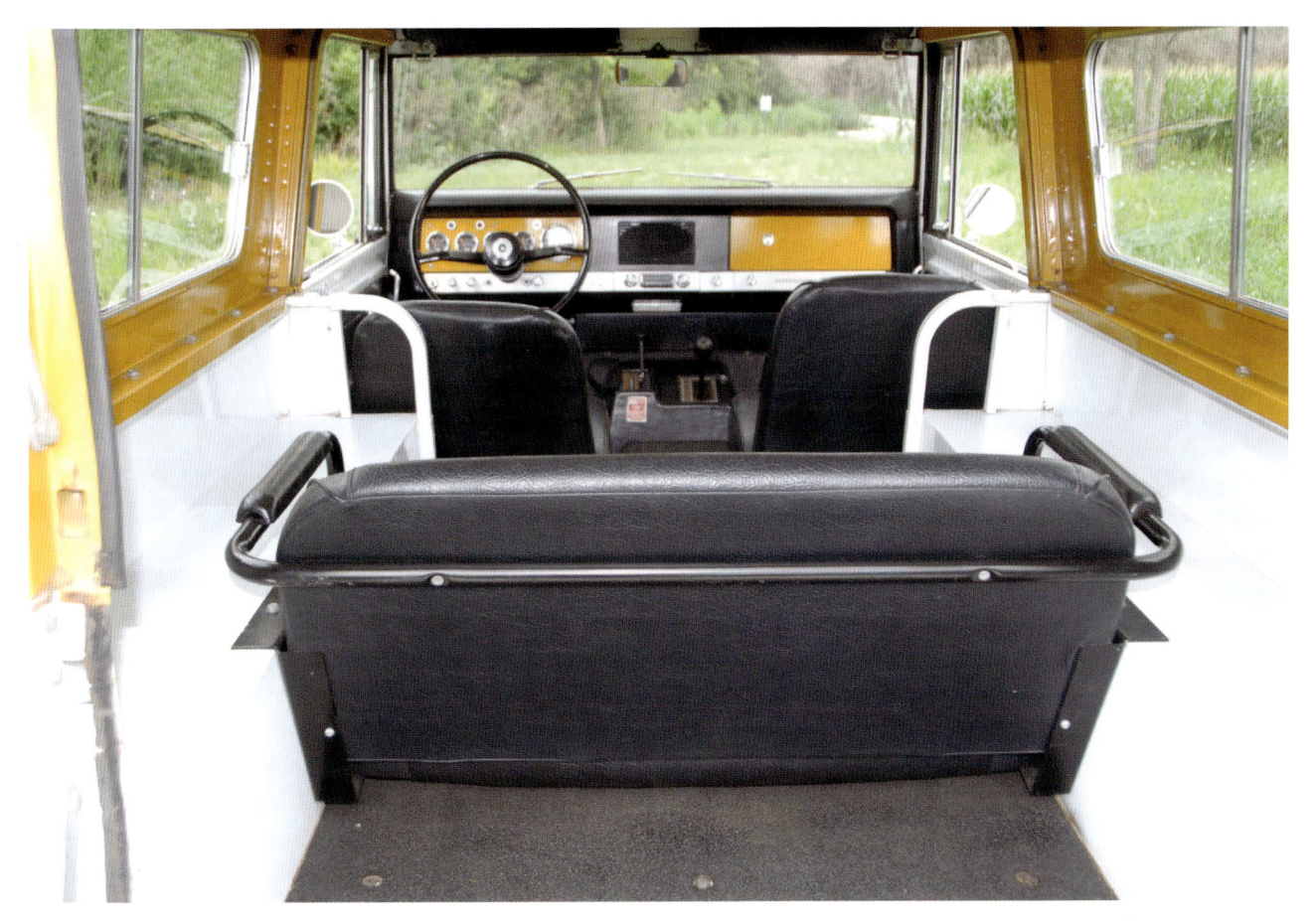

▲ Most of the Comanche interior was standard stuff. The headliner is the most difficult part to reproduce for a restoration. One of the more unique parts was the instrument cluster panel and glovebox door being painted Prairie Gold (4398) like the top and hood.

Beyond the array of available engine and transmission combinations, the Comanche were identically equipped with chrome bumpers, outside spare-tire carrier, FA-11 front axle, front locking hubs, courtesy lights, AM radio, undercoating, RA-28 limited-slip rear axle, dual fuel tanks, Traveltop Doll-Up trim, bucket seats, rear seat, and sliding rear quarter windows. The paint was a combination of Alpine White and Prairie Gold with hood and side stripes. Some confusion exists as to the correct paint code. It's listed in various places as Prairie Gold (4398), but some owners report not getting an accurate mix using that code. Collectors should supply their paint shop a good sample to double-check.

Various sources list 1,500 Comanche as having been built. That cannot be precisely confirmed. Comanche production started on November 17, 1970 (somewhere near VIN G4319740); research tracked the last Comanche (G445187) to March 26, 1971.

According to Price Bulletin MT-1419-B, MSRPs for Comanche Scouts were:

Version	MSRP
6-cyl. 883867	$4,222.91
8-cyl. 883887	$4,375.07
Automatic transmission	$306.24
4-speed manual transmission	$96.62

SCOUT SPECIALS ENCYCLOPEDIA

SCOUT 800B SNO-STAR

Model Year: 1971
Production Dates: August 1970–Mid-October 1970
Introduction: September 22, 1970
Number Produced: 474
Known VIN Range: G423249–G427008
Identification: Via LST, model code 884817, unique paint, equipment

In May 1970, with the SR-2 in production, Marketing began thinking about more promotional vehicles for the fall, anticipating they would mark the end of the 800A line in 1970. One of these would become the Comanche and the other was to be a snow-themed unit. Little did IH know that delays with the Scout 810 would lead to another stopgap model, the 800B, and that these new promotional Scouts would appear as 1971 models.

▲ International stylist Dave Higley did this Sno-Star rendering on May 20, 1970, and the production unit remained very faithful to his concept. Higley was the Styling Department's graphics specialist. Many of the appliqués seen from the 800 through the Scout II era came from his hand.

▲ Ready for duty! The Scout made a nimble plowing rig, better than the Jeep CJ because it was heavier and had a much stouter backbone. Built on September 15, 1970, Tom Thayer's Sno-Star is probably the best unrestored example known to still exist. Everything works according to spec, but it is unlikely to ever work for a living again.

The snow-themed project was different than any others because of a marketing tie-in to the light-truck line. Five hundred snowplow-equipped Scouts in a specific paint-and-stripe scheme were proposed, as well as 200 1210 4x4 pickups in the same general pattern. Both would be promoted as the "Sno-Star." International soon decided to have only one Scout model configuration (6-232 engine and T-39 automatic) and therefore only one model code.

The specifications for the Sno-Star Scout included: H78-15 mud and snow tires with chrome wheel covers, 6-232 engine with T-39 automatic, dual 10-gallon fuel tanks, heavy-duty front bucket seats, heavy-duty front and rear springs, RA-28 Powr-Lok rear axle with 3.73:1 ratio, locking hubs, heavy-duty rear-step bumper with hitch, heavy-duty 61-amp alternator, high-capacity battery, Meyer ST-78 plow with Hi-Lo power, and the full plowing-light package. Paint was to be a special yellow to match the Meyer plow blade of the era with a white Traveltop and hood. Research did not reveal an IH paint code for the yellow color, but various sources describe it as

NEW INTERNATIONAL SNO-STAR SERIES

SNO-STAR 800B SCOUT® 4 x 4
Features the Scout with bucket seats. Features a 145 hp six with "shiftable" three-speed automatic transmission.

SNO-STAR 1210 4 x 4 PICKUP
Deluxe model with 8-ft. Bonus-Load body, bench seat. Equipped with 197 hp high-torque V-8 teamed with a heavy-duty three-speed automatic transmission that can also be operated as you would a manual stick.

Automatic Plowing with All-Wheel Drive

BEST COMBINATION IN THE BUSINESS

Meet the finest equipment ever teamed up to best fit your snow removal requirements—the all-new International® Sno-Star series.

A choice of two models: a 1971 Model 1210 4 x 4 pickup or a 1971 Model 800B 4 x 4 Scout®—both specially engineered to earn you greater profits. Both equipped with the finest of snow plows—a Meyer with automatic power angling *and* lift systems. These all-wheel drive vehicles are also equipped with Powr-Lok limited-slip differential, "shiftable" automatic transmission, increased-capacity electrical system and other heavy-duty components. *All yours at a special "package price."*

PLUS THE MOST DISTINCTIVE COLOR-SCHEME OF ANY PLOW ON THE ROAD. Bright yellow and white two-tone with racy black striping and bright trim—exclusively yours on the International® Sno-Star series.

MEYER SNOW PLOW. Gives you 50 percent faster snow clearing with its automatic power angling. Without leaving the comfort of the cab, the driver can quickly set blade angle from 30 degrees right to 30 degrees left—in just three seconds—via convenient, positive-action in-cab controls.

The Meyer Monarch Hy-Lo lift system is also activated by in-cab controls. Monarch Hy-Lo is a reliable fan belt driven hydraulic lift unit consisting of an engine-mounted pump, tank valve, and front-mounted lift cylinder. In addition, a built-in safety system reacts instantly whenever the plow hits an obstruction. It automatically releases the angle setting and permits the blade to slide off the obstacle.

◀ The 1210 Sno-Star pickup was a similar package of options but on a three-quarter-ton chassis with an 8-foot bed, V-345, automatic, dual tanks, big alternator, and a 7 1/2-foot plow, versus the Scout's 6 1/2-foot unit. At this writing, only a handful of surviving 1210 Sno-Star could be confirmed.

▼ The Meyer ST-78 was the same unit that could be bought on the aftermarket. The blade (technically called the moldboard) was 6 1/2 feet wide and made so it fitted only the sector (the bracketry) for which it was designed. That prevented someone from installing a larger moldboard than intended. The two low cylinders are for angling the blade; the vertical one is for lifting it.

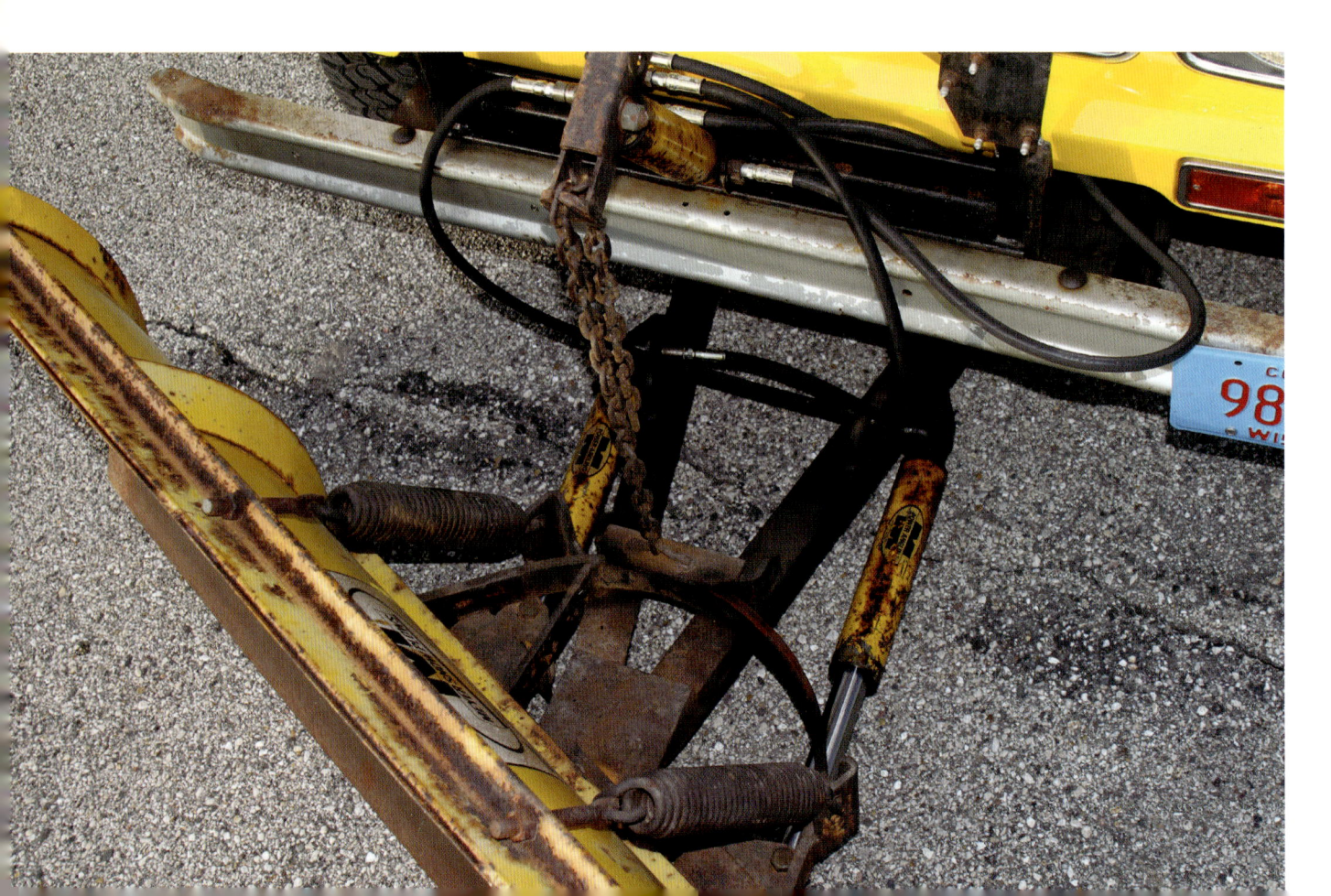

▶ The Sno-Star emblem was a last-minute addition. Stylist Dave Higley designed the Sno-Star graphics. Higley was responsible for most of the Scout appliqué designs from the 800 era on..

▲ The hydraulic controls were mounted to the left of the steering column (the two red knobs). The left knob raised and lowered and the right angled.

the standard Meyer Yellow of the era. A striping kit and door logo also was developed.

Meyer was set to supply training personnel to show line workers how to install the plows, and for the most part things went well from the engineering standpoint. One minor issue was interference with the plow brackets against the new steering-stabilizer bracket. That problem was ironed out by using stocks of the older axles with no stabilizer bracket. Target production date was eventually set for August 31, 1970, and the announcement date was for September.

The official public release was dated September 22, 1970, but dealers had been placing orders for about a month before. The announcement touted both

▲ A rear view shows the heavy-duty step bumper that was a part of the package. The spare was mounted on the interior bulkhead, presumably to allow unfettered access into the cargo section. Salt spreaders were available for the rear, but it's more than apparent this Scout never saw one.

the 1210 truck version and the Scout and showed an artist's rendition of each. There was considerable difference in the graphics and striping on the production models versus what was depicted in the artist's rendition.

Production records that indicate 474 Scout 800B Sno-Stars were produced, making it the least-produced 800-line variant. As snowplow rigs, their survival rate is lower than most Scouts and they are extremely rare today.

▶ No lap of luxury here. Heavy-duty seats and a heater/defroster were concessions to the Sno-Star Scout's intended purpose. It also had a bulkhead.

U.S. SKI TEAM, SPIRIT OF '76, PATRIOT, AND SNO-STAR SCOUTS

Model Year: 1976
VIN Range: FGD33458–FGD43230 (all Scouts with Spirit appliqué listed on LST)
VIN Range: ~FGD27976–~FGD32738 (Undocumented Patriot/Sno-Star)
Production Dates:
  U.S. Ski Team Traveler: September 10–12, 1975
  Documented Spirit, Patriot, Sno-Star: March 5, 1976–June 1, 1976
  Undocumented Patriot, Sno-Star: ~January 28, 1976–~February 26, 1976
Introductions:
  U.S. Ski Team Traveler: December 29, 1975
  Spirit, Patriot, Sno-Star: January 22, 1976
  Total Number Produced: 465 (all those with the Spirit appliqué not including U.S. Ski Team)
  Spirit of '76 Produced: 407
  Documented Patriot/Sno-Star Produced: 45
Identification: Visually. Via option code 10876 ("Spirit appliqué") on the LST. LSTs are commonly (but not always) hand-marked with "Spirit," "Patriot," or "Sno-Star" at top. Via other options as described below.

The Spirit of '76, Patriot, and Sno-Star Scout specials began with the decision to supply Scouts for the U.S. Ski Team. The Olympics were taking place on the nation's bicentennial and it isn't difficult to grasp the potential advertising benefits of a special for a U.S. team. The idea easily translated to bicentennial-themed specials eventually known as the Spirit of '76, Patriot, and Sno-Star and all used the same appliqué option, listed most commonly as "Spirit Appliqué," on a Winter White body (10876). The packages differed slightly, but they could be ordered with any available powertrain and many other à la carte options. A thorough search for all 1976 Scouts with this option code listed on the LST turned up a total of 465. The vast majority were Spirit of '76 Scouts, but a further breakdown of each is below.

As Scouts become more valuable and collectable, the undocumented Patriots have become a valuation problem. Value added to a collector vehicle usually requires a certain level of documentation which is not present with the TSPC rigs at this point. Plus, they could be faked and it's remotely possible a few had their stripes added by dealers back in the day. The authors, and others, have asked former dealers if this might have happened and to a person they say, "not at our dealership." Because of the difficulty applying the appliqué, most think no sane dealer would even try. At this moment in 2020, the authors think these are the real deal but a buyer paying extra for an undocumented Patriot is best served by careful scrutiny of the Scout. If restored, the seller should be able to prove the Scout originally had the appliqué and that it rolled off the line with the proper required equipment for the package.

U.S. SKI TEAM SCOUTS

International created a press gold mine by supporting the U.S. Ski team in 1975–1976 with a fleet of 17 Scouts just in time for the 1976 Winter Olympics at Innsbruck, Austria, held February 4–15. These included 10 of the new long-wheelbase Travelers and seven Scout II Traveltops. All had what amounted to the Patriot or Sno-Star package. Ten of the 17 had Warn electric winches mounted on Warn bumpers. A large "U.S. Ski Team" decal was added to front fenders just above the IH emblem, and some units are seen with driving lights. It was a striking package and probably the envy of other participants in the Winter Olympics.

A December 29, 1975, International press release stated that all U.S. Ski Team Scouts were 4x4s featuring V-8 engines, automatic transmissions, two-speed transfer cases, bucket front seats, ski racks, and AM-FM radios. The press release reported six of the Travelers were shipped to Europe to be used in Innsbruck and in World Cup events. The others were stationed at, or shuttled between, training camps in Cooke City, Montana; Thunder Bay, Ontario; Mission Ridge, Washington; and several places in Colorado.

Some digging uncovered the 10 Travelers in the LST files. All were built on September 11, 1975. The Spirit/Patriot appliqué option code was not found on the LSTs, likely because it had not yet been assigned a code number. They are, however, marked in the ship-to box with a team shipping address in Chicago. The

▲ The first appearance of the red, white, and blue appliqué came with the 17 Olympic Scouts built late in 1975 for use by the U.S. Ski Team before and during the 1976 Winter Olympics. One of the seven 100-inchers is shown here, one of 10 in the entire group that mounted 8,000-pound Warn 8274 electric winches on a Warn bumper mount. This is essentially the same package as the Sno-Star version of the 100-inch Traveltop, which would be virtually identical except for the U.S. Ski Team decal. Note the XLC decal on the rear fender. This is an indicator of very late 1975 or very early 1976 production. *WHS 110958*

LSTs indicate an order of 10, which does fit the release information, and all the options match: Winter White (9219), Wedgewood Blue Deluxe interior with buckets (16872, 16928, 16709), and Deluxe exterior (16835) and luggage rack (16903). In addition, the LSTs indicate the V-345, automatic transmission, two-speed transfer case, limited-slip rear, and 3.54:1 ratios.

The fate of most of these U.S. Ski Team specials was unknown until 2012 when one of the Travelers sent to Europe was found in Belgium in tough shape and acquired by a collector. The others may be hiding in plain sight. Their VINs, in order of their line sequence numbers, are: FGD14119, FGD14125, FGD14135, FGD14142, FGD14148, FGD14154, FGD14160 (the Scout found in Belgium), FGD14172, and FGD14178. LSTs were not located for any of the seven 100-inch U.S. Ski Team Scouts.

SPIRIT OF '76

The 1976 bicentennial inspired almost every American auto and truck manufacturer to build commemorative vehicles. An artist's conception of International's Spirit of '76 was done by the Styling team at Fort Wayne in the fall of 1975, about the same time that the U.S. Ski Team specials appeared. The pilot production model was built March 5, 1976, (VIN FGD33458) and was one of the minority powered by a 4-196.

▲ Ten of the 17 Ski Team Specials were Travelers. The appliqués were the same as the Traveltops, just a bit longer. Again, the Sno-Star or Patriot Traveler would be virtually identical in appearance except for the U.S. Ski Team decal. The Ski Team rigs were all V-8 automatics. Some, including this one, had the Warn 8274 winch package, which was also available to customers. Note the absence of the XLC sticker on the rear fender. This image was released December 3, 1975. All 10 of the U.S. Ski Team Travelers were built September 11, 1976, and came off the line identically equipped. *WHS 118242*

The Spirit option came only on a 100-inch Scout II in Winter White (9219) with the omit Traveltop code (16896). The package consisted of the Spirit appliqué, Wedgewood Blue Deluxe interior, color-keyed roll bar (painted blue to match the interior), Sport steering wheel, blue denim Safari soft-top, and 10-15 Goodyear Tracker A/T tires on seven-inch chrome Rallye wheels. Any drivetrain combination was possible.

The bulk of the Spirits were built April 12–June 1, 1976, the last being VIN FGD44250 with a V-345, T-407 automatic, rear limited slip, and 3.54:1 axle ratios. LST records indicate a total of 407 Spirit of '76 Scouts, 18 of which had the single-speed TC-143 transfer case. Only one was a diesel. There was also one 4x2 Spirit and one Spirit with a 4-196 mated to a T-407 automatic, and a very few were ordered with air conditioning. Both bench and bucket seats were available but most came with buckets, many with the center console.

In May 1976, a group of southern Illinois dealers held a "Spirit Driveaway" in which 33 Spirits (as determined by the resulting marketing photos) were driven to their respective dealers from Fort Wayne, forming a mile-long convoy. The LSTs found for 16 units produced April 12–13 are marked "Spirit Driveaway." A May 19 press release heralded the event, citing 40 dealers had participated.

The LST is needed to verify a Spirit. It will have a Scout 100-inch wheelbase with a 9219 paint code (Winter White). Also present will be the Spirit appliqué (10876), Omit Travel Top (16896), Deluxe Interior (16928), Wedgewood Blue interior (16872), Sport steering wheel (057090), seven-inch chrome wheels (29091), and 9-10-15 front and rear tires plus spare (925102, 925202, and 885102). Some, but not all, have "install roll bar," and "install convertible canvas top" written on the LST. Often, near the bottom, there's a typed "Spirit PKG" note. Most often, there is a hand-written "Spirit" in the ship-to box on top, and the option codes are typed with "TSPC, Fort Wayne, IND" also in the ship-to box.

▲ Dave Higley created the first rendering of the Spirit of '76, probably about the same time the U.S. Ski Team specials were being built. This rendering was created in the fall but not released until December 3, 1975. WHS 111085

▲ This image was in the first press packet announcing the Spirit, dated January 22, 1976. Knowing the very first LST with the Spirit appliqué shows up on March 5, it can be safely concluded this is a Styling Department Scout and very likely the Spirit prototype. Note the XLC emblem on the rear quarter panel. WHS 111083

▲ A production Spirit of '76 was photographed at the Styling Department in April 1976. LST research revealed a large batch of Spirits built that month. Note no XLC emblem is present, unlike the U.S. Ski Team unit.

On occasion some or all of the Spirit options are hand-written on the LST. In some cases, it appears a Scout in the right color and with the right options was purloined on the line to become a Spirit, even if it wasn't originally planned as one. In this case, a shipping address on top might be changed or added, and the side appliqué and special wheel and tires may be hand-marked to be installed at TSPC (Truck Sales Processing Center). It's still a "real" Spirit, as long as those notations happened at the factory.

PATRIOT

The Patriot was a hardtop variation on the Spirit of '76. Unlike the Spirit, it was available on both the 100- and 118-inch Scouts. The Patriot package came on a Winter White body with a Wedgewood Blue Deluxe interior, but it didn't include the roll bar, Sport steering wheel, or the wheel and tire package (though all that could be added except the roll bar on a Traveler). Any other option could be added as well, even the Deluxe Exterior, which included extra body trim and side molding. Two LSTs marked "Patriot" (no Traveltop omit code) included the 10-15 Goodyear tire and chrome wheel option, something that could be added to any Scout that year. This could lead to confusion or a claim of a "rare and coveted Spirit Hardtop," but the LST tells the true tale.

A documented Patriot has an LST showing the Winter White paint code, the Spirit appliqué (sometimes written "patriot appliqué" on the LST, though with the same option number), and the Wedgewood Blue interior. Strangely, one Patriot was ordered with the standard Tanbark interior. Most LST's found in the archive are also hand-marked "Patriot" at the top, though some aren't, having only the 10876 Spirit appliqué code. Some, mostly those built late in

▲ The rear three-quarter view of the Spirit gives a better look at the blue denim top made by Kayline. The material in this top was virtually the same as that used in Jeep's Levi's option.

the run, also list "Patriot package." All that were found had the sliding rear quarter windows, and there was a mix of engine and transmission combinations.

A number of apparently period Patriots have been found without accompanying notations on the LST."

A great deal more clarity on these has emerged since the 1st Edition. Quite a number of original, unmolested survivors have turned up and their patina leads us to believe they are period and original. They are found to share one feature, besides the right equipment; they were all shipped to the Fort Wayne TSPC (Truck Sales Processing Center). That's where unusual or special installations often took place. Our theory is that prior to the big Olympics/Bicentennial themed sales promotion, which was to take place in February and March, they produced a number of Patriots and Sno-Stars without any indications on the LST. Sales operated the various TSPC facilities around the country and we learned from former Scout Plant Manager Jim Poiry that there was a separate paper trail. Regrettably, those records are lost to time but Poiry confirmed it was not at all uncommon to have significant updates done at TSPC that manufacturing had no knowledge about.

The earliest of what we now call "undocumented" or "TSPC" Patriots still available for us to examine was built on January 28, 1976. A careful search could not turn up any undocumented Spirit of '76 Scouts. Documentation of the 10876 option on the LST started on March 5, and all Spirits, Patriots and Sno-Stars we know about from then on had the 10876 option documented on the LST. The latest undocumented Patriot we have found was built February 26, 1976. We suspect there may be other undocumented Patriots built later, but likely not very many.

In all, 40 Patriot LSTs were found. The first of these had a March 19, 1976, build date (FGD33989) and the last came on June 2 (FGD43271). Nineteen were 118-inch Patriot Travelers, the first of these coming March 26, 1976 (FGD35692). One diesel Traveler marked as a Patriot but which appears to be a special thrown into a fleet order, probably as a favor, with the appliqué code handwritten. Howard Pletcher reports finding a record of one Terra Patriot among the records at Navistar, but it could not be located decades later.

SNO-STAR

International announced a Sno-Star at the same time as the Spirit and Patriot. It's rare (but not so special)

◀ The Spirit Driveaway! Around May 19, 1976, 33 Spirits were assembled so a group of Illinois dealers could drive the new Scouts back to their respective dealers around the state as a publicity event. Sixteen LSTs from mid-April were marked "Spirit Driveaway" and presumably those Scouts are in this image. WHS

▲ This appears to be a Sno-Star (note ski rack) with a winch, a Patriot with a ski rack and a winch, or possibly one of the U.S. Ski Team Scouts at some early stage. What sets this apart is the Warn brush bar, which is not seen on any of the Ski Team Scouts but was available in the 1976 Scout accessories catalog. WHS

and difficult to tell from a Patriot, and it shouldn't be confused with the 800B Sno-Star of 1971. Basically, it was a Patriot package with the standard luggage rack and a ski attachment. The only other indication was "Sno-Star" written on the LST. There is no mention of the ski attachment on the LST and it's unclear whether that was installed at TSPC or at the dealer. Only five LSTs marked "Sno-Star" were found: four 100-inch Traveltops and one Traveler.

But is a Patriot with a luggage rack and a ski rack a Sno-Star? The luggage rack (16903) was present on many Patriots, so the dealer could attach the ski rack, which was available as an accessory. It's difficult to attach too much special status to the Sno-Star option, but it could be an interesting find if an LST is so marked. Technically, it seems the LST would have to say "Sno-Star" for it really to be one.

SSII

Model Years: 1977–1979

Production Dates: December 16, 1976–June 19, 1979

Introduction: December 22, 1976

Number Produced: 3,968

Identification: Model code prefix in VIN: G0052 (1977), H0052 (1978), J0052 (1979). Upside-down VIN tag on passenger side kick panel. Visual features (see below).

▲ This is the first evidence that an SSII-like idea was in the works. Dated October 1975, this Styling Department concept had the working title "Mountaineer." It's clear the idea was pretty fleshed out. One major difference between this and the final SSII is the side step. This concept was drawn and signed by Dick Hatch, who gets credit for the grille and much of the design work on the SSII.

In 1977, International confronted the mighty Jeep on its own turf: the hardcore off-road market. A well-conceived marketing plan included sponsoring four off-road race teams and creating an aggressive new version of the Scout. It took a bit of research, but it appears the idea for the race teams predated the SSII, which was then created to capitalize on race publicity. Larry Ehlers and Dick Bakkom got the race team up and running, but had a hand in creating the SSII concept too (see Chapter 7).

Looking back, one can see hints of the vehicle emerging, most noticeably in the 1968 800A-era Prairie Dog concept that predated the Aristocrat. Fast forward to October 1975 and the Mountaineer concept penned by Dick Hatch. The Mountaineer name was catchy and it's unclear why it didn't fly, but press coverage from 1976 called it the "Sundowner," a name that reportedly came from an IH source. Eventually, the working name became Scout Side Kick, or SSK. That might have worked too, but there already had been a Mercedes SSK. Calls went out within the company for a catchier name. Howard Pletcher, who worked in Product Reliability at International, recalls what happened next.

▲ This mockup of a Scout Side Kick used a Scout II body. Aside from some differences in the grille, this is substantially what the SSII became. The white spoke wheels had just finished going through engineering validation but may not have been available yet, so the mockup has the chrome wheels the Spirit and other previous specials used. *WHS 110232*

▲ The first running SSII prototype appeared in the summer of 1976. These images taken in June show a base SSII with only the barest of cosmetic options. Like the base model, it had only one seat, one mirror, one visor, and no rear bumper. Unlike the base model, it had a V-8, locking hubs, and mud and snow tires. The dual exhaust gives away its V-8.

"The naming of the SSII was a bit lower key than the major manufacturers' test marketing and focus groups," Pletcher explains. "One afternoon I was in Larry's [Larry Ehlers, sales engineer] office and he asked what we should call it. We began kicking around the names of other specials that had been used and tried to think of something similar. At some point 'Super Scout' came up and then "SS-2" with its similarity to the SR-2 and we decided that should be used. One moment in my fifteen minutes of fame at International Harvester."

As it happened, the Roman numeral II looked better and it became the SSII.

The earliest found mockup of the SSII appears in Styling Department images from June 1976. Images of a running prototype are dated the same month. It was a simple thing to strip off the roof, take out most of the interior, remove the doors, and add door cutouts, a roll bar, and a designed grille that added to the theme. The prototype was likely a 1976 Scout purloined by Engineering that reappeared in the advanced ordering information and announcement in December with the signature appliqué and a set of white spoke wheels with white-letter Goodyear Trackers.

SCOUT SPECIALS ENCYCLOPEDIA   249

RANCHER SPECIAL
4-196, T-427 4-SPEED
RA-18(4.09), H.D. CLUTCH
H.D. SPRINGS
POWER STEERING
STEP BUMPER, TOW HOOKS
2-BUCKETS
H78x15 MUD & SNOW

ELK

TERRA COTTA

BRUSH BUSTER
V-304, T-427 4-SPEED
H.D. CLUTCH, RA-28(3.54)
H.D. SPRINGS, MANUAL HUBS
POWER STEERING
REAR BUMPER, TOW HOOKS
2 BUCKETS
H78x15 MUD & SNOW

BUCKSKIN

SIAM YELLOW

SPORT
V-304, T-407 AUTOMATIC
RA-18(3.54), H.D. SPRINGS
AUTOMATIC HUBS
POWER STEERING
REAR BUMPER, TOW HOOKS
2 BUCKETS
REAR FOLDING SEAT
OFF-ROAD TIRE PACKAGE
FENDER EXTENSIONS
SIDE APPLIQUE

DARK BROWN METALLIC

BUCKSKIN

BAJA CRUISER
V-345, T-407 AUTOMATIC
RA-28(3.54), H.D. SPRINGS
AUTOMATIC HUBS
POWER STEERING
STEP BUMPER, TOW HOOKS
2 BUCKETS
REAR FOLDING SEAT
OFF-ROAD TIRE PACKAGE
FENDER EXTENSIONS
SIDE APPLIQUE

FIRE ORANGE

WINTER WHITE

▲ The 1977 SSII Gold Star Prototypes were packages selected by International to be generally appealing. Dealers were encouraged and incentivized to order from lists like this for dealer stock.

▶ An SSII doing what it was designed to do. This appears to be an image from the SSII dealer introduction in Parker, Arizona, in January 1977. A similar event was held in Reno, Nevada, a bit later. At both events, racers showed off the SSII's capabilities on a Baja-like course. The dealers were later allowed on parts of the same course. *WHS 110569*

Though the first grilles and door openings were made of fiberglass, the production units were made of vacuum-formed polycarbonate (a.k.a. Lexan) from 0.125-inch-thick sheets. Oakland Plastics of Troy, Michigan, was the original supplier.

Though the SSII was not a huge departure from the standard Scout Traveltop, the new body configuration underwent significant durability tests. Some 8,595 miles of off-road tests were done on a prototype SSII at Phoenix Proving Grounds from March 1 to June 9, 1976, and records indicate that approximately 200 airborne jumps were performed as a part of those tests. No particular mechanical issues were uncovered, but the standard seat structure was too weak for sustained hard use, so four types of seats were tested: the existing design and three reinforced designs. The final SSII had less than 50 percent of the deflection of the stock seat and was to be produced by Freedman Seating Company.

LST research didn't find any vehicles marked as an SSII prior to December 1976. VIN GGD29362, a pilot model marked as "Show Scout," was built on December 16. This rig was interesting in that it was spec'd for 15-inch chrome reverse wheels with white-letter Trackers. It was a Brush Buster package (7-SCSS-44-112) but in a custom yellow (code 4408).

The first dealer announcement and advanced ordering information came on December 22, 1976, in Product News G-674, but the reveal came to dealers in January 1977 at Chandler, Arizona, with two of the newly sponsored race drivers, Jimmy Jones and Frank Howarth, in attendance to show off its capabilities.

Virtually any drivetrain combination was available the SSII, as were most mechanical options. In fact, most everything beyond conflicting cosmetic or Doll-Up options was available. The base SSII came with a 4-196 engine backed up by a T-332 three-speed. Axle ratios were 4.09:1 for the four and the tire size was a modest H78-15 (a passenger tread, no less). It also came with standard power-disc brakes but manual steering. Among the standard off-roady concessions were the two-speed TC-145 transfer case, roll bar, fuel tank skid plate, and snapped-on defroster vent covers. The initial list price was $5,168 and it came without a top or a passenger seat.

Among the factory options were a soft top and doors, the V-304 or V-345 V-8, close- or wide-ratio four-speeds, T-407 automatic, power steering, heavy-duty springs and shocks, 61-amp alternator, an off-road tire and rim package (10-15 Goodyear

▲ Jeff Ismail put a lot of blood, sweat, and tears into restoring this '77 Gold Star Rancher Special back to its basic splendor, even down to procuring a full set of late-'70s General Gripper 780 tires. Gold Star information is found near the bottom of the LST. In this case, it's listed as "GOLD S 7-SCSS-44-110," which is the code found in the 1977 Gold Star book. Ismail's truck had but one option: the folding rear seat (16899, $160) and likely rolled out the door at under $6,000.

Tracker tires with or without raised white letters on 15x8 white spoke rims), rubber fender lip extensions, heavy-duty clutch, rear limited slip, and number of other small items. The optional top and doors were most often installed at the TSPC facility at Fort Wayne but could be installed by the dealer or owner. The tops were by Whitco and came in black or white, with or without a sunroof. The top option included a boot for when it was in the folded position.

Sales developed a package lineup in time for the first announcement and put together four Gold Star prototypes. For 1977, Marketing came up with some catchy package names: Rancher Special, Brush Buster, Sport, and Baja Cruiser.

The Rancher Special was the low-cost, four-cylinder package that included the T-427 wide-ratio four-speed, heavy-duty clutch, H78-15 mud and snow tires, painted rear bumper, front tow hooks, and passenger bucket seat with a sun visor to match. That configuration bumped the price to $5,557.50.

The Brush Buster package was almost the same as the Rancher Special except it mounted the V-304 V-8 and had power steering, a limited-slip RA-28 rear axle, and manually locking front hubs. The axle ratio was taller for the V-8 at 3.54:1, and the off-road tire and wheel package was offered (you could get it on the four-cylinder by special request, but it wasn't recommended). The 1977 Brush Buster package rolled out the door at $6,068.

The Sport added to the Brush Buster by tacking on the T-407 automatic the black sidewall 10x15 off-road tire and rim package with rubber wheel lip

▲ This is the SSII most people remember: the Baja Cruiser. This is a two-owner '77 SSII Baja Cruiser in Fire Orange (3195) that was owned by Mark Drake of Scout Madness and showed 7,200 miles when this picture was taken. The Gold Star code for this rig was 7-SCSS-44-116. The original owner ordered the rear seat, the gold wheels, and the dual horns. The gold wheels inspire a certain amount of controversy because technically they were not available until 1978 but the option is listed on the LST and this was a late-1977 build. The bumpers and lights are aftermarket but were installed at the dealer along with an ARE air conditioner kit. Aftermarket bumpers like these were also seen on certain prototype Custom Vehicles International (CVI) rigs. *Scout Madness/Larry Fintel*

▶ The aftermarket comes to the rescue! In 1978, T. J. VanDerBosch of Lake Zurich, Illinois, began offering a fiberglass hardtop kit for the SSII. It was good enough to be sold through the dealer parts organization and was available in white or black; it cost $695 with doors and $620 without.

extensions, AM radio and lighter, and the black and gold side appliqué. The Sport would set you back $6,843 in 1977 bucks.

At the top of the line was the Baja Cruiser—*the iconic SSII*. The Baja Cruiser added to the Sport by including the V-345, RA-28 limited-slip rear axle with

SCOUT SPECIALS ENCYCLOPEDIA 253

▲ For the 1978 model year, a SSII Special Edition was offered to jazz up units without the Baja stripe kits. It was a pretty basic kit, with a stripe similar to what was offered on regular Scouts starting in 1978 and gold-trimmed wheel covers with a blacked-out center. The kit was offered to dealers at $35 for installation on Scouts in stock. It could also be ordered on new vehicles and installed at the TSPC. These specials were available for order from November 18, 1977, to February 7, 1978. The LST for this rig shows it was built in February 1977 and had a 4-196, T-427 transmission, 4.09:1 ratios, and an open differential. It also had the second seat but little else.

3.54:1 ratio, raised white letter off-road tire and wheel package, a special appliqué, AM/FM radio, center console, and automatic front locking hubs. The Baja Cruiser cost $7,184. Overall, the Gold Star offerings were just a small part of the SSII lineup. To be a "true" Gold Star, it has to have the notation on the LST.

For '78, the Rancher Special remained largely unchanged, though the price jumped to $5,800. The bigger changes occurred in the middle of the line, including the elimination of the Sport. This ended the confusion with the similar-looking Baja Cruiser, though the appliqué was still optional à la carte at all levels. The Brush Buster was broken into two packages, one with the V-304 and the other with a V-345, both with the T-427 wide-ratio gearbox. Prices for a V-304 Brush Buster was $6,426 and the V-345 was $6,549.

The Baja Cruiser package had few changes but, like the Brush Buster, was split into two packages, one with a V-304 and the other with a V-345. An AM radio was standard for 1978 Baja Cruisers as opposed to the AM/FM for 1977. The prices bumped up to $7,216.50 for the V-304 Baja Cruiser and $7,339.50 for the V-345.

The Rancher Special package went away for 1979, though a base 4-196 SSII could still be purchased and optioned accordingly. The Brush Buster and Baja Cruisers remained virtually the same, but, of course, the prices increased. The Brush Buster leaped to $6,995 and $7,151 for the two V-8 models and the Baja Cruiser to $7,840 and $7,996.

Though mentioned in some preliminary 1980 information published in 1979, no '80 SSIIs were built. A number of leftover '79s were sold in subsequent years,

▶ Even with the high-end Baja Cruiser SSII, this was bare bones Scouting—a heater, four durable and semi-comfortable seats, maybe floor mats, and a radio. In 1977, the AM/FM radio was standard. In later years a good number of dealer accessories became available to spruce up the SSII, including weatherproof, washable Ozite carpeting and roll bar padding. This Scout was showing only 7,200 miles when photographed and is 100 percent unaltered. The first owner had the ARE air conditioning installed at the dealer, a common enough occurrence in Texas, even with ragtops. *Scout Madness/ Larry Fintel*

a situation that often confuses people into believing there were 1980 and later SSII models. Even if a truck is registered as a 1980 model or later, a look at the LST will confirm a 1979 build.

For many years, the exact number of SSII production by model year was unknown, though sales numbers were known. A recent hand count by Jim Allen gives us a better picture. Below are the annual sales followed by the counted production:

1977: 783/2,910
1978: 2,478/579
1979: 491/491
1980: 176/0
1981: 34/0
1982: 6/0

NO FAUX

Because it's easy to turn any 1977–1979 100-inch-wheelbase Scout II into an SSII using period or aftermarket parts, and since a "real" SSII brings a premium over an ordinary Scout II Traveltop, it pays to be able to tell them apart.

First, look at the LST, both the glovebox version and the one you should order from your Scout Light Line dealer. The VIN prefix is G0052 for 1977, H0052 for 1978, or J0052 for 1979. The top line of the LST will also indicate an SSII and usually reads, "SCT II 4X4 SS100WB."

Some of the visuals are obvious, such as the grille, door openings, and roll bar, but all these can be added. Next look at the VIN tag on the left kick panel as well as the International data plate on the right, which will be mounted upside down. Also, there will be snaps at the defroster vents for dust covers, even if the dust covers are gone. The door openings were thermoplastic in the original rigs, but International offered a fiberglass replacement. Aftermarket openings are also fiberglass, but the difference is that the original International items are gel-coated black while the aftermarket items are painted.

Other things to know include the fact that no SSIIs were factory-fitted with AC (sometimes added at the dealer, though), tilt steering columns, or bucket seats with folding backs, so be suspicious of any "SSII" that has them or signs that it once did. Also bear in mind that the SSII grille was available for any Scout from about 1978 on as an à la carte option.

SSII ID

▶ The left kick panel is where the federally mandated VIN sticker can be found. On a Traveltop Scout II, it will be on the door jamb.

▲ The International data plate is mounted on the passenger-side kick panel, conveniently upside down, so you can read it while bent over.

▲ Another key feature of the SSII is the snapped-in covers for the defroster vents. The covers often get lost, but the snap receptacles remain as an indicator.

256 INTERNATIONAL SCOUT ENCYCLOPEDIA

THE MIDAS TOUCH

Midas International of Elkhart, Indiana, was in its heyday when they contracted with International to doll up Scouts in the mid-1970s. Midas had several elements, including an RV Division and that very familiar chain of muffler and repair shops. Their RV Products Division started in the early 1970s and branched out to van conversions later in the decade.

Midas built some truly stylish and luxurious van interiors that translated well to Scouts, especially the Traveler. When a Scout was ordered with the Midas conversion, it was shipped to Elkhart, about 67 miles from Fort Wayne. The conversion was done according to order and the Scout went off to the dealer that ordered it.

"The Midas Touch," as it became known in advertising, came several ways. You could order just the interior or add even more luxury from a long list of Midas options.. The many interior options in the

▲ Approved in the summer of 1976, based in part on this Dave Higley concept drawing, was an the Off-Road Vehicle Package. It was introduced at the same time as the Midas rigs and included a front brush guard with a Warn mini 3,000-pound winch and two driving lights, a 23-channel CB with a magnetic-based antenna and lighter power cord, roll bar (for the Traveltop only, not available on Terra or Traveler), and the same Midas Z stripe appliqué as the Street Machine. The options codes were 10884 for the Scout II, 10885 for the Traveler, and 10886 for the Terra. The sales pamphlet recommended the two-speed transfer case, fuel-tank skid plate, heavy-duty springs and shocks, and the 10-15 tire package. This package disappeared after 1977.

Midas era included a folding second seat, a folding third seat for the Traveler, and even dual "His-n-Her" sunroofs. The interiors were particularly well liked, mainly because of the super-comfortable and stylish seats that came from the Freedman Seating Company of Chicago (which remains in business to this day). Midas' RV and van-conversion segment faded away in the early 1980s and the facility in Elkhart on South Nappanee became a plastics manufacturing plant in 1981.

MIDAS CRUISER

Model Years: 1977–1980
Production Dates: 1977–1980
Introduction: December 1976
Number Produced: Undetermined
Identification: Via LST and visually. Conversions will have a Midas sticker on the door jamb.

▲ The Street Machine package was available for the Traveltop or Traveler as shown in this page from the 1977 informational packet.

1977

The Midas Cruiser line was introduced to the regional managers on October 18, 1976, in two forms: the Family Cruiser, which was a luxurious people-mover edition of the Traveler, and the Street Machine, which was similar but left the cargo area open and was available for both the Scout II Traveltop and Traveler.

The Family Cruiser line had three option packages: Blue (code 10878), Green (10879), and Tan (10880). Each included a woodgrain instrument panel and shift console plate, color-coordinated shag carpeting with luxurious padding, two high-back front bucket seats that reclined and swiveled and had tweed cloth inserts, two nearly identical high buckets in the middle, and a rear jump seat that fully reclined into a bed. The door panels had a mix of tweed and vinyl that matched the seats, the headliner was the same tweed cloth, and the spare-tire cover matched the seat vinyl. The outside had a two-color stripe color-coordinated to the paint.

The Street Machine package used the same Blue (10881), Green (10882), and Tan (10883) themes as the Family Cruiser but had just the two high-back swiveling/reclining buckets up front. The cargo area was open but thickly padded and carpeted. The dash, door panels, and headliner treatments were the same as in the Family Cruiser, but the Street Machine had a sunroof and a black-and-white "Z" appliqué.

1978

The 1978 Cruisers were announced in Product News Bulletin G-755 dated September 22, 1977, and could be ordered for production starting November 1, 1977. The treatment changed for 1978 in that they were termed "Cruiser Packages" rather than catchy names with set package content. A Terra Cruiser package was introduced along with new optional accessories, while some of the items that had been included with

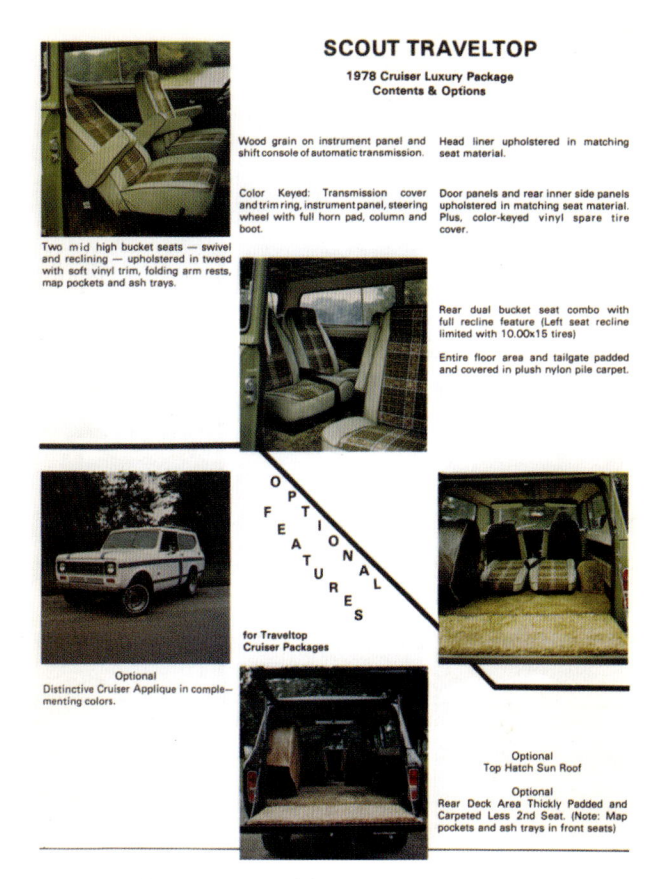

▲ The packages changed for 1978, as shown in this page from the dealer ordering packet.

the Family Cruiser and Street Machine were now à la carte options.

The tweed patterns changed for 1978 from the horizontal, Navaho-like pattern that covered the entire facing of the seat to a Scotch Tartan–like insert with vinyl on the bolsters. Again, IH offered Blue (10878), Green (10879), and Tan (10880) packages. The option code was the same for the Traveltop, Terra, and Traveler, but the prices were not. The Terra was listed at $950, the Traveltop $1,500, and the Traveler $1,575.

The Traveltop and Traveler packages included a woodgrain dash and shift console plate, front high-back buckets that swiveled, rear buckets that reclined but did not swivel, deep padded shag carpet, and a tweed headliner that matched the seats. The tweed material also carried over to the door panels, but the general layout didn't change much from 1977. The Terra's standard package was just the same as the other two but, obviously, without the rear-seat features. It included the center console/cooler with the padded lid that was optional with the Traveltop and Traveler.

Optional for all models was a one-piece sunroof (16497, $150). All models shared an optional exterior striping kit option (10711, $69). The range of stripe colors was not listed in the materials found, but are described as "complementary." Blue and brown are pictured. For some reason, a center console/cooler was not listed for the front of the Traveltop or Traveler, but one was listed for the rear of the Traveler (26498, $62). This was not available if the third seat (16499, $207) was ordered (an option for the Traveler only). Buyers could get a $158 credit with a "Street Machine-esque" rear-seat delete (code 16615), which left a flat, padded cargo area. A tonneau cover was an option for Terra buyers who didn't have the bed rails (16580, $124).

▲ The Midas interior radically transformed every Scout into which it was installed. For 1978, the Cruiser Luxury packages were just that—luxurious and much quieter than original. International cut the top away from this Traveler to show the arrangement in the blue color, including the optional folding third seat shown in its upright and locked position. *WHS 110653*

▲ Much of the Midas interior gear was translated from van conversions. The Scout had enough interior room to make it work and it gave International something unique and nifty to crow about in the SUV realm. As seen here in the Traveltop, there wasn't quite enough room between the wheelwells for individual rear buckets with armrests, but the seat backs reclined individually. This is a '78 Traveltop, but in 1979 the optional folding third seat could be installed in place of these buckets. *WHS 110475*

▲ The optional third seat folded down into a bed. By itself, it was only large enough for children or very small adults. The rear seats were a little too tall to fold down and make into a larger bed. WHS 110651

▲ The 1979 Midas lineup included all the Scout models and painted-on body graphics. The Rallye wheel/radial tire package on the Traveltop and Traveler were optional on the International side of the list. Ditto for the white spoked wheels and Tracker A/Ts on the Terra.

1979

The packages changed again for 1979 with more high-class options and features. A designated Family Cruiser package returned, alongside a new Baja Scout SSII package. The four interior colors offered for 1979 were Blue (108678), Green (108679), Brown (108680), and Russet (108681). Package prices went up to $1,682 for the Traveler, $1,600 for the Traveltop, and $1,065 for the Terra.

The pattern of the tweed seat insert changed again and it was made of nylon that covered the entire facing of the seat, eliminating the vinyl bolsters. The swiveling seats had folding armrests and a new over-center cam-latching mechanism for easier rear-seat access. The folding rear seat, previously available only as a third seat option on the Traveler, was now available as an optional replacement for the dual rear buckets in the Traveltop or Traveler. A new overhead console was optional, as were front and rear reading lights and one-way glass for the rear windows. Dual His-n-Her sunroofs were offered in place of the single panel units. Instead of appliqués, the graphics treatments were custom-painted.

The standard Traveler and Traveltop packages included the swiveling front bucket seats, reclining rear buckets (folding armrests on the Traveler only), and the tweed/vinyl accoutrements, as well as full shag carpeting with generous padding. The standard

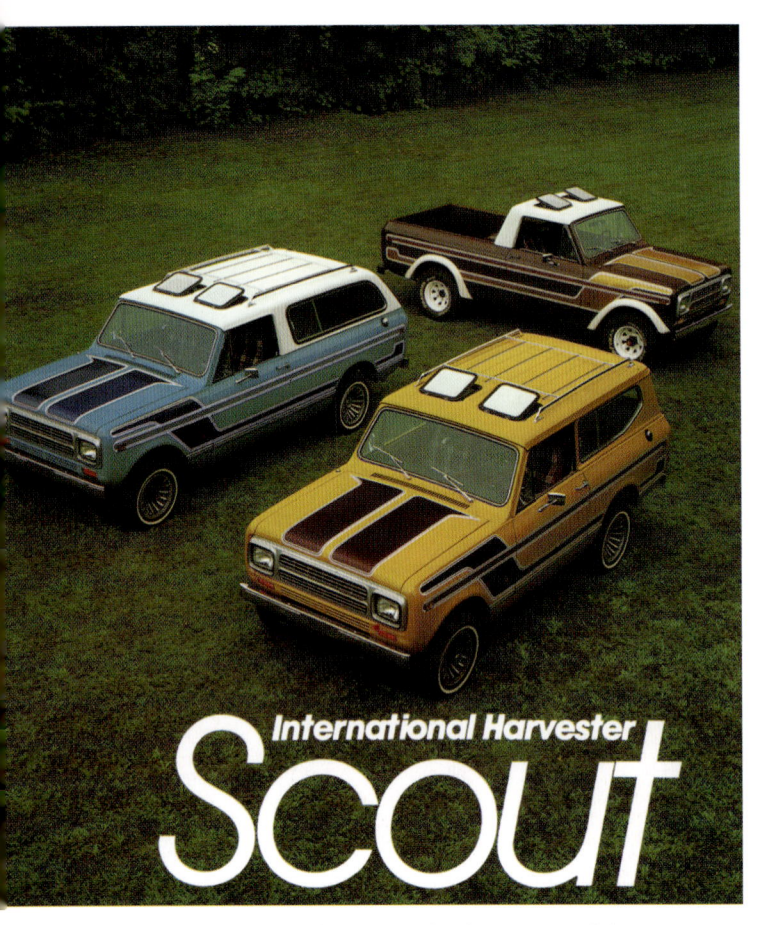

▲ The final year for all involved—1980. While advertising usually shows the full exterior graphics, it's not at all unusual to find a Midas Scout sold without them or with the factory appliques. The painted-on graphics seem to hold up better over time than the decals, but it's a lot less expensive to reproduce decals than it is to have a custom painter reproduce the graphics.

Terra package included no-swivel front buckets with folding armrests on the door side of the seat, the center console/ice chest with a padded and hinged armrest, and the rest of the interior treatment adapted for the pickup cab.

The options for all three models included the His-n-Her sunroofs (16497, $211), custom exterior paint (109963, $295), and the overhead console (08737, $76). The Traveler and Traveltop shared the one-way glass option (16495, $144), four reading lights (08701, $76), and the option of replacing the rear buckets with the folding seat (16494). That yielded a credit of $112 for the Traveler and $59 for the Traveltop. For the Traveler, the third seat (16499, $224) was an option, as was the rear-seat console/ice chest (16498, $66). For the Terra, options included a front brush bar and driving-light package (09581, $177) and a white tonneau (16580, $132).

1980

In the Scout's final year, the Midas packages changed only in small detail. An SSII version was initially listed in International preorder materials, complete with part numbers, but it was not pictured or mentioned in the Midas brochure and none are known to have actually been built. The tweed seat material pattern reverted to one very similar to 1978's but with the cloth over the entire facing. The armrests went from vinyl to velour. The same four interior colors were offered as in 1979 with the same codes (see previous). As in 1979, the exterior paint designs changed, but they were still custom-painted by Midas. White or black fender flares could be part of the package, and a personalized dash plaque read, "Custom Made for _____" with the name of the owner engraved.

The package contents were largely the same as in 1979, as were the options and codes. The white fender flares were coded 09591 and the black were 09592. The brush guard and driving-light package (09581) was offered on all models.

Midas continued doing Scouts right to the end, and perhaps even a few after. Among the last 35 Scouts built on the final day of production, October 21, 1980, was a well-optioned Scout II diesel with the Tan Midas Package, optional sunroofs, brush guards, exterior paint, and reading lights.

MIDAS BAJA SCOUT SSII

Model Year: 1979
Production Dates: 1979 model year
Introduction: August 1978
Number Produced: 82
Identification: Visually (see below). LST and package codes: 10678 (Blue), 10679 (Green), 10680 (Brown), or 10681 (Russet), with SSII model prefix K0052.

The Midas Baja Scout SII was a Midas edition of the SSII that shouldn't be confused with the Baja Cruiser, which was the top-of-the line SSII factory options package offered from 1977 to 1979. They are two distinctly different SSIIs, however, and sometimes the Midas version is incorrectly called a Baja Cruiser. The Midas is sometimes also called the Midas SSII, a generally acceptable nickname.

The 1979 Baja Scout SSII came in four interior color choices: Blue, Green, Brown, or Russet. Buyers

▲ This Midas Baja Scout SSII is the pilot model, built August 8, 1978, and is now in coauthor John Glancy's collection. It went to Midas first and was then sent to Melrose Park for preparation into a show rig. It did the rounds of some shows, the details of which are not clear. We are also reasonably certain it was used in the photo shoot for the '79 Midas brochure (see page 261). Besides being the pilot model and a show rig, this Scout was also a movie star. It had the dubious distinction of being in the 1980 grade B horror flick, *Humanoids from the Deep* starring Doug McClure, Ann Turkel and Vic Morrow. Filming started in October 1979 at locations in Mendocino, Fort Bragg and Noyo, California. It had been loaned to the production company and after shooting, it ended up with the Cincinnati IH dealer and was sold to an employee, from whom Glancy bought it. It's in unrestored condition and though storage-worn, it's complete and very solid.

▶ Steve Nieman's Baja Scout SSII was restored back to original configuration with a high degree of accuracy. This is one of ten in Black Canyon Black (0001). Nieman's Baja Scout has all the options: rear seat, striping, and brush bar with lights—a $2,395 set of options over and above the cost of the SSII. Equipped as shown, this Scout was around $7,000. Steve's also had a set of period fender flares that did not survive the years. Reproduction originals are not available but have been replaced with modern ones that look similar.

▲ The soft top from Whitco was included in the Midas package, along with soft doors. That differed from the SSII, where it was optional. The top was the same as that offered for the standard SSII models and had an optional sunroof.

could choose from any of the available stock exterior colors for 1979 or from the CT-399. The package came with vinyl-covered racing bucket seats that tilted forward for rear-seat access, padded roll bar with black vinyl covering, color-keyed snap-out carpeting, color-keyed cargo-area panels, black vinyl top with a zippered sunroof and black doors, a top boot, and a center console. An optional brush guard package included off-road lights with covers (09581, $177), custom-painted graphics (10963, $295), and dual rear bucket seats (16616, $363). The suggested list price for the basic package was $1,600. From there, any of the many dealer-installed accessories could be added.

The Midas SSIIs are among the more elusive Midas conversions and just 82 were built, including 22 Canada Dry versions for the Canada Dry's "Way to Go" sweepstakes giveaway in Canada in 1979. They were painted a special "Canada Dry Green" (CT-399 color 5018) with special striping. The drawing was held in August 1979. Another Midas SSII was given away in 1979 at the Toronto Sportsman's Show. Not much is known about it, beyond what was written in a 1979 *Scouts Unlimited* issue. The story clearly shows

▲ The V-345 and automatic was a popular choice for the Midas SSII, as well as the Baja Cruiser factory rig. Note the snap-out carpets, another handy Midas feature. The Midas SSII interior was typical Midas quality. The seats were nicely covered, stylish, and comfortable—compared to other racing type seats, at least.

▲ Jim Villerot's '79 Midas SSII is painted in a CT-399 color (6024) with the blue interior. He purchased it new late in 1979 and it retains the original paint, graphics, and fender flares. He still has the original Tracker A/T spare tire. It was ordered by an IH district rep with the full complement of options and Jim bought it from his local Michigan dealer with just a few thousand miles on the clock. The tube bumpers are not original, nor is the full roll cage, but Jim retained the original parts and is now considering reinstalling them. Villerot's was the only Midas SSII painted this color.

▼ Ten specially built Midas Baja Scouts were to be given away after an August 17, 1979, sweepstakes drawing. This Midas SSII is advertising the sweepstakes but is a bit of an anomaly. The striping does not match what was given away, so presumably it was not one of those as the "demonstrator" sign implies. *Art Stoyshin*

▲ Mike Grieble's Canada Dry is unrestored and nearly perfect. It's exact history is not known at this writing but Mike is probably the third owner. The original winner sold it fairly quickly to a gentleman in the railroad industry who used it to Scout abandoned railroad lines. He took good care of it and when he passed on, Mike bought it from the estate. Like all Midas of the era, the graphics are hand painted rather than being an appliqué. The Canada Dry Scouts were sent to Canada without graphics and the strongest theory is that they were applied in Ontario. Missing on Mikes are the original driving lights mounted to the brush guard.

a Midas SSII but with unique striping. The story mentions it was fitted with a V-345 and four-speed manual transmission.

As of 2020, 18 Midas SSIIs are known to survive, including coauthor John Glancy's, which was the prototype model, and two Canada Dry versions. Two Midas paint schemes are seen but it isn't known how many of each type were done.

A Midas SSII may be hiding in plain sight, disguised as a Baja Cruiser, so a copy of the LST is important. Clues are the option codes listed on page 263. Also look for shipping to Midas in the ship-to label on top. Most often, Midas conversions didn't go through the TSPC, though Baja Cruisers did, so there won't be any TSPC notations.

▶ The Midas SSIIs used special seats made by Freeman Seats. Though the SSII was a much simpler format than their normal Family Cruiser fare, Midas spent a lot of time making it top notch. The seats are very comfortable, durable and supportive. Take note of the original '79 vintage Goodyear Tracker A-T spare tire.

SCOUT SPECIALS ENCYCLOPEDIA

SELECTIVE EDITION

Model Years: 1978–1979
Production Dates: February–March 1978
 (1979 dates unknown)
Introduction: December 22, 1977
 (1978 model year; 1979 dates unknown)
Number Produced: Undetermined
Identification: Visually (see below).
 By LST and Package Code 10992.

The Selective Edition was another of International's efforts to spice up the line. When it was introduced to dealers in December 1977, they had from December 26, 1977, to February 15, 1978, to place orders. Production started February 20, and 51 were built into February 22. Amazingly, they were consecutive VINs and many of them ran down the line nose to tail. This first batch started at VIN HGD30301and ran to HGD30352. These were most likely a batch of the first orders plus some showroom eye candy. Production continued through 1978 and the Selective Editions were often built in batches of 50-70 units, many times with consecutive or nearly consecutive VINs.

▲ A Dark Green (5013) Selective Edition Terra shot in the mid-1980s. The stripe is almost identical to the '77 Special Edition SSII and similar to a stripe offered in 1980. This rig was for some years the personal vehicle of the owner of an IH/Navistar dealership but has recently been found in a rest home for old trucks—a.k.a. a wrecking yard.
Bob West

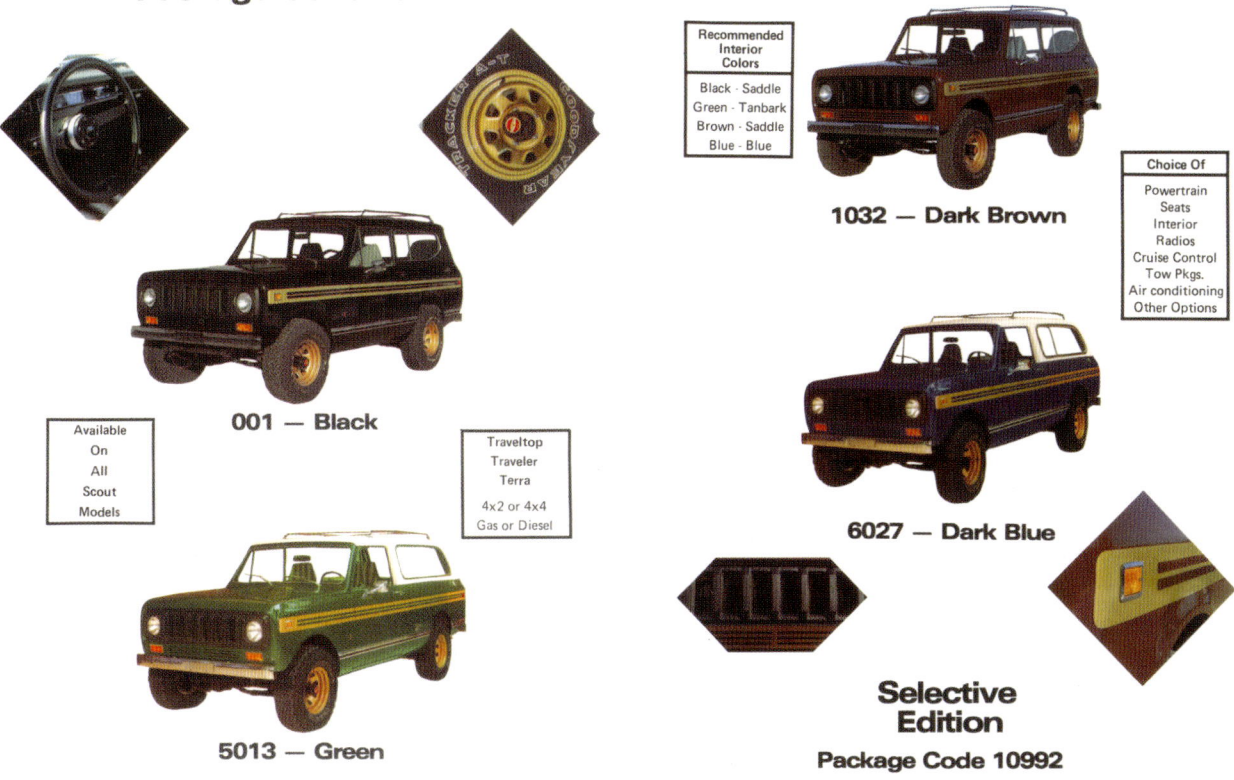

▲ Two pages of the 1977 announcement show the four color combos and the possible body styles, along with the options included in the package.

The package included narrow gold accent stripes on one of four body colors: Black Canyon Black (0001), Dark Blue (6027), Dark Green (5013), and Dark Brown (1032). In addition, a set of color-matched gold eight-inch spoked steel wheels with 10-15 raised white-letter tires were included, as well as a 15-inch Sport steering wheel. The grille was the same type used on the SSII.

The Selective Edition Package Code was 10992 and buyers had to specify the Chrome Exterior code (10898) or the Deluxe Exterior code (16835). Power-steering (05281 for gas, 05202 for diesel) and tire codes (88608 and 925608) were also necessary. List price for the package was $446. The Selective Edition was available for the Scout II, Terra, and Traveler, but not the SSII. The interior layout was left up to the customer.

SCOUT SPECIALS ENCYCLOPEDIA

SPECIALS POTPOURRI

MONTEVERDI

Peter Monteverdi, a German-speaking Swiss, was one of a handful of gearheads to put Switzerland on the automotive map. He started in the 1950s and early 1960s building race cars, but in 1967 he began customizing existing luxury and performance cars for the rich and famous in the Swiss and German markets. Eventually, Monteverdi's eye turned to the SUV market. He began importing SUVs, including International Scout IIs, for various conversions. His motivation for choosing the Scout is unknown, but it's been reported that he began customizing Scouts as early as 1974.

Monteverdi's Safari, which first appeared in 1976, was a heavily customized Scout. In fact, it was so custom that only a diehard Scout fan could tell it was a Scout. It was originally going to be a custom-chassis vehicle patterned after the Range Rover, but production issues forced a more practical solution. Enter the Scout II. The body was stripped of its top and most of its skin, then fitted with a new shell. The lines were

▲ The Rallye wheels are one of the giveaways to the Monteverdi Safari's Scout chassis. This is a 1978 with an unknown powerplant. It could be anything from a stock V-345A to a 440 Chrysler. *Charles01*

▲ The look at the back of the Safari is a little less clean and neat, but the Scout heavy-duty step bumper can be seen peeking out.

very Range Rover-esque. The base engine for the first unit was the Chrysler 318 two-barrel, which cranked out 152 horsepower and 255 lb-ft. Later, a Chrysler 360 was added to the mix (the International V-345 was also an option). The top option, however, was a Chrysler 440 big block rated at 306 horsepower and 477 lb-ft. It delivered 0–60 times of eight seconds and a top speed of 125 miles per hour.

According to 1979 literature, the Safari cost $35,000 for a base unit and was nearer $50,000 with the 440 and the other available goodies. Most sources state the Safari was available from early 1976 through the middle of 1980. Monteverdi built an experimental four-door on a Traveler chassis that was a work of art.

The 1978–1980 Sahara was more conventional and probably developed as a lower-cost model. It used most of the Scout's original body panels but with new trim and a totally redone front end. The interior, especially the dash, was heavily upgraded and new high-end luxury seating and trim replaced the IH-issue material. The powertrain was essentially stock. The 345 V-8 was the most popular choice, but small numbers of diesel 6-33-powered vehicles were converted. As with any Monteverdi, money talked, so a well-heeled customer could get anything he wanted. Occasionally one comes across upgrades not listed on the brochure, engine upgrades being one of the more common.

It is unclear how many Safari and Sahara conversions Monteverdi actually performed. Period magazines listed "expected" production in the thousands, but that is likely optimistic. Once source quotes Monteverdi himself as claiming 3,000 units.

One LST for a Monteverdi from late 1979 indicates a relatively standard Traveltop Scout with a V-345, T-407, TC-145, 3.07:1 axle ratios, air conditioning, and chrome Rallye wheels with Tiempo radials. Among the interesting items on the LST are EXPORT BUILD-UP (code 10520), METRIC SPEEDO-EXP&CAN (10427), and the two-tone color combo—a Black Canyon Black (0001) top and a Tan (1016) body. The black top was a standard Sahara feature seen on all Monteverdi images. The Tan color is more unusual and isn't on the standard color list but is found in the MT-90 catalog.

Monteverdi made a try at a military contract with the 230M, a barebones rig with a 90-inch wheelbase. It used a Scout 4-196, four-speed manual, and TC-145 transfer case but little else from the Scout stable. A few prototypes were built and that was that.

FELBER OASIS

Willi Felber (pronounced *FEL*-bare because he was a French-speaking Swiss) was a contemporary and competitor of Monteverdi. He too built performance and luxury cars and, seeing the profit potential in the wealthy SUV market, transitioned there. He started with the Scout II at almost the same time as Monteverdi, but unlike his competitor, he began with a more conventional conversion he called the Oasis. The base price in 1979 was $34,000 and went up from there.

Comparisons between the Felber Oasis and the later Monteverdi Sahara are inevitable because they are similar. The Oasis was also based on the Traveltop Scout II with enough changes to the body to hide its original identity. The powertrain was

▲ The Monteverdi Sahara was more recognizable as a Scout, though the front-end alterations were significant. Scout fans can decide if it looks better or worse than the '77 Traveltop it was built upon. This one appears to be in the original Fire Orange paint, though Monteverdi was known to do custom colors and treatments.

completely optional according to customer taste. A period magazine story stated one Oasis was retrofitted with a Rolls-Royce V-8. Conversely, it could also remain bone stock to maintain the IH factory warranty and receive service at an IH dealer. The baseline Felber was a V-345/T-407 combo. The Oasis body was more extensively altered than the Sahara's but not more than the Safari's.

Felber eliminated the standard Scout rear side glass and replaced it with large, two-piece sliding windows. The front end was altered to a forward-sloping design with rectangular headlights placed inboard of the standard location. The first Oasis grille had louvers that covered the headlights and radiator. The interior was gutted and replaced with whatever grade of over-the-top plushness was ordered. Felber built many fewer Oases than Monteverdi built Saharas (or possibly even Safaris), but his market ran more to the Middle East and less to Europe.

The Oasis was built more to order than in any standardized production package, but the outside look largely remained the same. The grilles changed, but there isn't enough information to verify if the three styles indicate model-year changes, customer preference, or random changes. It appears the Oasis was built, into 1980, though some sources variously list end dates from 1976 to 1979. It's unclear how many were built, but the number is likely in the very low hundreds or possibly under a hundred. Many ended up in the Middle East and so are not commonly seen.

REGIONAL SPECIALS

A number of regional specials were built to tickle the fancies of buyers in various regions of the United States, Canada, and elsewhere. Typically these were conceived, designed, and ordered by regional executives based on input from the dealer network in their areas.

▲ In 1979, Monteverdi produced a four-door Safari on a Traveler chassis. This was *exactly* what was needed to completely revitalize the Scout. It isn't clear how much Monteverdi influenced the Styling Department or the decision-makers at International, but they all had to have seen these things at some point. As it was, this was only an experiment that never went further than this one prototype.

Perhaps the best known Regional Specials were the Good Times specials of 1978 (see CVI Specials, page 278). Technically, these were designed and built for the Southwest region and did well enough there that the program was expanded to specials for nationwide consumption. Individual dealers were also known to do their own Doll-Ups. Likewise, the Shadow was built by CVI for the Canadian market. Among all of them, Scout collectors are sometimes left with a truck that can't be positively identified, leading to a mischaracterization of a vehicle as something rare. Most times that is not the case, but the regional and dealer specials are interesting in and of themselves even if documentation is nearly impossible to find.

▲ The Outlaw was a treatment for the Pacific Northwest region in 1974. It came in two version: the Outlaw I (top), with its custom paint job and the Outlaw II, which was a bit more conventional in its paint. The custom paint came from a well-known Portland, Oregon, painter known as "The Beard" (a.k.a. Jack Ruckman) who was legendary in the '60s and '70s for his custom paint. Both versions were built on a well-equipped Traveltop with a V-345 and T-407 automatic. They came in a two-tone with a white top, the Custom interior, sliding rear windows, and swing-away tire carriers with a gas can rack. Optional were the American Racing alloy wheels, bucket seats with a center console, and air conditioning. Specials like these turn up in all regions, most not as elaborate as this and often consisting only of some unique appliqués.

▲ Jason Wallace is the proud owner of one of the few remaining 1978 Van American Conversion Travelers. The exterior graphics were done with paint rather than decals and each one was slightly different. Acquired in original condition in 1999, by James and Wanda Setliff, the Traveler didn't need restoration, but it did require a lot of cleaning. The result was winning the Ted Ornas Award in 1999. It's Winter White (9219) with the V-345, T-39 automatic, and two-speed TC-145 transfer case. It has the Deluxe exterior package (16835) and the Radial Tire Chrome Wheel Package (10938), air conditioning (16923), power steering (05281), tilt steering column (05703), sliding rear quarter windows (16893), and luggage rack (16903). It was one of two similar Van American conversions sold at a Sioux City, Iowa, International dealer. *James and Wanda Setliff*

VAN AMERICAN

Van American is the most obscure of the IH-recognized Scout customizers of the late 1970s. Little is known about the company's origins, but it appears to have stayed in business into the early 1990s, primarily doing van conversions. Their involvement with International dealers began and ended in 1978. Their sales materials advertised painted stripes and custom exterior paint highlights rather than decals.

On the inside, the factory seats were reupholstered with a velour-type material in typical lurid 1970s colors. New vinyl door panels were heavily padded and an inch of foam was laid beneath the "high-grade shag carpet." Shag carpet was even applied to the headliner, with some areas covered in white, vinyl-clad ceiling boards. For 1978, the complete conversion was listed at $1,495, with a sunroof optional at $120 and padded sun visors at another $18.

It's impossible to determine how many were done, but the consensus is darn few. Even Van American literature is difficult to find. It isn't clear if they converted anything other than Travelers. Less than a handful of the small number built are known to survive, one notably winning the Ted Ornas Award at the Scout Nationals in 1999.

GLASSIC

Model Years: 1966-71 (1)
Production Dates: late 1965- mid-1971
Introduction: 1965
Number Produced: ~450
Identification: Visually. Glassic Serial number tag. On LST, "Half-Scout" SN prefix 71093, various deleted parts (see below) and "Glassic Industries" in the Ship-To box.

1) For Scout chassis. Glassic and derivatives continued into the '80s but not with Scout chassis.

International was no stranger to selling stripped chassis for conversion and those conversions ran the gamut of every purpose. None were more spectacular than the Glassic Model A Phaeton replica. Glassic development started in 1963 when Florida International Harvester dealer, E.V. "Jack" Faircloth (1905-1987), decided to recreate the iconic and beloved Ford Model A in fiberglass and mount it on a Scout chassis.

The first production run left the Glassic factory in late 1965 on '66 model year Scout 800 chassis produced by a crew of 20 in a 10,000 square foot building

▲ The Glassic prototype as it exists today. It was built on a used mid-'62 production Scout 80 4x2 chassis pulled off the Faircloth IH used car lot. It was completed in late 1965 or early 1966 (Faircloth is unsure of the exact date). This shows what a phaeton is all about—a tub-like open body with no fixed side windows, just snapped-in side curtains for emergency protection. The shelf in back is for a trunk—literally. A steamer trunk was used to carry extra gear. The prototype didn't have all the hardware found on the production models. *Rex Kirkingburg*

▲ Tom Thayer's Glassic is a later one and has the two-tone black and red scheme. The Glassic emblem on the radiator shell has been replaced by a Ford oval. The chassis rolled off the line at Fort Wayne on August 19, 1968, as a 1969 model year chassis.

in West Palm Beach. Glassic generally bought Scout 4x2 chassis in lots of 12 that were shipped together with crated parts, though there was a small run of 4x4s (some sources state 12) done for Abercrombie & Fitch in 1967. Why 12? That's the number that would fit on the semi trailer used to haul them to Florida.

The chassis specifications for the Glassic is largely the same for the model years in which they were produced. Glassic always ordered the basic powertrain; no 4-152T or 4-196 engines when they were optional. A V8 was tried but the small radiator needed to fit the Model A shell didn't have the necessary cooling capacity. Likewise, the optional 4-speed transmission was not offered for the Glassic, only the T-13 3-speed.

Glassic serial numbers started with 101 in late 1965 and when production moved away from Scout chassis in 1971-72, serial numbers were in the low 400s. Exact production of Scout-chassis Glassics is unknown, the company records lost decades ago when the business was sold, but the highest known first-generation Glassic (those built on Scout chassis) is listed on the Glassic Annex is 432. When asked why Glassic moved away from Scout chassis, Faircloth said the Scout II chassis no longer directly fit the Glassic body.

CVI SPECIALS

Model Years: 1979–1980
Production Dates: 1979–1980
Introduction: March 23, 1979 (1979 model year);
   October 18, 1979 (1980 model year)
Number Produced: Undetermined
Identification: Via LST, features, and Good Times or CVI tag on door jamb

The CVI story starts in 1978 with Good Times Inc., an Arlington, Texas–based van converter. Fate brought Good Times and International's Southwestern Zone Office together to make Scout history. Having debuted in 1976 at the peak of the van conversion era, Ed Russell's Good Times was well known in the industry and doing 10 to 15 conversions a day. The Southwestern Zone Office made inquiries to Good Times in 1978 about special editions, but Good Times was so swamped and understaffed it took a while to respond. Chief designer Dick Nesbitt and marketing manager Al Carpenter

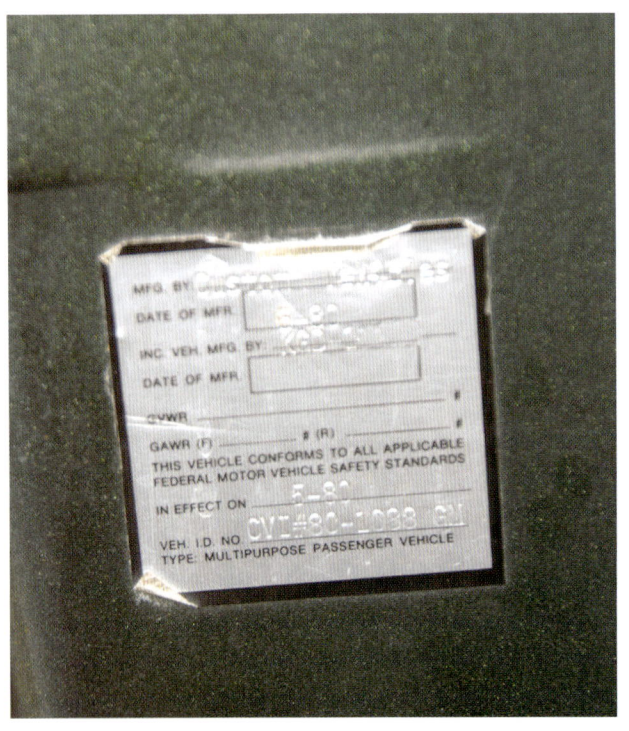

▲ The CVI-built rigs have a door jamb certification tag. Along with the LST, this is a key element in proving the identity of a CVI Scout. This one is from an '80 GMS built in May 1980. The tag is typical of CVIs from Fort Wayne. It's fragile and many are pressure-washed or sandblasted away during restoration.

▲ Here is an example of a Good Times tag from a '79 Midnitestar built in the Texas Good Times facility in June 1979. Rumor has it there are also tags from this era marked "CVI" rather than "Good Times." The "GT1H 136" gives an indication of how many conversions had been done to that time. Whether that was 136 Midnitestars or 136 of all types is not known. The Midnitestars, however, are the most numerous survivors of the 1979 rigs and Dark Brown the most common color.
Steve Nieman

were very new to the company but immediately realized the sales potential of doing specials for International.

Carpenter and Nesbitt offered design proposals for the four Scouts, all of which were accepted for prototype development. The prototypes were ready in August 1978 and named Midnitestar, Travelstar, Terrastar, and Trailstar. The only one that didn't get the go-ahead for production was the SSII-based Trailstar. Good Times began shipping converted Scouts by fall 1978, with the conversions done in the Good Times compound in Arlington. Russell owned most of the buildings in an industrial park on Peyco Drive and one building was dedicated solely to building the Scouts. Russell split the Scout business off into a separate entity called Custom Vehicles International (CVI). These Scouts were earmarked for dealers in International's Southwest district, which included Texas, Louisiana, Arkansas, New Mexico, and Oklahoma. If any new CVI Scouts were sold outside that district in 1979, it was very few, but many have migrated since.

The Southwest specials were enough of a success that in April 1979, Carpenter and Nesbitt were asked to Fort Wayne for a discussion with Jim Bostic, a new manager in the Scout product line, about building specials for the 1980 model year. Nesbitt penned 15 to 20 ideas and they returned to Fort Wayne in June. Two weeks later, CVI was informed they had a deal provisional upon locating an assembly facility as near the Scout factory in Fort Wayne as possible. Russell readily agreed and a building was purchased on Bremer Road, almost across the street from the plant and very near the engineering center and TSPC. Beyond the special models chosen by IH corporate, each sales region also had the opportunity to contract with CVI to make special models. Since the 1st Edition, we have found a number of individual specials that were concocted by the regional sales offices. For the most part, they were variations on the theme, taking various graphics and styling features already designed and mixing them with certain factory options to produce a new look. If these specials had names, like the ones established by corporate, we have not seen them in remaining documentation.

Most of the "mystery" specials we have seen came from the Southwest Region, which had led the charge to CVI in the first place, but we have found one so far from the Midwest region as well. The numbers of these were more than "onesies and twosies" but we don't have a clear picture just yet.

The CVI business in Fort Wayne went gangbusters at first. That success led to more ideas, one of the highlights being the Sunriser, an extremely stylish T-top version of the Scout. Like the other designs, it created quite a stir at IH and at least a strong hint of a go-ahead for 1981 or later, but storm clouds were on the horizon.

News that the Scout was doomed must have hit CVI and Dick Nesbitt like a gorilla backslap. Nesbitt said he first heard in April 1980. By then, he was only part-time with CVI but had been looking forward to a lot more work with International. The bad news started a new round of deal-making between Russell and International Harvester (see Chapter 6). Nesbitt barely missed a beat and carried on with a successful career in design that continues to this day. When Russell's deal to buy the Scout division fell through, CVI continued on for a few years doing van conversions, a few Scouts for individuals, and some dealer-special work on various makes, but the company closed up shop in the early 1980s. Russell died in 1990.

1979 SOUTHWEST CVI SPECIALS

MIDNITESTAR
▲ The Midnitestar was a blacked-out treatment of the Scout Traveltop inspired by the Trans-Am from the movie *Smokey and the Bandit*. The Midnitestar was originally announced on March 23, 1979, as available in Black, Dark Brown, or Dark Blue. CVI designer Dick Nesbitt reports that black Scouts were hard to come by in those early days, but the first prototype Midnitestar (shown here in a 1978 publicity photo) was of the black variety. Less than a handful of Black Midnitestars are known to have been built and only one survives.

▲ The Dark Blue Midnitestars were striking in their own right, though there might have been fewer of them had the black Scouts been more readily available. The package included gold accent striping, gold spoker wheels with 10-15 all-terrain tires (typically Uniroyal Land-Tracs), fender flares, faux hood scoop, rear quarter-window louvers, SSII-style grille, and a tailgate filler panel. The Scout was also given an Imron Clear Coat, a new thing back then that gave the paint a feeling of depth.

SCOUT SPECIALS ENCYCLOPEDIA

TERRASTAR

▲ The Terrastar prototype on its first photo shoot at Lake Arlington in 1978. The Terrastar was built on a well-optioned Terra pickup in Dark Brown Metallic (1606) with a dark brown top. Good Times added a light bar in the bed (faux roll bar) with a pair of driving lights. This light bar was angled forward to match the back angle of the cab and fitted with metal louvers. The side panels were painted tan (no paint code, but restorers report it's easy to match) with accent striping to complement. The black hood and hood scoop were also included. The spoked wheels were painted the same color as the side panels and mounted 10-15 all-terrain tires. Fender flares, the SSII grille, and special moldings completed the look. Like all the Good Times conversions, it was given a top coat of Imron clear for the "wet look." Aftermarket tubular bumpers were mounted front and rear, the front with driving lights and a brush bar.

▶ The production Terrastar had "Terrastar" across the lower body side, but was nearly identical to the prototype in most ways, including the SSII grille. Standard bumpers were used rather than the aftermarkets seen on the prototype. The graphics on the hood differed as well, including a Scout decal.

TRAVELSTAR

▲ The '79 Travelstar was based on a full-boat Traveler and conceived with a "luxury/sport" flair. Nesbitt's first concept shows it with the SSII grille, but this feature was dropped from the production models to understate it a little.

▶ Images of the '79 Travelstar are few and far between, but the first prototype done by Good Times late in 1978 was photographed at the Good Times facility while the paint was still fresh. Surviving Travelstars are almost nonexistent. One complete and mostly unmolested survivor has been found to not have the side louvers, nor the mounting brackets, leading to the possibility that not all Travelstars had them.

SCOUT SPECIALS ENCYCLOPEDIA

1980 FIFTY STATES CVI SCOUTS

HOT STUFF

▶ Hot Stuff, a 100-inch pickup revival that used a Terra top, was designed for the seven-state Southeast zone. This was probably the most elaborate CVI because it required a large number of unusual assembly line instructions. It started with the paint, which was a red CT-399 color (2024). In the "additional options codes" box on the order form, "Brown sport pickup top and bulkhead insert in lieu of Traveltop" was to be written. The special quote number was 430-693, which gave the production line a roadmap for building this rig. Other required standard options were black Deluxe interior (16928), omit cargo mat (16412), bucket seats (16709), and Sport steering wheel (05711). They wanted customers to order the standard wheels (they would be replaced), but they didn't want them to order the chrome or deluxe exterior, appliqués, rear seat, plaid interior, and a few other options. The package included the decals, Hot Stuff graphics, blackout treatment, rocker panel mountings, tailgate filler panel, light bar with two driving lamps, four red-accented alloy wheels with 10-15 BFGoodrich All-Terrains, tinted rear window, black fender flares, rear-body side steps, and wooden bed rails. *Howard Pletcher*

TRAVELSTAR AND TRAVELSTAR XW

▲ The Travelstar was the only CVI package offered on a long-wheelbase Scout for 1980. It featured a secondary brown metallic painted side panel with accent striping on a well-optioned Traveler. It had a blacked-out side molding and the standard CVI lower body moldings. There were two versions: the standard Travelstar and the Travelstar XW. "XW" signified "without the special wheels" and had the standard radial-tire package, which included the Tiempo radials on chrome Rallye wheels. The full-boat Travelstar used the Tiempo radials on gold-trimmed alloy wheels. The Travelstar package started with a Traveler in Winter White (9219) with a Midnight Brown top (10675), Sport steering wheel (05711), tan Custom interior (16834), rear seat (16915), bucket seats (16709), and chrome exterior package (10898). CVI added the special paint and striping, the moldings, alloy wheels (except on the XW), two driving lamps, a special hood ornament, headliner insert panels, upholstered tailgate panel, and cargo area carpeting, as well as the signature CVI center console. *Howard Pletcher*

CLASSIC

▲ The Classic came in three colors: Gray (8547), Saffron Yellow (4417), and the same CT-399 red used on the Hot Stuff (2024). They were accented in black (top sides and upper tailgate) and had the usual CVI lower-body molding and special black striping. The required packaging was modest, one of the three colors noted above, Sport steering wheel (05711), Tiempo radials on standard wheels, and rear seat (16899 or 16915). Included were alloy wheels and the rear tailgate filler panel. They were intended for the Western zone. Just as the book was going to print, John Glancy found a Classic with red striping. *Howard Pletcher*

CLASSIC SOUNDS

▶ The Classic Sounds was a very much upgraded version of the Classic intended for the Western zone. It could be ordered in seven colors: Green Metallic (5901), Tahitian Red (2300), Dark Brown (1032), Winter White (9219), Saffron Yellow (4417), Copper (3201), and Black Canyon Black (0001). All choices were compatible with the gold accent stripes, gold-accented alloy wheels, and black-accented top and trim. As usual, it got the CVI lower-body moldings and tailgate panel, as well as the center console. The big deal in this package was pretty much a one and only from International: a high-end stereo system. The maker of the system is not specified in the spec sheets, but it had an aftermarket AM/FM stereo cassette radio head with a built-in digital clock. A 40-watt amp was used with a five-channel equalizer with two tri-axial and two co-axial speakers, all topped off with a power antenna. Besides the color choices, the required options included Sport steering wheel (05711), automatic transmission (13407), tan or black bucket seats (16709), and Tiempo radials on standard wheels. *Howard Pletcher*

5.6 LITRE

▲ The 5.6 Litre was a sporty CVI that started with a Winter White (9219) Traveltop. First on that list was the V-345 (12114), automatic (13407), Sport steering wheel (05711), black Custom interior (16834), and rear seat (16915). Included were the tri-tone stripes, "5.6 Litre" graphics, black accents, CVI tailgate insert, center console, black fender flares, and red-accented alloys mounting BFGoodrich 10-15 All-Terrains. The 5.6 Litre was designed for the Midwest zone. *Howard Pletcher*

SPORTSTAR

▲ The Sportstar aimed to offer sporty looks without requiring any particularly sporty powertrain options. The package included two-tone striping, black accents on the hood and cab sides, Sportstar graphics, rocker panel moldings, black fender flares, black side moldings, front-bumper rub strips, quarter-window louvers, and the CVI center console. Buyers had to order a Winter White Traveltop (9219), automatic transmission (13407), Sport steering wheel (05711), Off-Road Tire Package (10964), black Custom interior (16834) with bucket seats (16709), and rear seat (16915). Though not listed in the original information, several have been seen with Russet Paid interiors. Likewise, some have been found without the rear window louvers, apparently omitted from the build. *Howard Pletcher*

TRAILSTAR

▲ The Trailstar package could be ordered on a Traveltop Scout in any color but black. Required options were automatic transmission (13407), Sport steering wheel (05711), utility rear bumper (01643), Off-Road Tire Package (10964), and bucket seats (16709). The Trailstar got the blackout treatment, the CVI console, lower body moldings, black side moldings, rubber fender flares, and twin gun racks. The '80 Scout pictured is in Saffron Yellow (4417). The "3.2 Litre" decal was an à la carte option. *Howard Pletcher*

GMS

▲ GMS stood for "Green Machine Sport" and was intended for the Eastern zone. It was available on the 100-inch Scouts with Green Metallic (5901) paint. Required options were the Sport steering wheel (05711), Tiempo radials on standard wheels, automatic transmission (13407), black or tan Custom interior (16834), bucket seats (16709), and rear seat (16915). The package included the GMS graphics and striping in gold, gold-accented alloy wheels, the CVI tailgate panel, heavily tinted quarter windows and rear window, the CVI rocker panel moldings accented in gold, black rubber side molding, and the CVI center console. *Howard Pletcher*

▲ GMS survivors are rare, especially in such original condition. Michael Buonocore's GMS was turned into a soft-top (he still has the original top), but otherwise remains very original. His was built May 1980. The LST indicates an Eastern zone distribution. The "ship to" section indicates "Custom Vehicles GMS Package." That is a typical notation for all CVI rigs, showing which special model was being ordered.

RAVEN AND SHADOW

▲ The Raven (rear three-quarter view) was intended for the Eastern zone, and the Shadow (front view) for Canada. They were very similar treatments, but the Raven had a few more goodies. The ordering spec differed slightly too. The Raven required a Scout 100-inch in Black Canyon Black (0001), Sport steering wheel (05711), radial tire package (10938) or Off-Road Tire Package (10964), Custom interior package in black or tan (16834), bucket seats (16709), and automatic transmission (13407). The Shadow differed only in that it didn't require the custom interior, and the Off-Road Tire Package wasn't listed as an option. The package equipment lists for the Shadow included the Shadow decals and the silver/red accent stripes, rubber moldings, CVI tailgate panel, quarter-window louvers, the extruded rocker moldings with silver accents, twin sport mirrors, and the CVI console. The Raven had all the preceding, with Raven decals in place of Shadow decals, but added a brush guard with a pair of driving lights, Polycast wheels, and tinted rear glass. The Raven did not get the sport mirrors. The Shadow here is pictured with Polycast wheels and the Raven is missing its decal. This likely indicates these are prototypes (though the Polycast wheels could be ordered à la carte).

THE LOST SPECIAL EDITIONS

1979 TRAILSTAR (SSII)
▲ The 1979 Trailstar prototype was based on the SSII. There is little information on what exactly was done, but the graphics, alloy wheels, rear bumper, and thermo-formed plastic tailgate insert used in all the CVI specials are all evident. This prototype was designed and built in the summer of 1978, and it isn't exactly clear why it wasn't selected for production. SSII sales were low, but they were on track for 1979. Likely, it had to do with the large number of SSII-based specials already.

BILLY KIDD SKI SCOUT

▲ The Billy Kidd Ski Scout was a promotional idea built to highlight Olympic skier Billy Kidd's paid endorsement of Scouts. It was built using the full array of CVI doll-up items but in a special color scheme and with Billy Kidd decals. Strangely, there seems to be no evidence that any were sold. One was built, that's for certain, and it made the cover of the January 1980 *Off-Road* magazine. Given magazine lead times, the story was likely shot sometime in October of 1979. The story and a few factory images are all that exist. *Howard Pletcher*

CVI POLICE SPECIAL

▲ CVI developed a police package for the Traveler, and while it was offered, it's isn't clear how many, if any, were sold. Pictures from late in 1980 show one on display at a convention in Poland. It may be the same truck photographed for the marketing shot seen here. The promotional materials list a number of accessories: prisoner-confinement cage including Lexan security shields, rear security and storage area, radio equipment access compartment, federal light bar and siren package, hood and lower-body graphics, lower-body side moldings, reflective stripes and police ID, rear-window defogger, side-window privacy appliqué, 80-amp alternator with high-output cable to rear compartment, fire extinguisher, reading lamp, billy club holders, hat caddy, shotgun mount, storage bins, and first-aid kit. Optional items included a second battery, spotlight, backup alarm, and roof numerals. *Howard Pletcher*

1981 SUNRISER

▲ Of all the work designer Dick Nesbitt did for CVI, this one makes him proudest. T-tops (a.k.a. Targa tops) were all the rage in the '70s. Having one in the lineup offered marketing cred on the hip side of the style meter. Nesbitt started work on the T-top Scout 100-inch eventually dubbed "Sunriser" in February 1979. When tentative approval took them to the next step, he worked with American Hatch engineer Lance Clark to prototype a vehicle. American Hatch was a T-top company acquired by CVI owner Ed Russell. Nesbitt was uniquely qualified to do the Targa design because in 1972 he had been the primary designer of the new Bronco, a vehicle that didn't appear until 1978. A T-top had been planned for the Bronco from the get-go but was not implemented. Nesbitt employed many of the same features in the Sunriser, utilizing the rear section of a Traveler top, complete with the hatch, to make the prototype. International execs were anxious to see it, and once a prototype was complete in February 1980, it was given tentative approval as a 1981 offering. *Richard Nesbitt*

▲ The Sunriser survives, though it's not as pretty as it once was and missing the Targa top. Coauthor John Glancy is its custodian now and hopes to bring it back to pristine condition. He may even reproduce the T-Top. When the Sunriser project was ended and Scout's fate sealed, the T-Top was removed and destroyed. Nesbitt says it was more a mockup than a functioning top, and that's why a standard top was installed. When IH began auctioning things off in 1980 and 1981, it was sold to an employee. The interesting back story is that the Sunriser was built from a Monteverdi conversion. It was a 1978 Scout II built June 2 of that year and shipped to Switzerland for conversion. How and why it returned is unclear but the Sunriser development started in late '79 and was completed early in 1980. As of 2023, the Sunriser can be seen at the Super Scout Specialists Museum in Enon, Ohio.

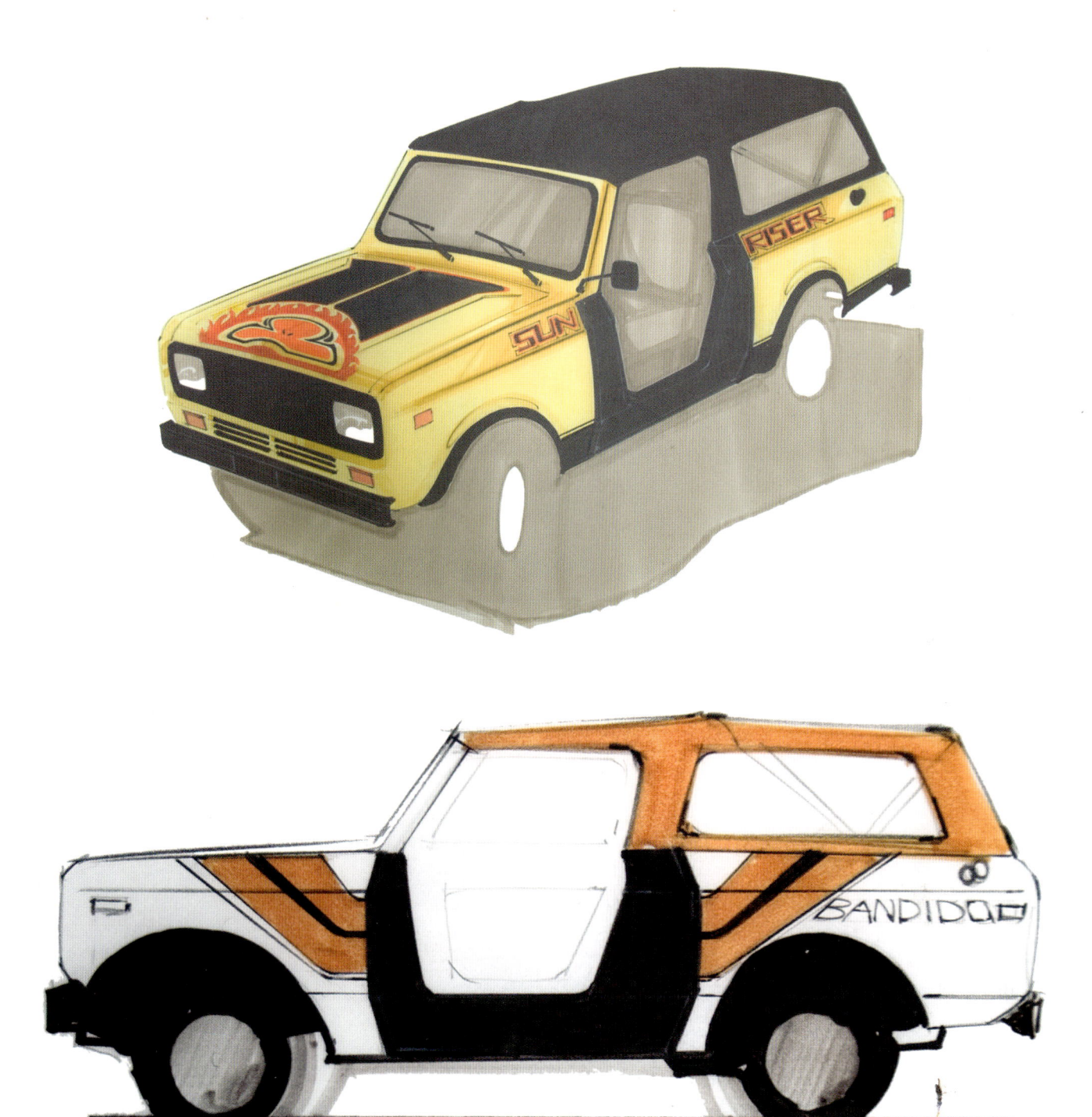

SSII PROPOSALS

▲ Dick Nesbitt's fertile mind turned out a lot of proposals for CVI Scouts—too many to show here. Often, design proposals don't make it past the point of a rough proposal and a quick look. While the Sunriser looked to utilize some Southwest/American Indian motifs, the Bandido was a much less developed idea. These were both done for the April 1979 meeting with Scout Division GM Jim Bostic in Ford Wayne. Visually, the Sunriser looks like it might have had some legs, but it went unpursued, likely due to very low SSII sales volume. Perhaps there was a sense that the SSII might not make it into 1980.

SCOUT SPECIALS ENCYCLOPEDIA

▶ The 434 and 844 are so rare even the leftover photos are difficult to find. Dick Nesbitt confirmed the striping kits came from CVI. Conceived for the 1980 model year, they were an attempt at jazzy economy models. The 434 (4-cylinder engine, 3-speed, and 4-wheel drive) was a base model with a couple of options, a stripe kit, and hubcaps. The 844 (8-cylinder, 4-speed, 4-wheel drive) was outfitted along the same lines. They were announced on March 21, 1980, and given Gold Star numbers 0-SCTT-44-502 and 0-SCTT-44-503, respectively. The 434 had a 3.73:1 axle ratio and was EPA estimated for 15.8 miles per gallon city and 19.1 on the highway. The 434 package included an AM radio, wheel covers, rear seat, a vinyl spare-tire cover, and passenger-area carpeting, along with the exterior décor. The 844 came with the 345 V-8 and was EPA estimated at 13.9 miles per gallon city and 21 highway. The package was identical to the 434 except for the V-8, a four-speed T-428 close-ratio transmission, and 2.72:1 axle ratios. Almost as they were announced, the writing began appearing on the wall. The 434/844 program was cancelled on June 4. The 844 was spotted in pictures of the 1980 Chicago Auto Show. It was thought only the two prototypes had been built but Nesbitt vaguely remembered more and after the 1st Edition came out, a surviving 844 was found, and it wasn't the one pictured.

SPECIAL EDITION RS

Model Year: 1980
Production Dates: ~June 1–October 6, 1980
Introduction: March 21, 1980
Number Produced: 8
VIN Range: Between KGD14656 and KGD23035
Identification: Options code 16838 on LST, visual features

The Special Edition RS (Rally Sport) Scout was the last notable Scout special from the minds at International and not many were built. This luxury rig signaled International's move toward more high-end offerings (see Chapter 6). A March 21, 1980, Product News Bulletin (G-878) announced the RS, along with the 434 and 844 fuel-economy specials. The RS package was designated option number 16838 and was available on gas or diesel Travelers (model codes K010203 and K010303, respectively). The RS interior was a burgundy International called Russet that had

▲ This Styling Department image was first seen in the March 21, 1980, introduction packet to dealers. It was apparent the RS was a stylish beast. Other than the pinstripe, the red-accented wheels, and the red top, everything on the outside could be ordered on any Traveler. This is the prototype RS, which rolled off the line on October 5, 1979, after which it was sent to engineering. This vehicle survives, albeit in rough shape, owned by Scout Connection, a Scout Light Line dealer in Iowa. It's a V-345A V-8 automatic with 2.72:1 axle ratios.

SCOUT SPECIALS ENCYCLOPEDIA

▲ A view through the rear hatch shows some of the over-the-top velour that was so popular in the '80s. The interior is where the whiz-bang resided—for fans of velour, that is! It's been said the RS was the most luxurious Scout built. Perhaps that's true, with the possible exception of the Midas conversions. One notable feature of this prototype was its reclining front seats, an option that did not appear in the production units.

been available previously, though the plush interior was new, with velour pillow-top front seats, velour headliners and door panels, deep carpeting in both the passenger and cargo compartments, a woodgrain insert on the automatic shifter console and instrument panel, and a Russet vinyl spare-tire cover.

On the outside, the paint was Tahitian Red (2300) and included a special pinstripe, a specially painted Tahitian Red top, color-keyed Polycast wheels with P-225/75R-15 Goodyear Tiempo tires, tinted glass, and the Deluxe exterior trim package (chrome bumpers, silver grille with bright trim, bright glass trim, black/bright side molding). Other goodies included dual horns, day/night rearview mirror, and lighter. To that, the buyer added the other options of their choice, including any of the available engines, transmissions, gear ratios, air conditioning, and a few others. There were a few restricted options with the RS, including right-hand drive, rear-step bumper, off-road tires, and a large number of cosmetic options that clashed with the package. The original list price for the RS package was $2,062.

No doubt there were high hopes for the RS, but not long after the late-March announcement, the Scout Division was up for sale. By June, a number of model-line changes and consolidations were made, but the RS survived the cuts. In a June 10, 1980, Western Region sales bulletin, the ordering deadline

▲ The nicest surviving RS currently known is Mark Drake's. It's a low-miles, one-owner, rust-free original that Mark found in New Mexico. It was built October 3, 1980, and is the third-to-last RS built. It's a 6-33T diesel with a four-speed. The interior is in nearly as-new condition, including an operational AM/FM eight-track. *Mark Drake*

was June 20, 1980, and deliveries based on orders were slated for the months of September and October 1980. The letter really pushed the diesel Traveler (International was trying to use up its significant stock of Nissan diesels) and all the production RS records found indicated diesels. The letter announced the list price of the options had been dropped from $2,062 to $1,462.

An RS LST will bear the signature 16838 option code, "RS SCT TAHITIAN RED," as well as several other unique codes. The ship-to box will generally have a handwritten "RS." Of the seven production RS LSTs found, five were marked for shipment to CVI. The significance of this is unclear, but speculation is that CVI was perhaps doing some of the custom-painting like the pinstriping or the color-matched wheels.

The prototype RS was built October 5, 1979 (VIN KGD14656), and it is the only gas RS on record. A diligent search revealed only seven more RS LSTs, all built October 2–6, 1980, with sequential VINs from KGD23029 to 23035. It is known with certainty that KGD23035 was the last RS built. One of these is in the possession of Scout Connection and the other was owned by Mark Drake at Scout MADness (a Scout Light Line dealer) but sold just as the 1st Edition was going to print. The prototype RS gasser is also owned by Scout Connection, making a total of three known to survive. Sightings of two more have not yet been confirmed.

SHAWNEE

Model Year: 1980
Production Dates: #1, unknown; #2, October 9, 1979; #3, October 9, 1979
Introduction: February 23, 1980
Number Produced: 3
VIN Range: #1, unknown; #2, KGD14795; #3, KGD14794
Identification: Visually

In 1980, Hurst had been in business 22 years and had a reputation making, or should we say *re*-making, cars to project a sportier and more exciting aura. Some of their work was strictly "show" while some was more on the "go" side of the equation. Either way, it was a pretty good bet a Hurst badge on any manufacturer's truck or automobile would inspire some excitement.

Little is known of how the Scout Division and Hurst hooked up. Jim Bostic, the general manager of the Scout Division, has long been suspected as the instigator. Bostic died in 1996, but an interview with his brother, Steve, connected some dots. Jim had a prior relationship with Hurst that dated back to his time as a marketing executive at AMC. After coming aboard at IH, he would have looked around for ways to generate marketing pizzazz and it's logical to assume he fell back on his past relationship with Hurst. A few documentary hints lead to late 1979 and one Scout went to Hurst in Michigan for development in that time frame. Beyond a few images, though, not much is known about what transpired there.

The next step came on October 9, 1979, when two special-ordered, identical 100-inch Scouts rolled off the line with sequential VINs and Line Sequence Numbers. They were given V-345A four-barrel engines, T-407 automatic transmissions, TC-146

▲ Shawnee #1 in an early stage of development. This is the rig upon which all the styling and product test-fits were done. There's no date for this image, but the foliage suggests it was taken in late 1979. The surroundings are not typical of Fort Wayne, so it may be near the Hurst facility in Michigan. The image's original source was Dick Bakkom, someone in the approval pipeline, and it was probably sent from Hurst as a progress shot. There is a lot to point out, the first being the extra-tall Targa top containing a set of driving lights—a design that was thankfully not used. Note the SS emblem is located on the Targa side panel, where the Hurst decal was later located, rather than on the rear fender. The GT-style outside mirrors are not yet fitted; instead, it has regular SSII mirrors. The mirror mounting holes distinguish #1 from the other two Shawnees. The CVI-style rocker cover is not in place, only one small, bright strip at the bottom. The Hurst shifters are visible, as is the digital dash insert, which may have been installed on only the one Shawnee. This Scout also appears to have a suspension lift (note the spring arch and the room above the tires versus the later pics) and Formula Desert Dog Xtra tires. A side step is mounted, something not seen on the other Shawnees.
Dr. Todd Sommer

transfer cases, and 3.54 axle ratios with RA-18 rear axles. They were ordered in Black Canyon Black (0001) with the Traveltop and doors deleted, SSII inserts and a Terra bulkhead installed, and a Scout roll bar. Both outside mirrors were omitted and only a temporary driver seat was installed. Street-tread H78 tires were mounted and no spare was included. The LSTs for these two Scouts are a showcase of unusual markups and seldom-seen option codes, some listed under the heading "Ordered by Description," a sure sign something unusual was about to happen.

The next destination was the Fort Wayne TSPC. The ship-to boxes don't list another location but are marked "Hurst Show Scout" and "Will Call." Does that mean the Hurst parts were installed at TSPC, or did they go off to the Hurst facility in Michigan? The documents are ambiguous and the timing unclear. Hurst was known to simply supply parts for installation by the primary manufacturer once the development process was complete, so it could have gone either way. No one at Hurst remembers, nor does anyone from IH who was directly involved.

The modifications designed or designated by Hurst began with the iconic Hurst Shifters (transmission and transfer case) and badging. A pair of Cobra Super Form racing bucket seats were installed, along with a small-diameter race-look steering wheel with a Hurst emblem.

▲ This is one in a group of styling shots showing configuration changes on Shawnee #1. It's seen here with the final-design Targa top and SSII-style soft doors. The Tracker A/T tires (versus the previous Formula Desert Dogs) indicate the shots were taken later in #1's life. This image, and the three others of #1 shown here, are Styling Department images dated August 1980, a very incongruous date. *WHS 116502*

▲ Hurst Shawnee #1 in open-air mode with the fiberglass Targa top and soft doors removed. The mounting holes for the SSII-style mirrors are visible; even at this late date, the GT mirrors used on #2 and #3 have not been installed.
WHS 116503

Outside, the signature fiberglass Targa top with a removable roof panel was the most prevalent feature, with an integrated fiberglass bed tonneau. The rear windows were either a clear vinyl piece that snapped in place for #1, and a glass panel on #2 and #3 as completed. Standard SSII soft doors were used. Up front, a brush bar was added that mounted a pair of rectangular KC lights. Chrome spoke wheels mounted the ubiquitous 10-15 Goodyear Tracker A/T tires. Early pictures show Shawnee #1 with a lift and larger Formula Desert Dog tires. A bright rocker panel cover was installed, and white and pale gold striping (the Hurst signature colors of the era) was added at the nose and tail. The rocker covers on #2 and #3 appear to be the same pieces used on the CVI conversions. They wore special Native American–themed Shawnee Scout "SS" graphics on the hood and rear quarter panels. A Hurst badge on the dash read "Hurst Special Vehicle Limited Edition #_" with the number stamped. The newly minted Shawnee SS Scout design prototype was designated #1 and the others were #2 and #3. Strangely, the Scout with the lower VIN became #3 and the higher became #2.

A February 28, 1980, Truck Group News release stated, "Inside, the Shawnee Scout features the first digital instrumentation offered on four-wheel drive vehicles, with digital readouts for the speedometer, tachometer, oil pressure, volts, engine temperature

▲ Most will probably agree this removable light bar is a lot more aesthetically pleasing than that in the previous shot of the extra-tall Targa. This image is the only time the accessory is seen. Close inspection indicates the light bar may be a mockup, but the preliminary price sheet lists a similar product. WHS 116500

and fuel level. It also incorporates a low fuel level warning and a headlights-on delay for illumination when leaving the vehicle." The date of this release coincides with the 1980 Chicago Auto Show (February 23–March 2), where Shawnee #1 was on display.

Shawnee #1 was evaluated on March 7, 1980, by W. J. Heidinreich in Product Evaluation not long after returning from Chicago and found to have many faults, several of them related to the digital dash placement and wiring. Neither #2 nor #3 had the digital dash. A price sheet was done up at the time of the Auto Show with a suggested retail price of $12,998.50, including most of the items shown on #1.

Another price sheet done later offered more detail and a package number for the Shawnee (10683) listed at $4,510. The light pod was given part number 08871 and listed at $275. The total list price was going to be $13,272.50, including a $7,748 cost for the Scout and the required options. Another listed option included an AM/FM stereo cassette (08855) for $1,014. The price sheet was in a 1980 data book, and until the decision was made to sell the Scout Division, the Shawnee was on track to be an optional package for 1980 or 1981.

The fate of the three Shawnee Scouts has long been a source of great interest in the Scout community. Their histories as known in 2023 (leaving out some

SCOUT SPECIALS ENCYCLOPEDIA 305

▲ A rear view of #1 shows the nifty fiberglass tonneau open. One notable difference between #1 and #2 or #3 is the snapped-in clear-plastic rear window versus solid glass or Plexiglas. *WHS 116501*

▶ Here are Hurst Shawnee #2 (foreground) and # 3 together in the Arizona desert, late spring 1981. At this time, Steve Bostic owned #3 and Don Painter #2. Bostic reports they were identical in every way, though when purchased, #2's Hurst horn button emblem had been pilfered. Both used the standard Scout rear-step bumpers, painted black, even though chrome rear bumpers were listed on the LST. *Steve Bostic*

▲ The Shawnee concept combined the sporty open-door aspects of an SSII with the utility of a pickup, which is made clear in this tailgate-down rear view of #2. While there was seating for only two, there was plenty of room for camping gear or what-have-you in back. *Steve Bostic*

▶ Shawnee #2 in the desert in 1981 with the top and doors in place. This is likely the configuration that would have been offered had the Shawnee been offered for sale in 1980 or later. Today, most would gladly pay the original 1980 proposed list price of $12,998.50. *Steve Bostic*

SCOUT SPECIALS ENCYCLOPEDIA 307

of the more speculative and unconfirmed reports, as well as the names of the current owners to protect their privacy) are:

#1 – In 1981, one Shawnee was reported as being held by International, pending a possible sale of the Scout Division. There are reliable reports of a Shawnee stored with other prototypes in the factory building in 1981. If the other two were in Arizona, then this must have been #1. From there, the trail goes stone cold.

#2 - This rig was sold to Don Painter, in Mesa, Arizona, early in 1981. Its early history is connected directly to #3 because Painter, a test engineer with Massey-Ferguson, was friends with Steve Bostic, brother of Jim Bostic, former head of the Scout Division, who arranged the sale of both Scouts. In early spring 1981, Painter and Steve Bostic flew to Fort Wayne and picked up and drove both Shawnees to Mesa, Arizona. It was early enough in the year that they encountered spring snowstorms, but by Steve's recollection, it was a fun trip. Painter sold #2 to Joe Flores in California in January 1982, and the Scout faded into obscurity, the location known only to a few. It was up for sale in 2015 with a $175,000 asking price but, not surprisingly, had not sold by the time this book went to print. It was advertised as having only 3,700 miles at the time. It was acquired in 2021 by noted Scout collector Mike Grieble and is the headliner in his stellar collection.

#3 - This Shawnee was purchased Steve Bostic in early 1981 through his brother Jim. This was around the time Jim's job as head of the Scout Division had ended and he had moved to Iveco Trucks. The date of the invoice was May 6, 1981, though the actual sale and delivery took place earlier. The invoice showed Steve paid $4,118.89. Bostic kept and used the Scout for a few months in 1981 and then offered it for sale locally in July 1981 for $8,800. There were no takers so he advertised nationally in October, finally selling it to a Phoenix car dealer. From there it was sold twice more and ended up in Iowa.

▶ The closest available thing to an interior shot, this one of #2 (note missing horn button). The Cobra seats and Hurst shifter platform are visible. Forward of the shifters is the switch panel for the off-road lights. It's also obvious this rig has standard Scout instrumentation. *Steve Bostic*

▶ Shawnee #3 in all it's unrestored glory in recent times. Shown at the 2016 Scout and All Truck Nationals, it was brought to the even by the Munson family in honor of the late Alan Munson who purchased it from a Prescott, Arizona, dealership in May of 1984. It's been in the Munson family ever since. Alan was an off-roader, so the Shawnee shows some minor battle scars but it's in remarkably good and complete condition.

▶ The 1980 Chicago Auto Show was likely the last big-money show for the Scout Division before it was offered up for sale. Shawnee #1 is in the background at the center of the shot with another aborted 1980 special, the 844 (in blue to the right). The #1 Shawnee is identifiable by its snapped-in rear window. Just beyond the Shawnee is the RS prototype, distinctive in its Tahitian Red roof.
Chicago Auto Show

SCOUT SPECIALS ENCYCLOPEDIA   309

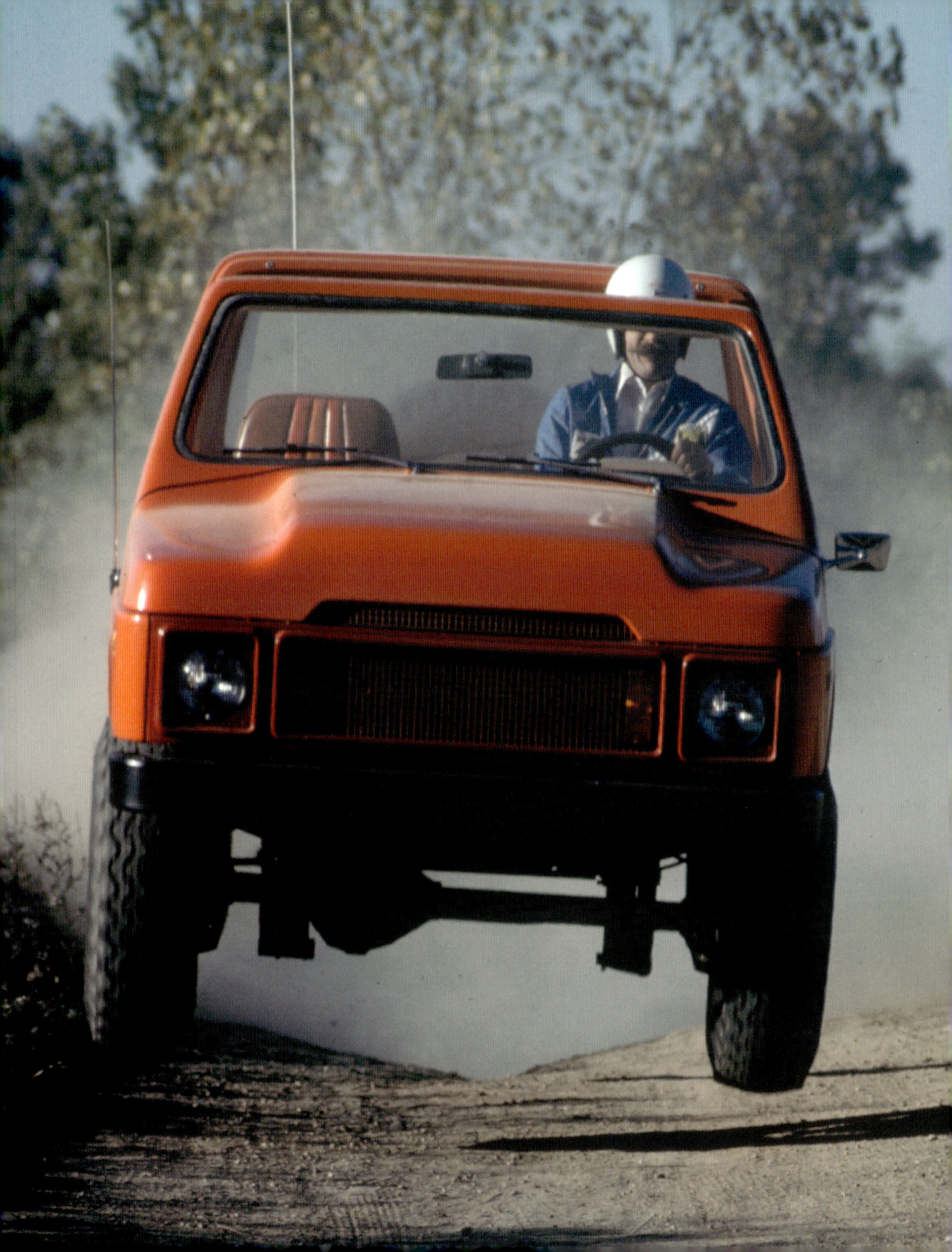

Chapter 6
SSV, HVS, AND 1981–1985 SCOUTS:
WHAT MIGHT HAVE BEEN

The 1981 model year planning was very complete at the time of Scout's demise in September 1980, with planning as far ahead as the 1985 model year well underway. All the work on the product was necessary but so, too, was expanding sales outlets onto Main Street, though little documentation exists to indicate a specific sales and marketing strategy. It is known International was expanding into car dealerships and there are examples of Scouts sold out of Pontiac and Toyota dealerships.

The *Scout Business Plan Overview* booklet from October 1979, extensively addressed how "premium" products were key to a successful Scout business plan, stating "margin improvement through substantial premium pricing—well above the competition (e.g., Mercedes)—is possible only when a substantially differentiated product is offered to justify the purchase cost difference to consumers—and such a differentiation is not readily apparent in the current Scout product."

Product planning for the 1981 and later Scouts focused on moving upmarket, with an obvious lean toward the Mercedes-Benz side of the meter. The British Range Rover, though not yet imported to the United States, was discussed in Scout business documents as an example. This proved a remarkable parallel, as history would show.

Range Rover started production in 1970 as an all-terrain station wagon for England and Europe. As it gradually went upmarket with luxury features and premium pricing, it became more profitable for its parent company, Land Rover. Brought to North America as a 1987 model, the Range Rover had a decent array of luxury features, European performance, a very carlike ride, and was extremely capable in the dirt. Working against it were appalling fit and finish, haphazardly applied luxury features, a tiny dealer network, and very little brand awareness—traits the

◀ The first running SSV prototype in June of 1977 at its first photo op on the back lot at the Fort Wayne test track. Of the next few years, it would get a fair amount of driving time during test when it made the rounds of the country at shows and races. *WHS 110570*

▲ Dating to 1979, some of the first visual representations of the '81 models included the 350S ("S" indicating Sport or Standard) and the base-model 100-inch wagon. The appliqué was new but not slated to be a standard part of the package. Note the absence of the Scout II badging on the rear.

Scout shared to varying degrees. Despite these shortcomings, in a few short years, a very astute North American marketing team placed Range Rover as the very profitable king of SUVs, and after setting up sales outlets inside existing high-end car dealerships, opened up stand-alone Land Rover Centres in hot markets around the country.

The Scout could have followed that same path to success. In fact, the division had taken the first steps in that direction, but the finances weren't there, nor, probably, was the corporate will to take some of the risks needed for such an unfamiliar course. And then there was timing. When IH pulled the plug on Scout, the SUV market was still stumbling. Not long after Scout's demise, the market regained its profitability and very soon became a focal point of the American automobile market.

THE 1981 MODELS

Based on the 1981 New Model Preliminary Information booklet and *Scout Business Plan Overview*, the 1981 Scouts would have remained an all-4x4 lineup with improvements including an all-new instrument panel, the availability of power steering with right-hand-drive vehicles, new interiors, a new suspension with tapered-leaf springs for an improved ride, an easy-start feature for the Nissan diesel, and improved soundproofing. A new tilt steering column would feature wiper controls

▲ Some of the standard interior that was to go with the 350S, highlighted by the new vinyl seat design. Not a great deal else to see in this August 1979 shot. This Scout does not have the new dash design or the other accoutrements being developed.

on a column stalk, with an intermittent feature. A new flow-through ventilation system would be incorporated into a new HVAC system. Cable-operated window regulators would improve operation and later open the door to power windows. On the safety side were new three-point seatbelts. A rear window defroster was planned for the option list along with an AM/FM cassette stereo with four speakers. At least one new engine was likely to be introduced and a 30-gallon fuel tank option was to be added. A new composite top was on the horizon for the 100-inch Scouts in 1981 or 1982.

New model designations were listed. Gone were Scout II, Traveler, and Terra. The short-wheelbase Scout became the Scout 350 and the 118-inch would

▲ This was more or less the new dash design selected for 1981. The Scout badly need a dash facelift and a great deal of time was spent on this upgrade. The SSV-100 used almost the same dash. This image dates to about April 1980.

▲ This April 1980 image shows a 450 model with a proposed woodgrain appliqué that beats the old Jeep Wagoneer at its own game!

▲ The proposed 350DS (Designer Series) in April 1980. It's an upscale, understated look from the outside and not a bunch different from the 1980-and-earlier 100-inch units.

▲ The '81 450LE (Limited Edition) was similarly understated, perhaps even underwhelming, but with different graphics than before.

be the Scout 450. Additional suffix designations were to be added. In the 350 line, the S (Sport) was the base model. The SC (Sport Custom) was the mid-level offering, and DS (Designer Series) was the high end. In the 450 passenger Scout line, S (again, Sport) was the base, the 450DS was the mid-level model, and the 450LE (Limited Edition) was the top dog. It's unclear how the brand name "Scout" would have been portrayed in the new model lineup because the work on badging was not finalized.

Two pickup models were listed as well: the base S (Sport), which was also tentatively named Apache, and the SC (Sport Custom) called the Aztec, which would have had the upper-end features in some measure but not be equivalent to the top dogs in the wagon lines.

International was working toward replacing all its legacy gasoline engines and the new powerplants would not likely have come from their own Indy Engine Plant. The IH engines were too heavy for the drastic weight reductions required, underpowered in the market for their displacements, and, most importantly, not up to upcoming emissions regulations without a great deal of expense. Further, the profitable International medium-duty truck lines were following industry trends by moving away from gas engines and toward diesels. The legacy gas engines in the medium duties, also used in

▲ The proposed Coachman interior was an overstated upscale option in the style of past American luxury cars. A folding third-row seat shown in a series of images from April 1980 is a possible giveaway the idea came from Midas, as are certain other small accoutrements. The limited info available indicates it was rejected. It's pretty clear International was going for the more subdued Euro look, so it's no wonder this one didn't make the cut.

the Scout, were slated to die off without replacement in fairly short order.

Engineers were test-fitting outside-sourced engines at least as early as 1978. Finalized blueprints for a Chrysler 225-ci slant-six installation in Scouts are dated early in 1979, and two slant-six-powered test Scouts that survived were sold in one of the many engineering department "yard sales" from 1979 to 1981. There could have been as many as 20 slant-six Scouts built as early as February and March 1979, and product-planning documents discussed the Chrysler sixes replacing the V-304 and possibly being the base engine. Documents from the Indy Engine Plant indicate Chrysler 318 and 360 V-8 engines were also being tested as replacements for the venerable V-345, and at least one GM 350 V-8 was tested in a Scout with the same idea in mind.

The Nissan turbodiesel would have played a bigger part in 1981, as it had in 1980, possibly past 1982. One product-planning document speaks of replacing the 6-33T in 1983 with a 2.5-liter Nissan 4-25T four-cylinder diesel. A Peugeot four-cylinder diesel was also on the list for consideration and plant employees confirm testing a Perkins diesel four in a Scout.

▲ The 350SC (Sport Custom) was a nice-looking rig with a silver appliqué. It appears to be a middle- to upper-level package.

▲ The 350SC interior in blue was reminiscent of the 1980 models and their plaid cloth upholstery. Again, the new dash design had not yet been fitted. The VIN tag is enhanced in this image. This Scout was built in September 1979, so it was a 1980 production rig being used by Design as a test mule.

▲ The Terra pickups hadn't sold well in the late-'70s, but there was still a place for them in the 1981 lineup. This image was labeled as being the base "S" Apache with an optional appliqué.

▶ A series of images was shot to show the adaptation of a Traveler top to a 100-inch Scout. This is the end result and dates to the latter part of 1979. There are indications this could have appeared as early as the 1981 model year. The difficulties encountered revolved around getting the rollover protection built into the top. With the '80 grille and a new interior, the '81 lineup would have been greatly upgraded.

▲ In July and August 1978 this group of renderings by Larry Nicklin represented a look ahead to 1981 and would even form the basis for ideas looking as far ahead as 1985. One of these is wearing a shorter version of the Traveler top and rear hatchback. The Euro influence is evident; there's even a slight resemblance to a two-door Range Rover.

A 1979 Future Scout Proposals document discussed the merits of eight engines, including the existing IH engines but also a Peugeot V-6, a Peugeot SDZ4 turbodiesel, and a turbocharged 225 Chrysler slant-six. A DV-345 diesel, based on the V-345 gas block, was being discussed as well, both for Scout and the medium-duty lines, and a turbocharged 4-196 had been discussed too.

Two new transmissions were listed: a T-37 four-speed from Tremec with an overdrive fourth gear, and a T-410 automatic, which may have been the smaller Chrysler TorqueFlite automatic with a lockup torque converter.

A short-wheelbase Scout with a drastically revised roofline was on the agenda for 1981 or later. The answer was as simple as redesigning the Traveler top to fit the 100-inch chassis; it's amazing how it improves the look. Photographic evidence of at least one prototype built in 1978 was shown in the *Future Scout Proposals* booklet from 1979. It hints the top was on the agenda for 1981, though the 1981 Preliminary Model Information booklet doesn't mention it.

One thing is clear: development of the 1981s was not 100 percent done, but it essentially stopped the moment the decision was made to sell the Scout Division. The available information is like a movie

▲ Larry Nicklin's rendering of a possible 1985 design is dated April 1980 and definitely shows a redefined Euro direction.

▲ This full-sized clay model was called "Design A." The image is dated April 1980 and is obviously a conventional steel-body-on-frame design that shares some lines with the Scout II. A version with single rectangular headlights was also done.

▲ Design B differed substantially from Design A in the front end and rear end treatments, with a band of color contrast along the base of the windows. The body shape remained substantially the same as Design A.

▶ In looking for a precursor to the composite Scout, one need look no further than this cart-like 4x2 buggy on an 800A chassis shortened to an 85-inch wheelbase. This rig was a part of a styling exercise instigated by Ted Ornas and built in 1970, but little more is known about it other than it had connections to the Scout 85 program.

▲ In March 1970, with the Scout II on the horizon, members of management expressed a desire to keep the 800 line in production as a low-cost or commercial product. Sales only saw this as a way to dilute sales of the new Scout. The idea morphed into an 85-inch-wheelbase version of the new Scout 810 chassis with a composite body and was codenamed Scout 85. The project was approved in September 1970 with the go-ahead to produce drawings and a clay model. The tentative plan was to introduce it as a 1973 model with as many as 5,400 units. Uniroyal Plastic Products produced four prototype bodies by March 1971 and one was installed onto a shortened chassis built by the Prototype Shop. By March 30, 1971, the SPC had viewed the complete vehicle. Sales disapproved, citing lack of a market. After discussions, production was rejected. There was some talk about producing it in Mexico for export, but that idea went nowhere.

that stops mid-frame just as the plot is developing. Scout Division's general direction can be seen in that early-1980 time frame, but not its final destination. Using the first announcements of the intention to sell the division in May as a marker for when development stopped, it's clear IH had plenty of time to refine the choices. There just aren't enough connectable dots to reliably make predictions on exactly what the "final answer" would have been, or to make any predictions on the success or failure of those changes.

SSV: THE SCOUT CORVETTE?

The SSV (Supplementary Scout Vehicle) was slated as a 1981 or 1982 limited-production, fiberglass-bodied coupe and designed to do for International what the Corvette did for Chevrolet. The SSV morphed out of a 1973 program to streamline commercial truck development and reduce manufacturing costs by developing composite bodies or body parts.

"Composite" is a loosely defined term encompassing various ways to use fiberglass and plastics. The touted advantages were lower weight and manufacturing costs and easier and less costly styling changes. The biggest downside at the time was the costly transition from traditional methods. This was not International's first look at composite bodies, but a decade later, the technology had advanced and become less costly to implement. Ted Ornas, among others, instigated another look, the initial focal points being the medium- and heavy-duty truck lines. Outside-sourced fiberglass front ends were already in use at IH and it was a natural progression to a composite cab.

▲ One of the first times a composite Scout rendering came to light, if not the first, was this October 1975 drawing by designer Chuck Harris. Because virtually any design was possible with composites, Harris went a little crazy with it, though some elements translated to later ideas, such as the integral roll bar.

▶ It's difficult to keep track of design changes from 1976 into 1977, but it's recorded this quarter-scale model was built in the February–April, 1976, time frame and photographed in June. It gives an early look at the general appearance of the upcoming SSV.

▲ Here is the first full-sized SSV clay model, shot in September 1976. Compare this to the scale model picture nearby and note the lines of the hood as it meets the fender near the firewall. This complex feature was altered in the later designs.

SSV, HVS, AND 1981–1985 SCOUTS: WHAT MIGHT HAVE BEEN

▲ The rear view of the first SSV clay model with doors installed. Note the removable top and squared-off door edges, as well as, again, the lines of the hood where they meet the fender and firewall.

▶ An exploded view of the early SSV-95 body and its seven major components. The pieces were resin and glass bonded together. Each had composite structural members for strength.

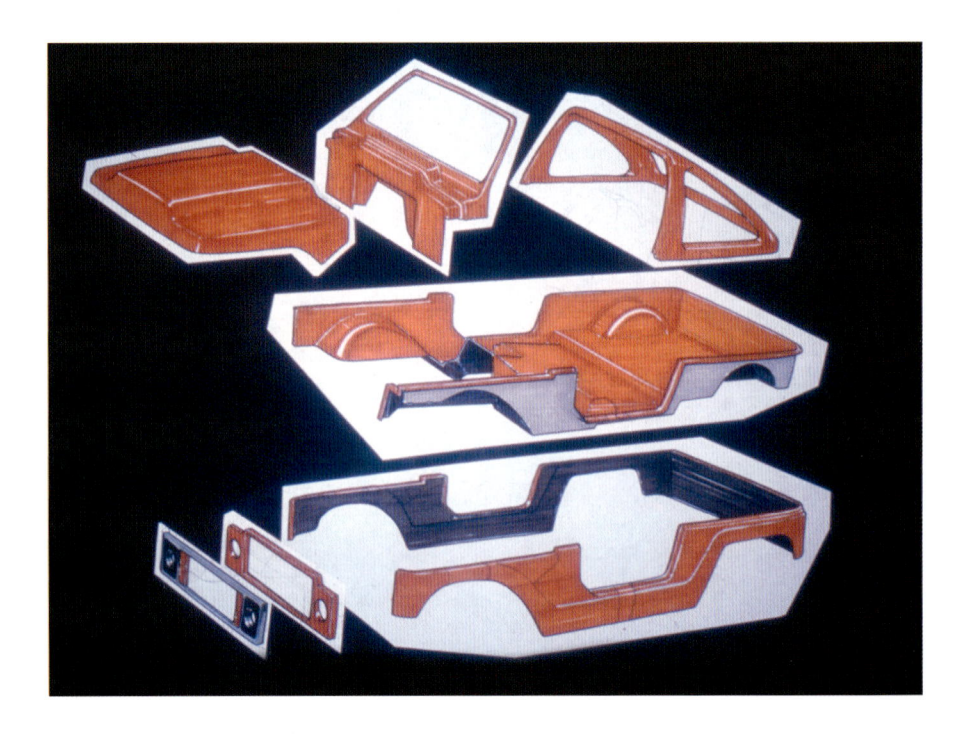

▲ The first running SSV prototype under construction. Note the unique front and rear bumpers, which became a signature feature of the SSV line. This is the first complete body delivered by Leo Windecker in February 1977.

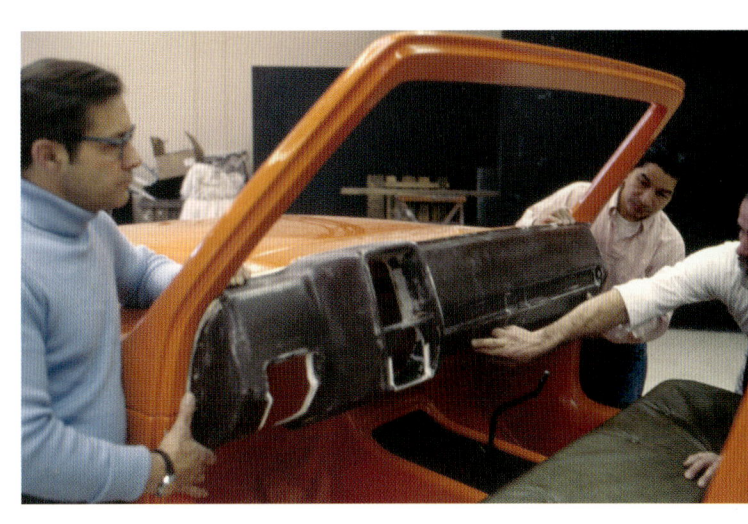

▲ The dash design for the SSV was intended to be entirely different. Here, engineers and designers test-fit an early version to the prototype.

The project started with discussions with companies developing composites, including GE Engineered Plastics, Dow Chemical, Research Plastics, and Windecker Industries. This gave IH engineers a better understanding of the technology and implementation costs. IH even had discussions with John DeLorean, who was involved in acquiring the rights to a composite developed by Royal Dutch Shell.

By 1975, Dr. Leo Windecker's company, which existed primarily to build composite aircraft, had risen to the top of the list. A dentist by training, Windecker designed and built the first FAA-certified composite-powered aircraft, the Windecker Eagle. Dr. Windecker's designs were based on a lightweight, but very strong, honeycomb material and methods for strengthening the material at attaching points. Windecker also demonstrated a knack for solving issues with new applications of the technology.

In spring 1975, the Truck Division decided to move ahead with a composite program. In May,

▶ May 3, 1978, at the International Test Track in Fort Wayne. The first SSV-95 crash test is ready to start with the first of many SSVs to be expended. On the left are Dr. Leo Windecker to the outside and Ted Ornas near the Scout. The other man may be the plant manager. The crash-test SSVs were finished only as much as necessary to meet testing requirements.

▲ The first SSV runner as it was equipped for testing, more or less. This image dated June 1977 shows the SSV with a set of white spokers and Goodyear Tracker A/T tires.

▲ After being dolled up later in 1977, the first SSV did a roadshow routine and appeared at many International publicity events all over the country. This image is dated June 1978, almost a year after being photographed under rigorous testing conditions. It has acquired a Targa roof with a sunroof and a set of soft door, some pinstriping, and color-keyed alloy rims mounting BFG A/T tires.

IH made Windecker a job offer, which he initially refused. At that moment, he was trying to restart his aircraft program, which had failed due to funding disputes. By August 1975, the restart had had fallen through and Windecker was hired by International as a full-time consultant to develop composite technology with a contract lasting into 1981.

When Windecker came on board, Truck Engineering was in the early process of designing a more aerodynamic cab for their cab-over line, which became a test project for the composite technology. Windecker worked with the design team on the cab and solved a number of complex structural issues. By the early part of 1976, aero testing on the cab had shown the improvements were not significant enough to warrant further development so the research moved in other directions.

By February 1976, Windecker's involvement bore possibilities in the form of an armored Scout Traveler prototype that featured lightweight Kevlar armor (see Chapter 4). The Scout platform was chosen as the new testbed for developing composite materials. From that came the idea for a sporty, 95-inch-wheelbase Scout with a composite body.

By June 1976, a quarter-scale model was built and the greenlight given for a first prototype by November 1. The Design Studio also produced concepts that included a roomy four-wheel-drive called the Family Cruiser, a four-wheel-drive, four-door "Mini-Van," and a couple of compact box vans on the 118-inch chassis.

A prototype body arrived at Fort Wayne from Windecker's Midland, Texas, research facility on February 9, 1977, and was mounted to a Scout chassis shortened to a 95-inch wheelbase. The SSV body weighed 348 pounds (a comparably trimmed Scout

▲ The rear three-quarter view shows a completed vehicle and how the first SSV was shuttled to various dealerships and shows to be evaluated by International customers and dealer personnel. There are literally hundreds of images of this rig all over the country, some showing customers with clipboards giving it the once-over.

body weighed 900). It became the first SSV prototype to be photographed, measured, tested, and evaluated. Eventually, it was sent to the Prototype Assembly Shop, where it was dolled-up to go on the road and promote the SSV all over the country.

An early goal was to crash-test the composite structure and the first of these tests occurred on May 3, 1978, with Windecker, Ted Ornas, and others in attendance. A number of other SSVs were also crash-tested subsequently, perhaps as many as 13 units. Predictably, weak links were discovered, primarily in the areas where the body mounted to the chassis and where the body sections joined. The failures were analyzed and led directly to the last bodies having 16 assembled components versus just seven at the beginning. The standard Scout body, incidentally, had 112 individual stampings that were welded or bolted together.

▲ *Bajaro!* Another dolled-up SSV-95 was built late in 1978 with copper bronze paint, color-keyed wheels, and an earth-tone interior. This image is dated January 1979. Like the orange SSV-95, this rig was used for show and promotional purposes and appears in various publicity shots from 1979 and 1980. This rig is known to have survived and was recently photographed in storage. It was reported to have been given to the owner of the J. B. Hunt trucking company at or near the time Scout ended production. Recent images show it stored and dusty in an unknown location but still substantially in the condition shown here—less the Bajaro decal.

By June 1978, testing had progressed enough for a "go" to develop the SSV with a target production start of November 1980 and a production run of approximately 4,000 units for the 1981 model year to supplement regular production. SSV was the working name, but the project had not progressed into discussions of an established trade name. "Scout III" had been bandied about, as much in periodicals as anywhere, but it was mainly used by International to describe the upcoming all-new Scout. Plus, the SSV wasn't a major evolution of the Scout, merely a side development, so the name "Scout III" didn't apply.

The SSV would see four distinct body evolutions and two wheelbase lengths. Most were on a 95-inch wheelbase. Toward the end of the project, a small but unknown number of new bodies were built for the standard 100-inch Scout chassis. The 95-inch

SSV-95 (SSV, 95-inch wheelbase; these are our designations, not from IH) were built of seven pieces with removable tops and soft doors. There is documentary and anecdotal evidence of about 30 SSVs, most of them SSV-95 used for crash tests and other engineering purposes.

The body evolution culminated with the 100-inch-wheelbase SSV-100 Coupe (again that is our designation). The Coupe is very different, with a full roof and hard doors. It's unclear exactly how many SSV-100 Coupes were built, but it it's clear they were near the final evolution intended for production in 1981 or 1982. We have found documentary evidence of 11 SSV-100s being built.

A good deal of the SSV program is laid out in the internal IH publication *Composite Materials: A Technical Evaluation*, dated February 1979. The composite design team estimated the production line as originally envisioned could produce 20 SSV bodies per day with about 60 workers per shift. That did not inspire a lot of confidence at International HQ. It was vital to either increase the daily production or more fully automate the process, but the group was optimistic they could improve the numbers.

The original plan called for 19 fiberglass bodies. Six would be crash-tested, one would go to the

▶ The first SSV Coupe in 1979, shortly after completion, with the team from the Fort Wayne Prototype Shop that built it and all the other SSV prototypes. Left, from the front: Don Sprandel, Ron Landrum, Chris McCombs, Al Sprunger, Gaylord Warner, Walt Domer, and Bill Freeman. Right, from front: Susie Melchi, Don Geradot, Dallas Dewey, Harl Donley, and Keith Conner. This is the SSV that currently resides in the Auburn Cord Duesenberg Automobile Museum. After this image was made, it acquired some color-keyed alloy wheels and pinstriping, and lost the driving lights. *Dallas Dewey*

▲ By October 1978, the composite Scout had evolved into the SSV-100 Coupe. Built on a 100-inch wheelbase, it featured a hardtop coupe body with full hard doors. This is the finished clay model of what would become the final evolution of the SSV.

▲ The SSV-100 Coupe today. The Auburn Cord Duesenberg Automobile Museum was kind enough to bring the Coupe down from its second-floor display to be photographed. The VIN reveals the chassis was built on March 20, 1979, as a Winter White Traveltop Scout with a V-345, T-407 automatic, TC-145 two-speed transfer case, and 3.07:1 axle ratios. Because it was destined for engineering to have its body removed, it was ordered without any cosmetic accoutrements, though it was given air conditioning, tilt steering column, cruise control, and 10-15 Goodyear Trackers on white-spoke wheels, all of which would be reused.

▲ Given the lack of practical features, it's pretty clear this was going to be a sports-oriented Scout and not a practical family rig. The rear-step bumper with an integral tire carrier was a signature element of the SSV. It isn't clear if this feature would have been retained in production, but it seems likely. The glass hatchback lid opens for rear access.

Phoenix Proving Grounds for hot-weather and off-road testing— one would be tested at the Fort Wayne test track, three would be used for emission testing, one would be used to test seat and shoulder belts, and seven would be held in reserve. In an interview, the group leader of the prototype shop where the SSVs were built, Dallas Dewey, recalled the initial order he saw was for 23 units but remembered as many as 30 built. Most, he said, were rough-finished and used for crash-testing, or other engineering purposes, but he remembers the shop turned out three nicely finished ones.

Dewey remembers all the SSV-95 chassis had to be shortened in the prototype shop before the bodies were fitted and an extra set of rear-body mounts added. The 100-inch chassis that were slated for crash testing were modified with straps welded to the top of the front frame rails. With a few exceptions, Dewey recalled they came to the shop as rolling chassis with complete powertrains. Of the surviving chassis, none of those slated for crash testing have VINS, but the units intended for display do. Those with VINS were originally built out as complete

▲ The SSV-100 shared the dash that would have appeared in the standard 1981 Scouts. The air conditioning from the original Scout was reinstalled into the Coupe—without the usual roll-up windows, it would have been sorely needed. The AM/FM stereo cassette player was the same type offered in Scouts at the time with four speakers fitted. The odometer of this SSV shows a mere 209 miles.

▶ Under the hood, the SSV-100 Coupe is all Scout. The V-345A four-barrel has only a few hundred miles on it, virtually none since it was donated to the Auburn Cord Duesenberg Automobile Museum on November 22, 1983. It was driven to the museum but didn't run again until 2020, when it was brought back to driving status.

▲ The SSV-100 Coupe in its customary place on the second floor of the Auburn Cord Duesenberg Museum in Auburn, Indiana. It lives in a part of the collection that highlights cars and trucks built in Indiana. On the wall behind is one of Larry Nicklin's renderings of an SSV-100 Coupe. Nicklin lived nearby and was a big supporter of the museum until his death in January 2015 at age 88.

Scouts, according to their original LSTs, and later had the steel bodies removed.

By the end of 1979, IH's financial problems slowed SSV development down as investments in the program were reduced. A 1981 introduction was looking iffy. Some thought was given to having Coachman Industries (an RV manufacturer in Indiana) manufacture the SSV bodies and assemble the complete vehicles on chassis supplied by IH. When the UAW strike began in late 1979, SSV development was largely suspended and never ramped back up. Had the Scout Division been sold, it may have been revived, but that would have been up to the new owners.

Two complete and nearly perfect SSVs are known to survive: a 100-inch coupe held by the Auburn Cord Duesenberg Automobile Museum, and the 95-inch "Bajero" SSV that was given to one of IH's biggest truck customers and is known to be hidden away in storage. A third gussied-up SSV is also thought to survive and was reportedly sold in an East Coast auction not long before this book was written but has since disappeared. If this third SSV exists, it's almost certainly

SSV, HVS, AND 1981–1985 SCOUTS: WHAT MIGHT HAVE BEEN

▶ Here are two SSV-100 chassis owned by John Glancy, along with a large number of other small SSV parts. The nearer has a VIN, so it was intended to be a driving vehicle. Evidence points to this rig having been a complete and fully assembled vehicle at some point. The farther chassis does not have a VIN, an indication of it being a crash-test vehicle. Further evidence of that is found in the reinforcement straps on the top rail of the front chassis, something found *only* on crash-test chassis. Finally, written on the fuel tank (an adapted Loadstar tank, no less) is "do not run–crash test only." The rear bumpers and tire carriers are further SSV giveaways, as are extra body mounts on the rear that are not found on standard Scouts.

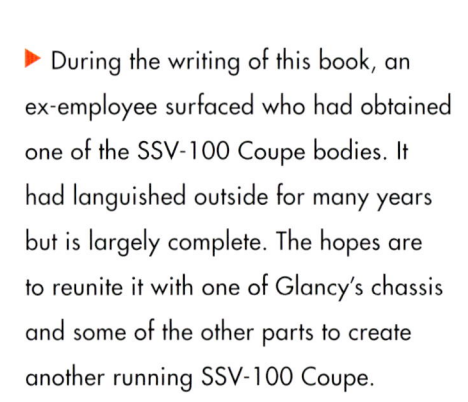

▶ During the writing of this book, an ex-employee surfaced who had obtained one of the SSV-100 Coupe bodies. It had languished outside for many years but is largely complete. The hopes are to reunite it with one of Glancy's chassis and some of the other parts to create another running SSV-100 Coupe.

the very first running SSV prototype that was eventually dolled-up for shows. There's also a semi-derelict early prototype that came from Windecker's Ranch in Texas. There are only a few pictures to speculate from, but it appears to be the very earliest iteration of the body design. Was this unit assembled by Windecker from leftover bits and pieces or is it an actual International test vehicle? That remains to be discovered.

John Glancy has pieces of several SSVs in his possession, including two SSV-100 chassis, one with a VIN. There are many remaining SSV-unique parts on them and it's clear they were once assembled vehicles. They came from the Engineering Test Department and were sold in some of the many auctions held in Fort Wayne after the Scout was discontinued. A surviving SSV Coupe body and crash-test chassis was also located in Fort Wayne that survived the crusher by being buried in the corner of a warehouse. The hope is that it will be reunited with an SSV chassis in the coming years and again become a complete SSV.

The SSV in the Auburn Cord Duesenberg Automobile Museum is the closest representative example of what the final vehicle might have been. Up to 11 finished SSVs very much like it were built for tests in a variety of venues.

FUTURE HOPES: THE HIGH-VOLUME SCOUT (HVS)

The SSV was intended as a sideline, a means to perfect the manufacture of composite bodies while earning money and bragging rights in doing so. At the time, a totally new Scout was also in the works and composite body ideas were being considered for them as well. The proposed implementation date for an all-new "Scout III" commonly focused on a 1984 or 1985 model-year introduction, the years a large number of federal standards on emissions, fuel economy, and safety would be implemented. The composite ideas in this program resulted in the designation High Volume Composite Scout, or HVS. Some product planning documents from late 1979 refer to an all-new Scout as having "the crisp and clean lines associated with European luxury sedans."

The composite HVS Scout wasn't going to be just a Scout in the traditional sense. The tentative lineup was to feature more than one model and include an off-road version for sports enthusiasts, a more civilized four-wheel-drive coupe on the standard 100-inch wheelbase, and a very stylish 118-inch-wheelbase four-door with three-row seating for up to eight. The working name for the four-door began with "Mini-Van" then became "Scout-Van" before evolving into "Trava."

The most radical HVS idea mentioned in *Composite Materials: A Technical Evaluation* was to build the new rigs with a unitized composite body structure with steel subframes for the suspension. These rigs would feature a front-drive powertrain with a transaxle, and the weight savings would have helped its fuel economy. The front wheels would be the primary drive system, but the transaxle would feature an optional two-speed integral transfer case with an output to a rear axle with locking hubs. This cutting-edge concept had been produced successfully in Europe.

The 1984–1985 models were pursued on three levels: the composite HVS mentioned previously, a more conventional composite body on a standard steel chassis, and a newly designed steel body on a steel frame. The bets were being hedged on the composite idea because it wasn't at all certain the technology could be developed to production status in time and on budget.

Design studies and clay model development for the composite HVS went right to the end. Some of the last renderings are dated August 1980—about when hope for the fate of the Scout Division had fully dimmed. Given the financial climate of the era, the state of IH finances overall, and the labor situation, the cost of implementing the HVS by 1984 or 1985 was very doubtful, even if the Scout line had been continued by International or another entity. It would most certainly have been a very innovative Scout, but we will never know if it was a viable idea.

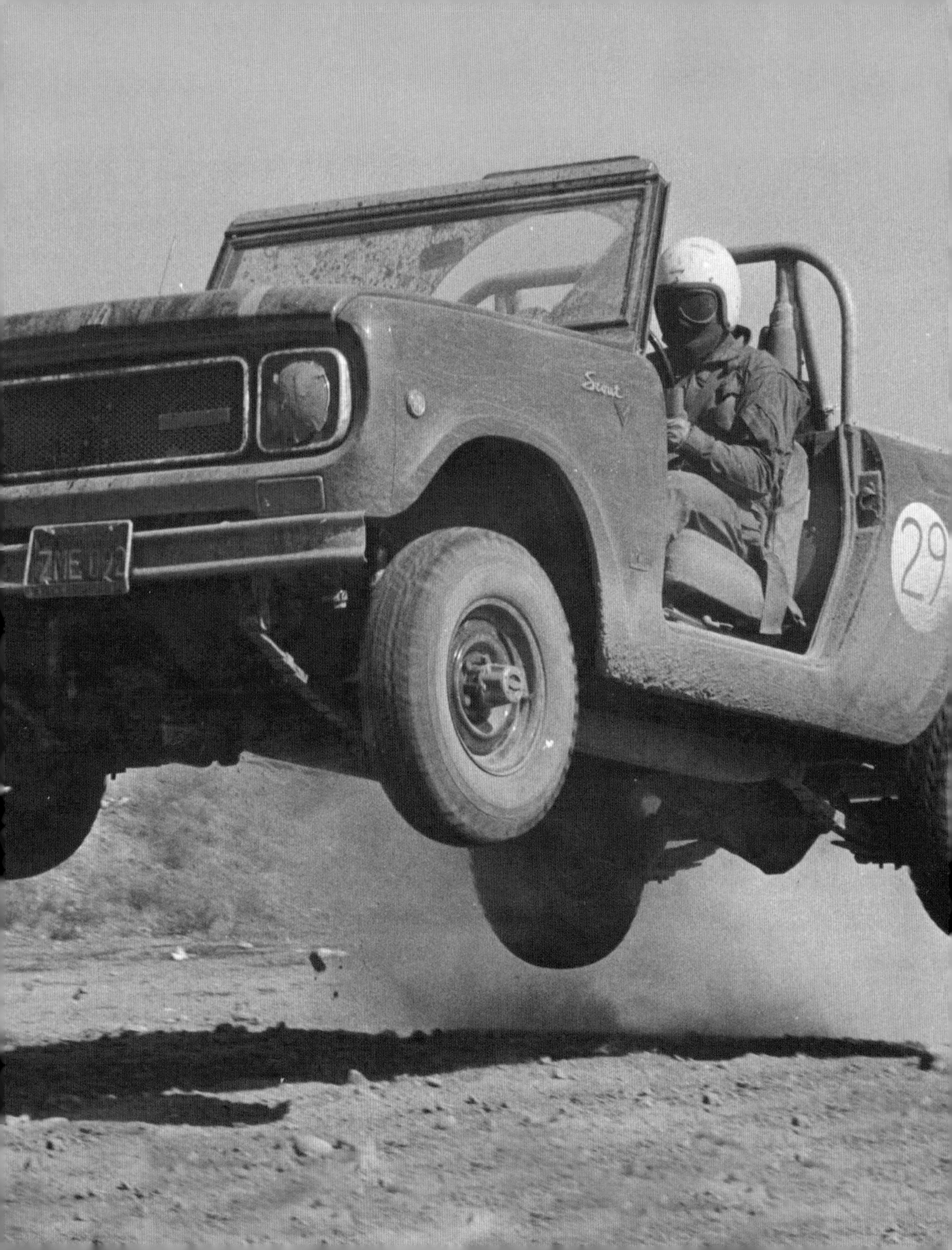
Scout
29

Chapter 7
RACING SCOUTS

In the last years of the 1960s, off-road racing had grown to national prominence and 4x4 manufacturers, including International, were getting involved. Though there were, and still are, many types of racing for 4x4 vehicles, the spotlight was on desert-style racing, where speed and rough terrain were combined into one event. NORRA (National Off-Road Racing Association) was one of the first sanctioning bodies and organized the first big races in the Mexican Baja and Southern California deserts. On-the-spot coverage of the Baja 1000 by ABC's popular *Wide World of Sports* in 1968 cemented that race as a cornerstone motorsports event.

When the first oil crunch hit in 1973, NORRA canceled the race in Mexico to run an abbreviated race north of the border. The regional Mexican government hastily organized a race that didn't go well, so they contacted the professionals at SCORE (Southern California Off-Road Enterprises), a Mickey Thompson creation, about running the Baja 1000 race in the future. It was not run in 1974, but SCORE ran the 1975 race and ran it well. Under the guidance of Sal Fish, SCORE rapidly expanded, overwhelmed NORRA, which went dormant, and soon began sanctioning races all over the country, becoming a multi-million-dollar organization that continues to this day. (NORRA restarted in 2010 as a body that organizes desert races with vintage vehicles.)

Desert-style endurance racing combines high-speed runs over 100 miles per hour with very slow, technical sections where achieving 5 miles per hour is a struggle. Today there are more than 25 classes that run the gamut from almost bone-stock to the wildest modified motorcycles, 4x4s, cars, trucks, and buggies.

◀ This '69 800A was the first Scout to race in a national event. Jimmy Jones bought it as a family vehicle early in 1969, took it on a few four-wheeling trips, and found it very capable. Soon, he was running it in local competitions, this one in Mira Mesa near what is now the Miramar Marine Air Station. By the end of 1969, it had been completely converted into a race machine. It had a V-304A and T-39 automatic. *Lance Jones*

▶ Larry Ehlers (left) and Dick Bakkom were the two people at IH most responsible for the race program and were instrumental in the development and production of the SSII. This image was taken at the Mint 400 Race in 1977. Both gentlemen have passed on, car guys and Scout fans to the very last. Some of the images in this book, especially this chapter, come from Bakkom's collection, which he passed on to Dr. Todd Sommer. *Dr. Todd Sommer*

The classes have changed over the years since Scouts raced, but back then, they most often were seen in Class 3, the production 4x4 class with limits on the extent of modifications.

In the early 1970s the bean-counters at International were convinced to look at off-road racing as a marketing tool. The story of how this came to be is a bit vague because the principals at IH who put it together are deceased and few records remain. It is known that a couple of key players were Dick Bakkom, IH Light Truck Marketing Manager, and Larry Ehlers, who was a Marketing Department engineer at the start of the process. When the race program went into gear, Ehlers was given a new title and a second job as Special Equipment Marketing Manager. Bakkom and Ehlers are remembered as being the prime movers behind establishing the Scout racing program, but they had a lot of enthusiastic help within the company, and most of those names are lost to time.

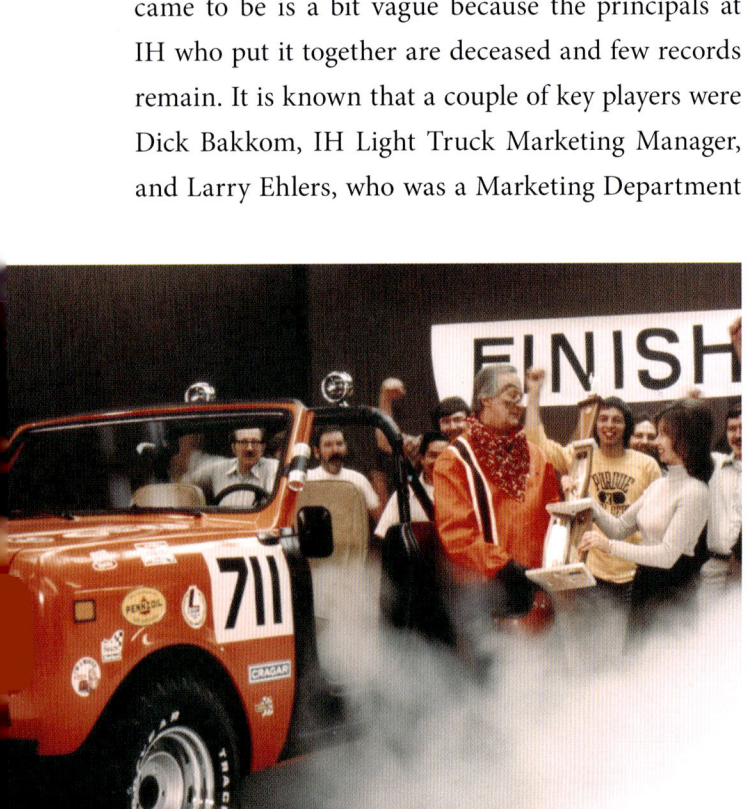

The foundation was laid in 1972 when a desert racer named Jimmy Jones won his class at the NORRA Baja 1000 Race in a near-stock 4x2 1972 Scout Cab-Top. Jones had been the first to race a Scout in the

◀ In order to promote factory support of Scout off-road racing teams and push the SSII into production, Ted Ornas set up a December 1975 shoot with the SSII mockup to mimic a Scout driver winning an imaginary race and getting a big trophy and a kiss from a buxom lass, all with an admiring crowd looking on. For obvious reasons this became known as the "Walter Mitty Ornas" shot and it worked! The SSII went into production for 1977. Most of the remaining Design Studio guys remember the event. The racing number, 711, was no accident—it just happened to be the same one on Jimmy Jones' Scout that won its class in the 1972 Baja 1000. Jones' truck was also orange, so this orange paint may not have been an accident either. Jones' 1976 Scout would be nearly a clone of this after it was "SSII-ized."

▶ A desert-racing Scout lived a hard life. The Mint 400 near Las Vegas, Nevada, was one of the hardest desert races due to extremely rough terrain. Half the competitors finishing was a milestone. Here, Jerry Boone's team conducts repairs in a dust storm. The drivers interviewed gave high praise to their support crews, which often consisted of family members. These were the days before multimillion-dollar teams with support trailers equipped better than many shops became the norm. *Dr. Todd Sommer*

Baja 1000 in 1969, placing thirteenth in the race with a 1969 800A. Most of the old IH hands interviewed think Larry Ehlers instigated the idea of a race team by informing the Marketing Department of the Baja class victory and the potential publicity value. By all reports, Dick Bakkom then took the idea upstairs but had an uphill battle getting the conservative, tightfisted, ag-centric board at International to run with it. Then Ted Ornas got involved and staged the now-famous "Walter Mitty Ornas" shot in December 1975 with the SSII mockup. Legend has it this event finally pushed the unconvinced to approve the sponsorship program.

The rubber met the road 1976, when the first contracts were signed with three proven drivers to run Scouts starting in 1977. Reportedly, Jimmy Jones provided input on choosing the other drivers: Frank Howarth, Sherman Balch, and, later, Jerry Boone (Jones was the shoe-in, of course). The sponsorship deal included supplying Scouts at $1 each and covering all the racing bills, including parts and travel. The drivers got to keep any prize money. International got the marketing benefits, including access to the drivers and vehicles for events, publicity, and promotions, and the teams had to run as many races as possible. The drivers were connected to dealers near their home bases. Boone was part owner in Truck-n-Tractor, a Parker, Arizona, IH dealer.

Howarth was already building Jones' Scout, and both were connected to Earley Truck Center in El Cajon, California. Balch was associated with Bayshore International Trucks in Hayward, California. Each had other sponsors as well.

As a result of the race program, International garnered a lot of ink across many marketing spectrums. Remaining documents and interviews with surviving personnel indicate the marketing department saw it as a successful venture. In a letter to Balch, Dick Bakkom wrote:

> Our awareness level with first time buyers was practically nil and the demographics of second time buyers was almost entirely commercial. After that initial first full year, our awareness level for first time buyers increased 27 percent and the majority of our buyers bought their Scout for personal transportation, which then represented approximately 85 percent of the sport utility market.
>
> During the last full year of racing support, we received what we estimated to be

between 2.5 and 3.5 million dollars' worth of free media editorials, articles, photos, and mentions. This support cost us $150,000.

The racing program brought the Scout Division into the mainstream and could have been part of the Scout's salvation. Unfortunately, too many larger factors were driving events and the race program was much too little and way too late.

Available records don't indicate exactly when the plug was pulled on the race program. International continued to pay the bills into 1980, but none of the surviving principals remember exactly when they stopped or when, exactly, they got the word it would stop. All four drivers raced into 1980 under the IH flag, but the press coverage of the March 1980 Mint 400 offers a notable clue. Only Howarth showed up and magazine coverage cited the impending sale of the Scout Division and the end of factory sponsorship as the reason.

Balch raced a Scout on his own as late as the end of 1982, the last of the factory-sponsored guys to do so. In the big races at least, Scout was mostly gone from the scene not long after that. The lack of sponsorship was one reason, along with the Scout being out of production and no longer viable in a production class. In the hands of Dick Sasser, Jimmy Jones' 1976 Scout raced until 1997, where it was seen still plugging along in Class 3 at the Baja 1000. New technology would eventually have put any leaf-sprung, old-school 4x4 on the non-competitive list in Class 3, but Scout raced on more competitively in Class 4, which imposed fewer limitations on the level of modifications needed to remain competitive.

JIMMY JONES

Jimmy Ray Jones was the inspiration for the Scout racing program. Born in 1939 and an electrician by trade, he lived in the San Diego area and was an avid off-roader. Jones began racing in local Southern California events in the 1960s and, according to a late-1970s bio, he "raced 'em all . . . minitrucks, sedans, bugs, and modified four-wheel drives."

In 1969, he bought a Scout 800A V-8 as a family vehicle but thought it capable enough to run in local

▶ Jimmy Jones in January 1977, just as the Scout race team was starting in full force. By this time, Jones, Sherman Balch, and Frank Howarth had been signed up. Jones was 38 and the oldest of the drivers. This image, taken at the Chandler, Arizona, event that introduced the SSII, was likely Jones' first big publicity event for International. WHS 110474

▲ In 1971, nearing the end of racing with Jones, the 800A was barely recognizable as a Scout. It had been extensively lightened, including at the signature Scout front end. Jones raced it like this into 1972 when it was sold and raced by another individual. Its current fate is unknown. *Lance Jones*

▲ Despite rolling—twice—Jimmy Jones put Scout on the racing map by winning the 4x2 utility class in the 1972 NORRA Baja 1000 with this '72 Cab-Top he named the *Baja Binder*. The '72 Scout was scarcely a year old at this point in November 1972. Note the caved-in side and roof, plus the damaged roof lights. This rig was 6-258-powered and had the T-407 automatic. Besides the required safety gear, the suspension was modified and larger, sturdier tires were added. The *Baja Binder* mounted an Offenhauser four-barrel intake and Hooker headers. It wasn't the fastest rig, but it had staying power.

competition. According to Jones, "It beat the pants off the local competition." He was impressed enough to try the Scout in the big leagues and raced in the 1969 NORRA Baja 1000. He was pleased with thirteenth place in a race where to finish was a victory in itself. This was also the first time a Scout had raced in the Baja 1000. From there Jones was hooked on Baja racing—and the Scout. The 800A soon became a fully built race rig and was run into 1972, but it was sold soon thereafter.

When the new Scout II was announced, Jimmy ordered one and ended up with an orange 1972 two-wheel-drive Cab-Top with a 6-258 automatic to run in the NORRA two-wheel-drive utility class. Why a 4x2? Perhaps he figured a 4x2 would be lighter and faster, with less stuff to break. In 1972, against all the odds, Jones won the class and unknowingly caught International's attention.

Jimmy raced this Scout II for several more years during which time he began working with Frank Howarth. Frank would do much of the fabrication and suspension work on Jones' vehicles, even though they would eventually be competitors in racing. The Jones Scout soon acquired the nickname *Baja Binder* and continued to place well.

By late 1975, representatives from International had chosen Jones as their go-to guy in the desert racing world and, for a dollar, set him up with a new V-8-powered 1976 4x4 Scout. This rig also was orange and became the new *Baja Binder*. He raced it for the first half of 1976 more or less configured as a 1976

▲ Repaired after the 1972 Baja 1000 win, the *Baja Binder* raced the 1973 Mexican 1000. Reportedly, this vehicle still is owned by the person who purchased it from Jones in the late '70s, who converted it to a 4x4. *Lance Jones*

▲ The new *Baja Binder* at its first big race, the April 1976 Mint 400. It was fresh from Frank Howarth's ORD shop, where it got extensive modifications, including a front shackle reversal, a lift for more suspension travel, and room for 35-inch Desert Dog tires. It had a full cage and axle and chassis beefing, as well as all the other safety devices required by race rules. At this point, it was still running the original doors and even the windshield. It would race like this into May and June, when it was converted to SSII configuration.

Dr. Todd Sommer

▲ As soon as prototype SSII body parts were available midyear 1976, the Jones team converted the '76 to SSII configuration. It raced this way through the rest of 1976 without most people knowing they were getting a sneak peek at a new Scout. Notice Frank Howarth is listed as co-driver on the door insert. The number is from the Baja 500, which took place in June 1976, possibly the first time it raced as an SSII. This is around the time the first press releases from International were produced regarding their interest in Jones and possibly sponsoring a race team. Most likely, it was already a done deal at this point, awaiting only the public announcement. WHS 110383

Cab-Top. When the SSII accoutrements had been designed by the Styling Department, Jones was given all the relevant pieces for a conversion. This was well before the SSII was formally introduced, thus giving race fans an early look at the SSII without actually telling them what it was. According to Jimmy's son Lance, the SSII parts were early prototype pieces made of fiberglass, not the thermoplastic used in production. He remembers getting them in the summer of 1976 and the first time they appear in period photos was in the June Baja 500 race. The end result looked a lot like the "Walter Mitty Ornas" SSII mockup (or vice-versa) and the *Baja Binder* raced as an orange SSII in the latter part of 1976 and into 1977. According to Lance, that Scout saw several cosmetic revisions, most notably white paint and the addition of SSII appliqués in 1977, even though *Off-Road* magazine called it a Sundowner in a 1977 article. The Scout remained like this to the end of its time with Jones and after it was sold to Dick Sasser, who raced it into the late 1990s.

In 1979, Jones had an opportunity to do something very unique in the Baja 1000. After some arrangements with Sal Fish, head of SCORE, Jones was allowed to "race" a Scout Terra turbodiesel pickup. With only 120 horsepower, and only mild modifications, it wasn't competitive, even in Class 3. Nor did it really fit into a racing class. It was purely and simply a publicity stunt for International to highlight their new turbodiesel, but Jimmy took it seriously. Howarth added appropriate safety devices and made a few suspension and durability modifications so the truck could be pushed hard in the outback.

To prove a point, Jones drove the barely street-legal Terra from his home in La Mesa, California, to the starting line at Ensenada, Mexico. From there, the Terra completed the 975-mile race with only minor suspension trouble and made a calculated 13 miles per gallon during the race on fuel that cost

▲ By January 1977, coinciding with the dealer introduction of the SSII, Jones brought the refreshed '76 to Chandler, Arizona, wearing the numbers for the 1977 Parker 400, which was just days away in Parker, Arizona. By this time, the truck had the full SSII treatment, including the appliqué, and was presented as an SSII even though it was a hard-worked Cab-Top. Other notable differences included the absence of the factory windshield and frame. Not only did it save weight, but the glass collected dirt and obscured vision, especially when it shattered. A helmet visor could be cleared with a quick swipe of the hand.

▲ The Jones SSII at speed at the 1977 Riverside SCORE championship race in August. By this time, all four factory teams were up and running, and Jones was facing competition from his own camp as well as others. This Scout was potent. The engine was built by Wayne Tharp, who blueprinted it using a mostly stock lower end. A Schneider 280H cam was used and Tharp ported the heads to increase airflow, adding Schneider dual springs. The Jones team got one of the prototype IH aluminum manifolds, which was massaged by Tharp, and a Holley dual-feed carb was added with a remote air filter. Hooker headers and a fabricated dual exhaust eliminated the used gasoline. A stock electronic ignition was used, but the distributor was recurved. Art Carr built a special TorqueFlite automatic and the axle ratios were 4.27:1. Lance Jones tells us Jimmy preferred to run in four-wheel high range, liking the handling in that configuration.
Dr. Todd Sommer

▲ The Jones '76 racer resurfaced just as this book was going to print. An inquiry by the new owner about the history of an old race Scout led to the authors' discovery that this rig, raced for many years by Dick Sasser, was Jimmy Jones' long-lost Scout. This was verified by its VIN and by Lance Jones, who knew it well. Apparently, it was raced into the late '90s and also used as a pre-runner. The new owners, Ron and Lori Anderson, were ecstatic to find such a historical Scout. Beyond getting it up and running, future restoration plans were uncertain at press time, but this rig will no doubt be making the rounds as Scout shows. *Ron Anderson*

▶ Big plans were afoot when Larry Nicklin penned this rendering of the upcoming turbodiesel racer in February 1979. The graphics on the actual vehicle remained very much the same as rendered here. The appliqué was very likely done by Excello, the company that did all or most of this work for International.

▲ The turbodiesel Scout at the start of the 1979 Baja 1000. It was driven from La Mesa to the race start in Ensenada and given the special number "IH." Little is known about the actual Scout other than it was a very early production unit, possibly a prototype used for tests. The engine was largely stock except for an updated air filter. The hood was replaced with a fiberglass piece, and the dash was replaced by a sheet of aluminum to which the appropriate gauges were added. The doors and fiberglass top were retained, but a roll cage was added and the door glass and mechanisms were removed. The headlights were replaced by halogens, and four Hella driving lights were added. Howarth added a mild lift to fit 11-15 Uniroyal Laredo tires and triple-shocked it at both ends with Gabriel shocks. It proved to be over-shocked—by the end of the race, it had only one shock per corner and was doing fine. Jones and his co-driver, journalist Brian Brennan, probably had a few loose teeth after the first few hundred miles.

13 cents a gallon. The fuel bill for the race was a mere $13.50. From the finish line at La Paz back to La Mesa, the now well broken-in turbodiesel managed 18.25 miles per gallon on average. Calculating the entire trip, including the race, the Terra used 127 gallons of fuel over 1,975 miles for a 15.5 miles per gallon average. The turbo Terra raced Baja again in 1980, and, as far as is known, that was its last race. According to Lance Jones, the turbodiesel Terra is now owned by one of Jones' old racing buddies but has been "de-dieseled."

When the hammer dropped in 1980, it didn't quite end Jimmy Jones' Scout racing career. He raced his SSII a few times in 1980, including the Parker 400, and then drove the diesel in the Baja 1000 late in the year. From there, Jones eventually started racing with Nissan and Jim Connor but more or less retired in the late 1980s. He passed away on New Year's Day 2013 at age 74.

▲ Frank Howarth (right) and Jimmy Jones had a long friendship and collaboration. At the Mint 400 in April 1976, Howarth showed up to help out as needed since he had done many of the modifications to Jimmy's new '76. Howarth co-drove with Jones in the latter part of 1976, reportedly in three major races. He also continued to do most of the chassis and suspension work on Jones' rigs, even after they were competitors. Howarth also built a race Scout for Mike and Pat Mulligan from the East Coast.

▲ By February, Howarth's long-wheelbase Scout was prepped and ready for the 1977 Parker 400, where it is shown here wearing number 303. At this early stage, it was still running a V-345, had a Dana 44 semi-float rear axle, and was without the extensive suspension modifications Howarth later designed and built in his ORD shop. Howarth ran the Parker and the Mint 400 in March with the truck configured like this. By the time the Baja 1000 came along late in the year, it was vastly improved but still was a DNF in that race. Oiling problems led to a hybrid 345/392—a 392 block running the extensively ported 345 heads built for the first engine. *Dr. Todd Sommer*

FRANK HOWARTH

Frank Howarth's first trip to Baja came in 1954 and from then on he was hooked on the picturesque and sparsely traveled desert. In the 1960s, Frank was working in the aerospace industry as a machinist/fabricator when he began weekend desert racing. He first raced the Baja in 1971 in a single-seat buggy he built. By the time of his first serious race, he had left the aerospace industry and was working for race teams as a builder and fabricator, eventually opening his own shop and retail operation, Off Road Distributors (ORD). It was a well-known supplier of Scout buildup parts for a few years but closed down not long after Scout production ended. In the early 1970s, Howarth's work attracted the attention of Jimmy Jones and he began doing prep and fab work on the *Baja Binder*. This put him on Jones' good-guy list, which likely had something to do with the offer of sponsorship from IH that came later.

Howarth's choice of a long-wheelbase Scout has long been a source of interest and debate. While the other three Scout drivers ran both short- and

▲ The Mint 400 is a tough race and Howarth shook out some weak links in his Traveler in the April 1977 event, notably the original Dana 44 rear axle. At about 3,800 pounds, Howarth's rig was a little heavier than the 100-inch Scouts. This was the point at which the Dana 60 axle came to be on the team's radar.

long-distance events, Frank focused on the longer races and thought the longer wheelbase had stability and comfort advantages over the short Scouts. On the short track, of course, the extra length was a disadvantage, which kept him out of some of the coliseum races of the late 1970s. His record speaks for the choice, however: Howarth was usually at or near the top of the SCORE points board in his class.

Howarth ran a lot of races—10 in 1977 and 12 in 1978—but didn't achieve any firsts in the big races. In 1978, he was driving in "the zone" and headed for a very spectacular class win in the Parker 400 race but ended up with a different flavor of spectacular by crashing into an oak tree. That put the Scout down for the count and required a new chassis and most of a new body.

The Howarth Scout initially had a very modified 345 that was hand-built at the IH engine plant by IH engineer Terry Hankins. It was replaced with a 392 that Hankins estimates produced 450 horsepower. The 392 raised a few eyebrows because it was not offered in the Scout, but since it was in the same long-stroke engine family as the 345, it met class rules. Another controversial element to

▲ By the 1978 Mint 400 in April, Howarth's truck was sporting a full-float Dana 60 rear axle, a built-up V-392, and a very capable suspension. The truck was rebuilt in mid-1977 to improve performance and durability. Early in 1978 at the Parker 400, Frank was in "the zone," running in second place with a win not out of the realm of possibility. Then, in some whoop-de-doos, the right ball joints on the front axle failed. The spindle, hub, and wheel separated from the vehicle, and it spun off the course and wrapped itself around a tree. The Scout had to be completely rebuilt, with a mostly new body and chassis, but it raced in the Mint just a couple of months later, as shown here. Howarth lamented the loss of points because he missed a couple of races but said the Scout was a little better after the rebuild. *Frank Howarth*

▲ In 1980 one of Howarth's sponsors was Gott Coolers. They used the Scout for a spectacular ad campaign in which it jumped through a wall of coolers. *Dr. Todd Sommer*

▲ The Howarth Scout at the 2016 Scout Nationals. After the late Terry Hankins restored it to racing splendor, it was much in demand for display. It spent a number of years in the National Auto & Truck Museum in Auburn, Indiana, but it 2019 it moved into coauthor John Glancy's museum at the new Super Scout Specialists location in Springfield, Ohio. It's still hale and hearty enough that, had there been enough financial backing, it might have competed in a NORRA race for vintage desert race rigs.

the Howarth Scout was the rear-axle upgrade. The factory Dana 44 rear was a bit on the light side for racing so Howarth opted for a Dana 60. Considering the Scout never came with one, this seemed a bit iffy too, but the rules left enough room for interpretation to claim that since the Traveler was the successor to the Travelall, and the Travelall did have an available D60, it was a legal upgrade. Howarth also dared to be different by choosing a TC-143 single-speed transfer case. This unit was lighter, and since four-wheel drive was needed only occasionally, it worked better with his driving style.

Howarth ran the same Scout to the end of his career, the Penrod Mexicali 250 in 1982. Howarth ran ORD for a while longer, finally selling out and taking an early retirement. That only lasted a few years before Howarth became bored and, because he was fluent in Spanish, was "drafted" by Navistar to be a technical representative to their plant in Mexico. That job expanded to include plants north of the border. Frank retired again in the early 2000s and today divides his time between a home near Ramona, California, and one in Baja, Mexico.

Howarth's Scout languished in storage until 1990, when it was handed over to longtime friend Terry Hankins. The Scout was essentially unchanged from its last race and Hankins used that as a template to restore it to fighting trim. It made the rounds of the Scout shows in the early 2000s and now rests in the Super Scout Specialists Museum in Enon, Ohio.

SHERMAN BALCH

Sherman Balch began his racing career in 1958 at age eight, racing quarter-midgets in Fresno, California. In 1967, he bought and began rebuilding a basket-case CJ-2A flat-fender Jeep in high school auto shop. He began competing with it in 1968 right after high school while also starting a career as a carpenter and contractor. His primary off-road training ground was an event called the Georgetown Gold Rush in the Sierra, an all-around event that featured both high- and low-speed sections. Balch reckons he learned a lot there. As the opportunities to race grew in number and profile, he transitioned to the more elaborate ones and grew with the sport.

In 1971, Balch went big-time, winning his class at the Riverside Grand Prix. The following year, he built a full-blown production-class 4x4 (later known as Class 3) CJ-5 and became the youngest driver to win the class at the Mint 400. He was often called the king of the short courses and was known for beating modifieds in his production-class rig. He had a great 1975 season and a spectacular 1976 year, not only winning the 1976 SCORE Baja 1000, but the SCORE Points Championship, SCORE Driver of the Year, and the Riverside World Crown. All that had a lot to do with him being approached by International.

Like most of the other IH-sponsored drivers, Balch signed his contract in the summer of 1976 and was offered a Scout for $1. The SSII was not in production yet, so for that princely sum, Balch became the proud owner of a white-over-tan 1977 Traveltop built

▲ Sherman Balch (left) and Dick Bakkom at a 1978 race. Balch recalled Bakkom approaching him for the first time at the 1976 Riverside Championship, where he had just won the class title in a Jeep. The two remained in contact long after the end of Scout, judging by the friendly letters Balch shared for this book.

▲ The paint was barely dry on Sherman Balch's new race Scout when it debuted in the February 1977 Parker 400. The engine was largely stock that first race, with only a DeLong cam and Hooker headers listed as modifications. Chassis Works built the suspension, cage, and chassis. After this race, Balch was very satisfied with the ride and handling of the Scout, but a load of small failures plagued it through most of 1977. When the truck was running, Balch was at the front of the pack. *Dr. Todd Sommer*

▲ By the end of 1977, the bugs were wrung out of the Scout and Balch handily won the SCORE Championship Race at Riverside in August. Even when racing Jeeps, short courses like Riverside and others were Balch's cup of tea.

on November 16, 1976, with a 345A, T407, TC-145 transfer case, and 3.54 gears. Oddly, it was a Gold Star Deluxe Model (7-SCTT-44-604), so it actually had a few luxury items. The LST says it was shipped to an IH dealer not far from Balch's San Mateo home. It was then race-prepped by Dennis Harris and Billy Brooks at Chassis Works in Roseville, California, in a lightning-fast job completed in only six weeks, just in time for some shakedown before the Parker 400 in February 1977. The Scout did well in that race for as long as it ran, but did not finish (DNF) due to a blown tranny seal.

What followed must have chafed Balch's patience. A succession of DNFs followed that year as various items broke. Balch said the Scout was easy enough to learn and plenty fast, but teething trouble kept it out of the money. The first big win came in the SCORE Championship race at Riverside in September 1977, where the Scout held together and Balch kicked butt. Between Riverside and the Baja 1000, the Balch Scout

▲ By the time the Baja 1000 rolled around in November 1977, Balch had decided on a more colorful look for the Scout. By this time the engine had been further modified with oversized 10:1 compression pistons, head porting, and a Holley four-barrel carburetor. On the suspension side, the front shackles were reversed and the spring packs were scragged (prestressed). Shock tuning was dialed in and the steering box was reinforced. The steering linkage was beefed up too, and the tie rods replaced by Heim joints. Balch was running very fast and a class win was not out of the question, but a collision with a buggy at 100-plus miles per hour damaged the rear axle and put the Scout out of the race.

acquired its signature green paint that carried the rig through 1979.

Balch's driving record in the Scout from then on was a series of high places wherever he raced, including various championship wins. He was always high in the points. The short courses were his best venue, and while his aggressive driving style didn't always earn hugs and kisses from the competition, it got him a lot of respect.

Among the Scout bunch, Jerry Boone was his biggest competitor, especially on short courses. One highlight came at Mickey Thompson's 1979 La Coliseum Race, when Balch endo'd and landed on the fence, halfway into the stands while Boone won.

At the peak of the Scout racing story, Balch partnered with some of his technical people to produce a line of Scout upgrade parts under the

▲ In March 1978, Balch was flying high at the Mexicali 250 with the Scout dialed in and running well.

Tuff Terrain name. They designed a built a line of suspension and engine performance products and fleshed out their mail-order lines with products from other manufacturers. Like Howarth's ORD, Tuff Terrain didn't last long after the end of Scout.

When the bad news came in 1980, Balch had just replaced the tired 1977 Scout with a new 1980 and missed a few races while he was building the new truck. Expert builder Byron Rexwinkle went to great lengths to build a very light Scout by using aluminum, titanium, and fiberglass wherever possible. It was also 392-powered. Balch raced the truck from 1980 to 1982 and racked up an impressive series of wins. In 1982, he won the class in several races, including the Baja 1000 and Riverside, as well as Driver of the Year.

In 1983, Balch started racing in Class 7 with Nissan, where he remained into 1990, driving both pickups and Pathfinders. In September of that year, while helicopter scouting the route for the Nevada 500 (Las Vegas to Reno), the chopper crashed, nearly killing Balch and leaving him with a permanent back injury that ended his racing career.

Balch's first Scout was sold to one of his old sponsors, Archer Brothers, in the early 1980s and despite having changed hands, somehow remained largely unmolested and survives to this day. The

▲ A little the worse for wear, Balch parades the Scout around the Riverside track after a first-place in the August 1978 race. The beating taken by the Scout in that race probably led to the new paint and graphics seen in the Baja 1000 a few months later. The short courses usually resulted in multiple collisions, so body damage was expected.

▲ The first Balch Scout as it looked in 2011 while owned by Scout collector Wes Brent. It has since found a new home and may one day appear in fully restored race splendor at a Scout show near you. This is the color scheme it wore from late '78 on. As far as we can tell, the last time it raced under Balch was the '79 Baja 1000. In the latter part of 1979, Balch scored what we think was the last big racing support, a $1 1980 Scout. There are no available images of that second Scout, but it was built to very high standards, wrecked early in '80, rebuilt, and raced into 1982. *Wes Brent*

exact fate of Balch's second Scout is unknown, but the last reports have it in a private collection in Texas.

Balch became a very successful contractor in the San Francisco Bay Area and is well known for his charity work, which has included using his company assets to help reconstruct Haiti after the 2010 earthquake. When interviewed in early 2015, he had just returned from there.

JERRY BOONE

Jerry Boone was born in Downey, California, but moved to Parker, Arizona, in 1969 at age 25 to run a hay-hauling business. With his in-laws, the Rexwinkles, he opened an International truck dealership in 1976, which they called Truck-n-Tractor. While attending the introduction of the SSII in January 1977 in Chandler, Arizona, he met two of the drivers just signed for the race program and was hooked on the

▲ In just 10 days, the team at Truck-n-Tractor took a bone-stock '77 Scout SSII and turned it into a race Scout. Boone bought the Scout direct from the '77 SSII intro at Chandler, Arizona, and was the only of the four factory-sponsored teams to race an actual SSII, not a clone. In fact, this was the first time Boone had raced anywhere in anything. The Scout was still very stock at this point February 1977.

SSII and the idea of racing. Jerry bought one of the early production SSIIs at the event and decided to take it racing. After only 10 days in the Truck-n-Tractor shop, it went from a bone stock Scout SSII to a race rig and was entered in the Parker 400 race with the number of 327L ("L" for late entry). It was literally the first time Boone raced anywhere.

Jerry is quick to point out that he would have gotten nowhere in racing without the help of his in-laws, Wayne, Byron, and Bruce Rexwinkle, all of whom Jerry describes as "mechanical geniuses." The Rexwinkles would become well known in Scout racing and in off-road racing in general. Two of their major accomplishments were developing leaf-spring

▲ By the 1977 Riverside race, the SSII had lost its appliqué but acquired a good number of modifications to the chassis and suspension. It still had a more or less stock engine.

packs for Jerry's Scout that allowed him to travel faster over rough terrain. These thin-leaf packs were eventually adopted by some of the other Scout teams. Front hub failures were a common problem for all the Scouts and Bruce Rexwinkle developed a heavy-duty front hub and built enough to supply the other factory Scout teams.

Jerry Boone was well into his first racing season, at his own expense, when Dick Bakkom approached him to become a factory-sponsored team. Boone had won some rookie awards in his first year and his unexpected (except to Boone!) rookie win at the 1977 Baja 1000 attracted a lot of attention. As Boone described it, "After I kicked some of the factory guys' butts a couple of times, they thought I might be a good addition to the team." That proved more than true.

What is very interesting about that first season is that it was run on a bone-stock engine. That included the Baja 1000 win! The Boone team had no trouble with the engine and Boone credits the Rexwinkle spring packs for allowing him to maintain a higher speed over rough terrain, thus fully or partially negating the power disadvantage. It wasn't until the 1978 season that the 345 got some upgrades in the form of cam, pistons, and breathing. Ditto for the front-shackle reversal. Ironically, the only engine troubles they had were *after* the engine was modified. In 1978, the team received one of the IH engine plant's prototype aluminum four-barrel manifolds. It was only partially machined, but the Rexwinkles took care of that. While the new manifold may not have offered a big power gain, it did save about 40 pounds.

A back-to-back win at the Baja 1000 was within his grasp for 1978 when, just a few hours short of the finish, Jerry ended up stuck in a sand pit and took hours getting out. There's no shame in being second behind Rod Hall, but there would have been much more joy in making *him* take second.

Boone started off very uncomfortable on the short courses but learned he had a knack for them.

▲ The desert racing world was astounded when Jerry Boone walked away with a Class 3 first-place in the 1977 Baja 1000. Some of that was luck, but Boone had a lot of innate skill and cunning that served him well. Though the Scout was still running the stock V-345, the suspension and chassis had been given the first of many upgrades.

This was proven out when he won Class 3 in Mickey Thompson's 1979 La Coliseum race, an event everyone acknowledged was a tough nut.

Boone reports a good relationship with IH, but when the strike came very late in 1979 and ran into early 1980, all the factory support stopped along with promotional events. The end of the strike was followed by a total end of support from International, but Boone does not recall how, when, or from whom he got the news. This slowed the Boone racing schedule but didn't stop it. Jerry placed second in the 1980 Parker 400 with his wife, Cleone, riding shotgun as an anniversary present. Boone raced again at Parker in 1981, but that was his last year in the saddle. That year, the Boone Scout lost its signature Grenoble Green paint and acquired a white finish with a Christianity-inspired rainbow appliqué on the sides.

The Boone Scout raced its last in 1982 with Jerry's son Rick behind the wheel. It was the year after his 1981 high school graduation, so the Scout was adorned with "Class of '81" across the hood. Jerry had more or less handed over the keys and said, "Have fun, but on your own dime." With a little help from his uncles, Rick prepped the Scout more or less as he thought fit. The tired Scout broke a number of times at the

▲ Boone had the crowd on its feet at the 1979 Mickey Thompson LA Stadium race, the first time an off-road race was held inside a sports stadium. There was no doubt Jerry was pleased to take home a big win, snatching the Class 3 trophy out of Balch's hands.

◀ Once the factory sponsorship was over, Jerry went back to the IH dealership and trucking. The Scout languished a short while, but in 1982, after his son Rick had graduated from high school, Jerry gave him permission to race the Scout on his own dime. He scraped up the money to enter, found a few sponsors, and managed to get his tech-savvy uncles to contribute their expertise one last time. He entered the 1982 Parker 400 but logged a DNF due to a succession of mechanical issues, the last being a major axle failure. That was the last time the Boone Scout raced and it hasn't run since. It's still in their possession and the family talks about bringing it out again—for shows at least.

RACING SCOUTS 361

▲ Mike and Debra Ismail's (owners of IH Only, a Scout Light Line dealer) race Scout Traveler is very likely the last Scout seriously campaigned in sanctioned professional desert racing. Built in 1994 and first raced in the 1995 Barstow 250, it was raced into 2010, mostly in MDR (Mojave Desert Racing) Class 400 or Class 850. Brothers Mike and Jeff Ismail teamed up to race it, sharing expenses and running the races together. It's shown here at a jump called "The Wall." It has raced 29 times and DNF'd only three times. Mike took a break after 2010 but plans to update the truck and race again. The Scout is powered by a very potent 392 backed up by a built T-407 and has a Dana 60 rear axle.

beginning of the race, but even with his uncles' help, it could not be resurrected in time to finish. That was the Scout's last race.

Jerry Boone still has the Scout. Most of its breakage from the 1982 race has been repaired, but the truck remains partially disassembled at the Boone hacienda in Parker. Boone went back to running the IH dealership, and when it shut down, he went back to hauling hay—something he is still very good at, both on and off the road.

OTHER DESERT RACERS

There was no shortage of teams racing Scouts. Most of them raced on a regional level, but more than a few competed on nationally. One of the standouts in the national realm was Jerry Colton of Colorado Springs, who did well whenever he appeared and often raced in Class 4, with some sponsorship from the dealership and corporate levels of IH.

The Mulligans raced an ORD-built long-wheelbase Scout with some success. Larry Thew was another racer who competed nationally when finances allowed.

There are still a few Scouts racing where class rules allow. Mike Ismail, of IH Only, is very likely the last person to regularly campaign a Scout in a sanctioned SoCal desert race. He ran until 2010, then took a break, but may campaign his Scout yet again in Southern California and Nevada desert races. As a revision of this book was being completed in 2023, Anything Scout of Ames, Iowa, was setting up to campaign a Scout Terra in the 2023 NORRA 1000 race. There has also been talk of resurrecting Howarth's racer for the NORRA 1000.

Appendix 1

LINE SETTING TICKETS

If, after reading this book, you still don't get the importance of the LST, here's one more chance to be beaten over the head with it. The LST lists all the components for one particular Scout and was the "recipe" line workers followed to build it. All these years later, it provides owners and restorers the same recipe to bring their Scout back to original. An abbreviated version of the LST was attached to each vehicle, but decades later, they are often deteriorated or missing. That's where the Wisconsin Historical Society and your favorite Scout Light Line dealer come in.

During Scout production years, and following the demise, the buyer went to a dealer to get their LST. The order was made to the firm in Illinois (later Michigan) that maintained the records. When Scout Light Line entered the scene in 1991, they became a part of the game, though the ultimate source was still Navistar or one of their designated recordkeeping vendors. Navistar chose the Wisconsin Historical Society to house the remaining records from the IH days, and in the mid-2000s, WHS acquired one set of the microfilm that contained the Line Setting Tickets for most IH trucks. After a bit of organizing, they offered a lookup service so owners could get copies of their LSTs, either directly or via a Scout Light Line dealer. The funds generated go toward maintaining the McCormick collection in which all the IH records are kept, so consider your purchase a donation for that purpose.

The location of the LST in a Scout varied over the years and few early Scouts still have them. After all, how long can paper and tape last? Surviving early LSTs are likely to be found only on a well-preserved Scout or if it was removed by an early owner for safekeeping. On a Scout 80, they seldom remain, but when found, they are glued to the back of the glovebox. On an 800, they are taped to the inner firewall (right behind the glovebox insert). On 1971–1976 Scout IIs, Travelers, and Terra, they can be found in the engine compartment, on the cowl near where the hood latch is located. From 1977 to 1980, it was attached to the glovebox door. These very commonly survive because they are better protected and made of a less vulnerable material.

The on-vehicle LSTs are somewhat abbreviated but list the major options and sometimes the build date. The dealers had a third source of LST material in the form of microfiche with the major component numbers listed with a VIN. Those component numbers need to be decoded, however, and certain things, such as paint codes, are missing. These dealer microfiche do not cover the entire range of Scouts, only from '73-on.

ANATOMY OF A LINE SETTING TICKET

The LST can be considered the DNA profile for a Scout. Like a DNA profile, parts of it are very arcane and not immediately useful to those outside the lab (or in this case the factory). That said, they are rich with historical and technical detail for those who know how to read the important parts.

Component/Option Codes: These are numbers indicating the individual options ordered for the vehicle. They may vary by year and model, but often the code for something such as a chrome bumper on a 1962 Scout 80 might be the same as for a 1979 Scout II. They all share a common convention, the first two digits indicating the parts group.

01 - frame and bumper
02 - front axle
03 - springs
04 - brakes
05 - steering
06 - driveline
07 - exhaust
08 - electrical
09 - front-end sheet metal
10 - misc (décor items, speedometer, undercoating, etc.)
11 - clutch
12 - engine
13 - transmission and transfer case
14 - rear axle
15 - fuel tank and related
16 - cab, body, and interior
17 - wheels and hubs

Going down the column, the options are listed by group, so the 01 bumper stuff will be first and the 17 wheels group last. Tires used a different numbering convention.

Over the years, there were a lot of options codes. Some partial lists have been done up over the years. Most of the powertrain codes are listed in the Encyclopedia sections and many of the other codes are found throughout this book in pertinent locations. The more obscure codes (e.g., the various codes for lighters in the three major Scout eras, or one-off delete codes) are probably a bit too arcane for most of us.

Paint Codes and Paint Notations on the LST: Unlike some other makes, International did not put the paint code or any other paint information on the body tag. The *only* place it is found is on the LST, either the one on the vehicle or the one purchased from WHS. The paint code can be found at the bottom of the Line Setting Ticket and there are several ways it might appear, depending on era.

In the early Scouts, there's a simple one-line notation such as "PAINT CHART 102S 4393GO." This is from a 1974 Scout II LST that indicates the 4393 Burnished Gold color. The 102S is the Paint Instruction Code (read on) and the 4393 GO is the paint code and a two letter abbreviation for the color, in this case Gold. Similar codes go back to the first Scout 80.

Starting later in 1974, the codes got a little more detail. For example, in an LST that reads "PT4 9219 WH SINGLE," the inscriptions PT1, PT3, etc. (up to PT6) are line markers for paint instructions. In these cases, the paint code is usually found under PT4, such as: "PT4 9219 WH SINGLE." Here, 9219 is the later 1975-on Winter White and WH is the color abbreviation. Sometimes there is no space between the number and the abbreviation. SINGLE indicates the body and top were painted a single color (TWO-TONE was used for a two color combo). If the colors were two standard colors for that year, it would read STANDARD TWO-TONE. If a CT-399 catalog color, it would be indicated by CT-399 COLORS.

The LST also mentions Paint Instruction Codes. These were the painting instructions for particular models and body styles. Some of the different types of bodies required painting in different areas, a particular combination of colors, or even the use of paints that required special application (metallics for example), and this code told the paint shop what to do. They are very arcane and there isn't much information available on what exactly they signify. The 100 code is most common in early Scouts into the 800 era and 101 is the most common in the Scout II era. Two-tone treatments were commonly seen as well, and the LST will indicate the colors. The Paint

Instruction Code always used 201 for two-tones and listed the two colors. In the later Scout II era with PT line indicators, PT4 was usually for the body color and PT5 for the top color.

There is a method to the madness of the three- or four-digit paint codes. The first digit loosely defines the color group (e.g., 0 = black, 2 =red, etc.). You can find the paint codes (and names) listed by year in the Encyclopedia sections of this book. Many early colors were only three digits long, but, with only a few exceptions, they were replaced by four-digit codes starting in the late Scout 80 era. For the most part, paint manufacturers can match a standard color from the IH paint code or a sample from a Scout.

The CT-399 was a book of special colors from which the dealer could order.

As mentioned, for a fee, the buyer could order any color in the CT-399 from International. They would also do custom colors for a greater fee when supplied samples according to their guidelines. For 1977, a single special color from the CT-399 cost $68. A custom color was $130. More common in later years was the two-tone option, which could be special-ordered in a couple of ways. The least expensive two-tone option was to choose from the standard colors available that year; in 1977 that was $40 extra. Prices were reduced in the cases of fleet orders over 50 units.

We haven't seen many LSTs with custom notations, but most have some version of "SPEC PAINT IN LIEU OF STD" and a three- or four-digit code. The first is the color group (read below). For example, in a late Scout 80 with special paint, we found the code 501GN. The 5 indicates a green color and an MT-90 chart shows three 501 code greens and indicates they are all custom colors and not in the CT-399. Another example comes from the 1966 800 era and shows "211RD." The MT-90 shows two custom reds with that code. Yet another 800 had a "6452TQ"— which was a turquoise (which is technically a blue color but was listed as a CT-399 color in those years).

PAINT GROUPS

0-1 - Tan, Beige, Brown, or Black
2 - Red
3 - Orange
4 - Yellow or Gold
5 - Green
6 - Blue
7 - Purple
8 - Gray
9 - White

KEY TO LINE SETTING TICKETS

This key applies to all three Line Setting Tickets shown on the following pages. Because they are from different eras, some have elements that do not appear in others or are in different positions.

1. The shipping company and method used to transport the vehicle. Among the most common trucking shippers were KAT (Kenosha Auto Transport) and SOBER (Howard Sober Trucking). If shipped by train, the box would list "RAIL" and the station destination, usually a city name.
2. The IH sales district in which the ordering dealership resides.
3. Order quantity. Sometimes more than one order was made of a particular combination of options or was grouped by who ordered them. The LST might list the number of the particular vehicle in the total, "2 of 6," for example.
4. Order number. These come from the order sheet sent in by the dealership or entity making the order.

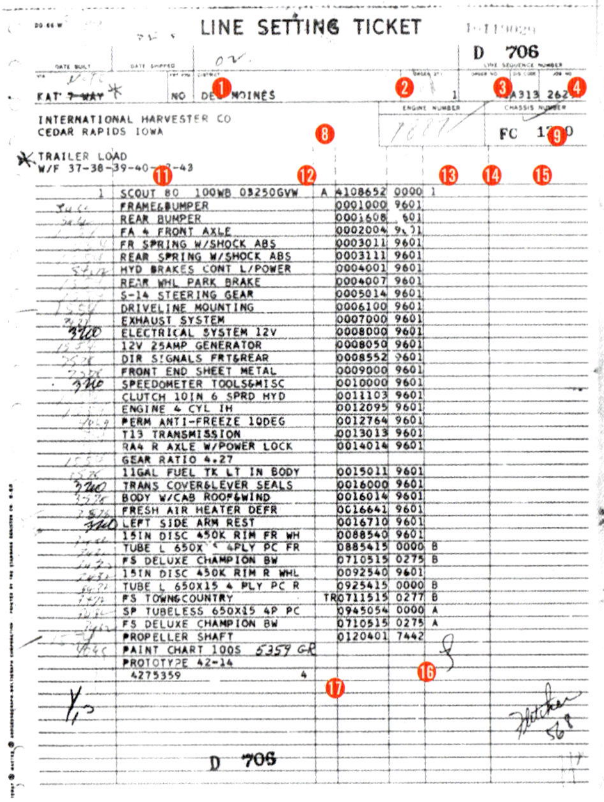

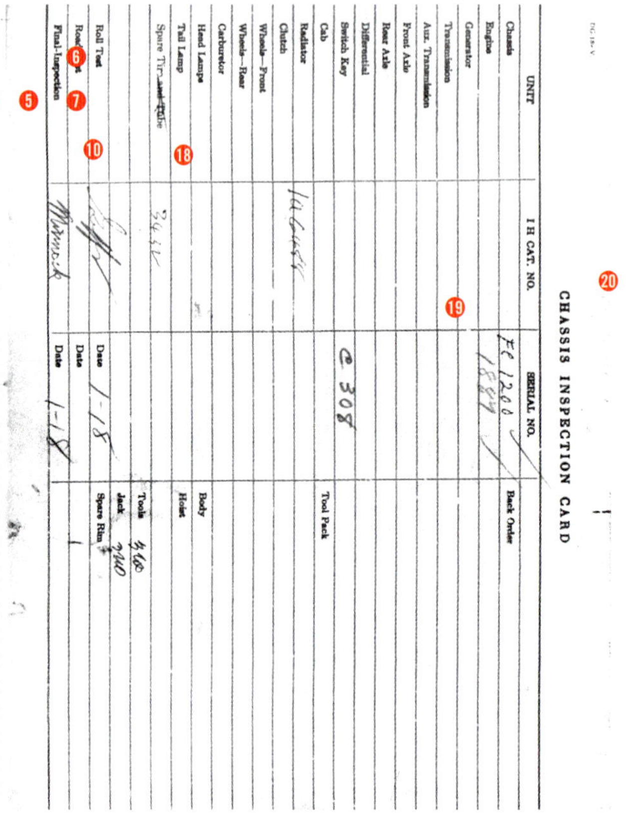

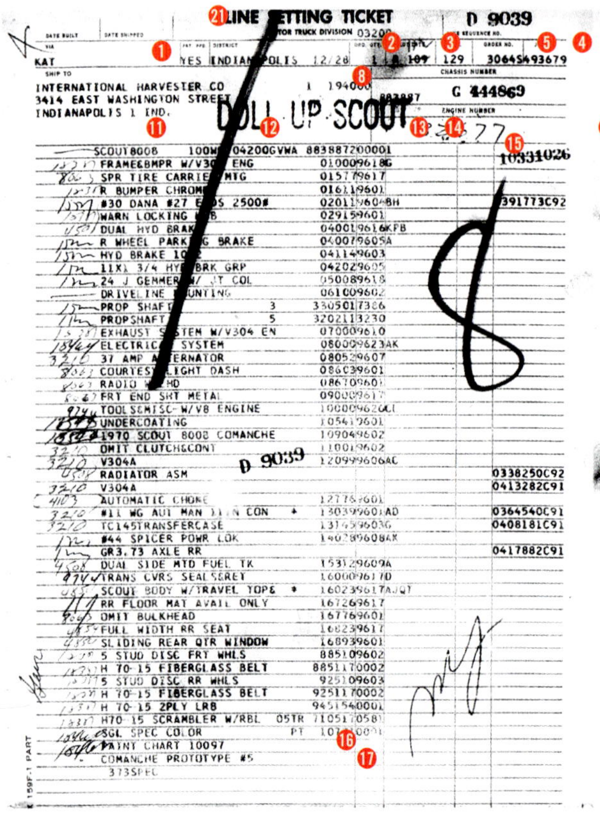

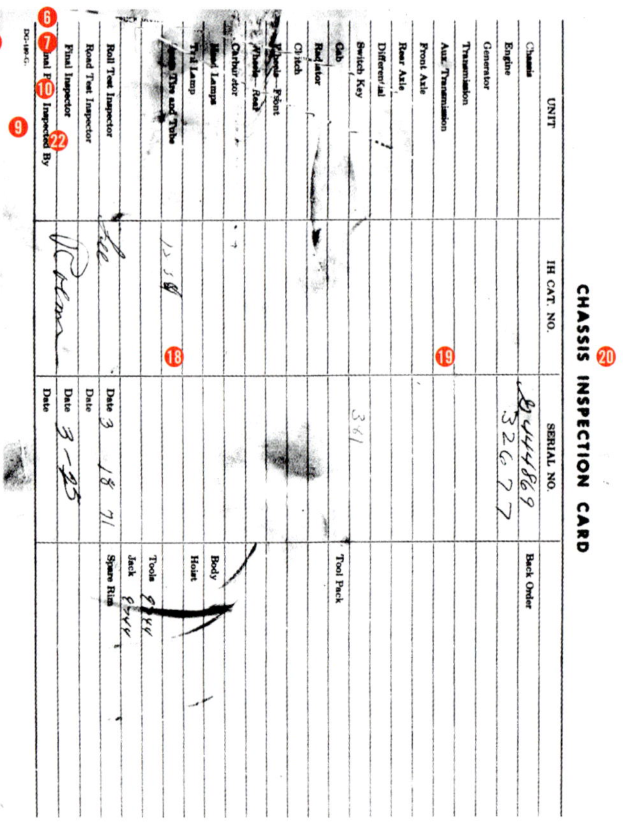

5. District code.
6. Line Sequence Number (LSN). Very useful when tracking batches of Scouts. This number would reset at 9,999. Typically the sequence started with a totally new model, such as the Scout 80 and the Scout II. The numbers did not reset for the Scout 800 transition, nor the 800A or 800B. For the most part, Scouts ran down the line sequentially according to LSN. Often, but not always, the same was true of the serial number.
7. Job number. This arcane number came from the scheduling department and what they indicate is unclear.
8. Ship-To. When you see "International Harvester Co." here, it indicates a factory-owned dealership. A franchised dealer will be listed by name. Many other interesting notations are sometimes found here. If the Scout is a special model, there is often a notation stamped or handwritten in the box. Some that we have seen include:

    6 Cyl - Seen on some of the earliest rigs when the 6-232 was first introduced.
    Arist - 800A Aristocrat.
    Camper - Scout 80 Camper.
    Diesel - Many, if not most, diesel-powered Scouts.
    Doll-Up - Usually stamped, sometimes handwritten. Starts in the 80 era with the Red Carpet and Champagne series and carried through the 800 era and into the Scout II era on any special model and for an early Custom trim level model; occasionally, may include a further notation on which special, e.g. "Red Carpet."
    FC - Presumably a derivative of "Family Cruiser" and seen on Scouts ordered with the Midas Package; sometimes seen with "Midas."
    Gold Whl - Seen starting 1978 with the optional gold painted spoked wheels and also for the gold accented alloy or polycast wheels. .
    Hot - A priority build.
    Midas - Any Midas package. Alternate to *FC* but sometimes seen with both notations.
    Patriot - 1976 Scout II Patriot edition.
    Pink Paper - No sure information on this, but appears often on LSTs for prototypes, special models, rush jobs, or show Scouts.
    RHD - Right-hand drive.
    RS - Limited edition 1980 Scout RS Traveler.
    Show Scout - Scouts were regularly built for shows and the LST are so marked; usually have some interesting special features.
    Sno-Star - Seen only on the 1976 Scout II version, not the 1971 800B.
    Spirit - Spirit of '76; sometimes written "Spirit '76"
    Sport Top - 800 Sportop; sometimes just "Sport" but occasionally seen with "Conv"
    SR2 - 800A SR-2

9. Engine number. Seen here and on the back (item 20). This is usually handwritten according to what engine was next on the rack. No effort was made to install engines sequentially by number, only to make sure the correct type of engine was installed. In the early days of Scouts, you see the least amount of "skew" between engine number and serial number.
10. Chassis number or VIN. Usually, but not always, stamped or typed. Most often, Scouts went down the line sequentially by serial number or VIN. In later LSTs, the model code is included here. Handwritten numbers are

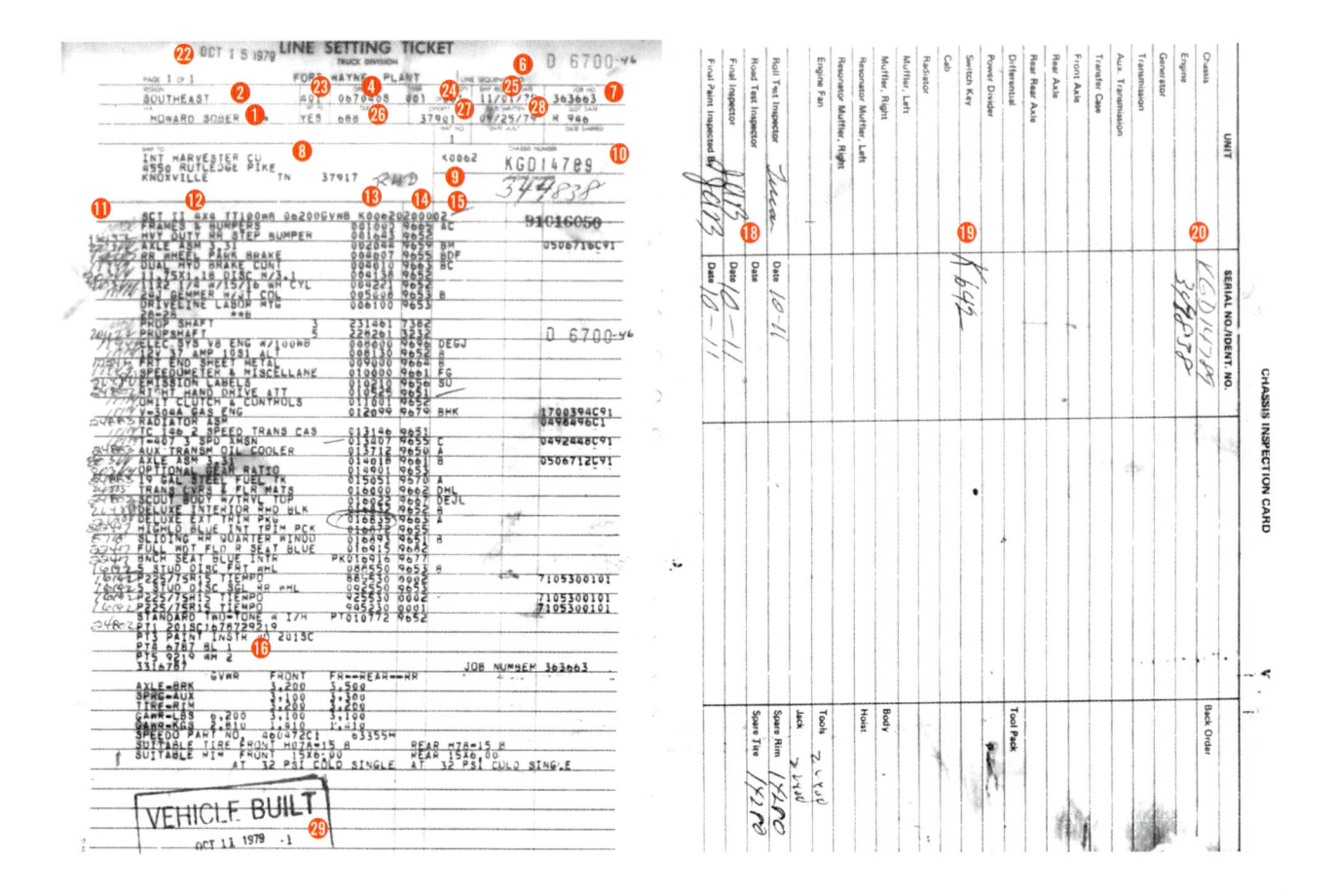

also seen, and sometimes one is crossed out and another written in for a bevy of reasons. LSTs so marked usually provide interesting reading.

11. First column. It is unclear what these always handwritten numbers indicate, but they are often the same as the component code so perhaps they are a line check.

12. Second column, top line. This is the model, the wheelbase, and the GVW.

12. Second column, all but top line. The component/option descriptions, listed in order according to parts group.

13. Third column, top line. This is the model code. This will be found here and on certain internal documents. See the Encyclopedia sections for how to decode these.

13. Second column, all but top line. Component codes, listed in parts group order. Some are options, some are standard or mandatory items by Scout model. For example, a 100-inch chassis was the same for all 100-inch Scouts and there were no dealer or customer options, just a component code. Many of these codes will show up as options in the price book, but many won't. They vary in length and the first two digits are parts group numbers. In older LSTs there are some extra zeros at the front; these are placeholder digits.

14. Unit division variation. A certain component code may indicate the basic component, but this number indicated a variation. These numbers are important because a component code may be used for many generations of Scout but the

Unit Division Variation would highlight the differences. Regrettably, we do not have a Rosetta Stone for these, though we have figured a few of them out..

15. Fifth column. Any notations here indicate a change, variation, special instruction or upgrade to the standard part or installation process. Line workers would have documentation to indicate what these meant at particular times or with particular models. These changed often and we have seen some for specific moments in time. They can be lists (some very informal), instructions, or even wiring diagrams.

16. Paint information. See "Paint Codes and Paint Notations on the LST" (page 369) for more information.

17. Prototype notation. It is not always present, but many get confused when it is, thinking they may have a prototype Scout. Not so. International had standardized build configurations they called "prototypes." Generally speaking, the prototype system was an ordering expedient for both the dealers and factory production planners. Sales put together a configuration that was deemed, or had proven to be, a popular one and that combination could be easily ordered by listing the "prototype" number assigned to it. Later, this evolved into the Gold Star system in which dealers got a booklet with a large number of option combinations listed under various descriptive categories such as "Deluxe Town Traveler" or "Sports Wagon" that described how the customer might use the Scout.

18. On the backside of the card is listed "Roll Test" and "Final Inspection." The roll test was done as the Scout exited the line (usually). It was placed on a rolling road that simulated a highway load and tested for a period of time. If any glitches were found, they would be fixed right there, if possible, or the Scout would be taken to another location for repair. Final inspection was complete when the Scout had passed all its tests and was essentially ready for shipment. Matching Roll Test and Final Inspection dates indicate the Scout was more or less trouble-free at assembly. A difference in dates either indicates quitting time came as the roll test was being done or that Scout had a problem and the final inspection date came only after it was resolved. Sometimes the line would be short certain parts and that could also account for delays in the final inspection.

19. Ignition key code
20. Engine and serial numbers again
21. Date Shipped. This is most empty in the early 80 to 800 era.
22. Part number column for the component listed. Usually numbers are only shown when the part is something outside the standard build or something new that has not yet been fully integrated into production. The part number is usually not the same as that issued for the same part sold through the Parts Department.
23. TSR (Truck Sales Routing), an internal code for how the Scout was to be shipped
24. Number of vehicles in this particular order (e.g. "#1 of 2")
25. Ship Request Date. Seemingly often ignored.
26. Dealer number
27. Destination zip code
28. Date the sales order was written
29. Vehicle-built stamp. Appears in the Scout II era. Sometimes the date is handwritten.

▲ The location of the LST on a 1977 and later Scout II. These are the most complete versions to be found and are generally photostats of the original tickets printed on a durable material.

▲ In the 800A and B eras, the LST is found on the passenger firewall, just above the footboards.

▲ The Scout 80 LST is the most fragile of all, being just paper glued to the firewall inside the glovebox. To find one intact and pristine like this is rare. Owner Patrick Caraher is the second owner of this '65. Obviously the first owner took very good care of his rig. Interestingly, this Scout originally mounted a 4-152T turbocharged four. *Patrick Caraher*

▲ On Scout IIs through 1976, a photostat was found attached to the cowl near the hood latch.

▲ The CT-399 book was issued to dealers and contained paint chips for about 200 optional colors applicable to all IH trucks. The number of colors, and the colors themselves, varied over the year. John Glancy's CT-399 is from the late 1960s and contains 205 colors. The chips gave customers an idea of what the paint would look like on the Scout body.

370 INTERNATIONAL SCOUT ENCYCLOPEDIA

Appendix 2

SERIAL NUMBER AND VIN LOCATIONS

The Scout serial number or VIN (serial number into 1966 production with the VIN after) is stamped on the chassis and onto a plate attached to the body. The location for each varies according to the Scout era.

▲ The serial number/VIN is stamped into the chassis on the driver side, forward of the axle, just behind the bumper attachment on both the Scout 80 and 800. The stampings are usually fairly light and sometimes don't survive the ravages of time and rust. Heavy painting can also obscure them. This is an 800 chassis. The 80 chassis does not have the lifting hole.

▲ The early Scout 80s had the serial number plate screwed to the firewall on the passenger side.

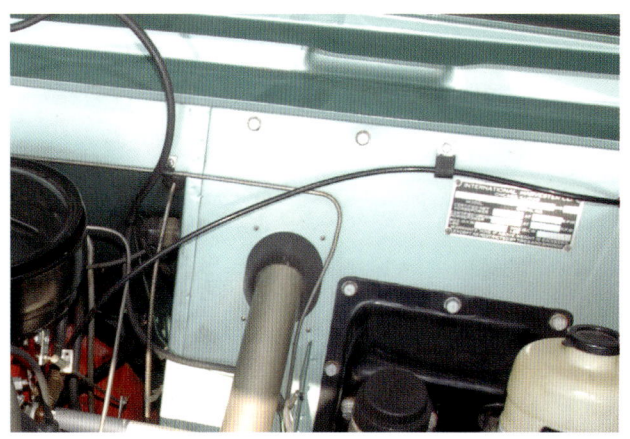

▲ The serial number/VIN tag is screwed to the firewall above the brake and clutch master cylinders on the driver's side (left-hand-drive Scouts) on Scout 80s from about the '62 model year and into the 800 era through 1968.

▲ Starting in 1969, the metal VIN tag was moved from under the hood to the passenger-side door opening.

▲ In 1969, new federal regulations required a Federal Safety Certification Sticker, which in the 800 line was placed in the rear edge of the door opening, below the door latch.

SCOUT 80 AND 800

▲ The chassis VIN stamp is located in the area shown in white here. It's just forward of the trailing front-spring hanger on the driver's side, usually near the top edge of the frame. The stamping is usually better on Scout IIs than on 80s and 800s, but with a lot of rust, it can still require a fair bit of work to bring out.

SCOUT II

▲ On the 1977–1979 SSII models, the metal VIN plate was located on the passenger kick panel, conveniently upside down so you could bend over, lean in, and read it.

▲ For the 1971 model year Scout IIs, the VIN plate was located on the cowl. This is the only year Scout II with this feature. *Michael Onstad*

▲ Except for 1971, the Scout II metal VIN plate was located in the leading edge of the passenger door opening. The exception being the SSII, as noted nearby. Through most of 1971, the VIN plate was under the hood near the hood latch facing up.

▲ The Federal Safety Certification Sticker was in the leading edge of the door opening on the driver's door throughout the Scout II era.

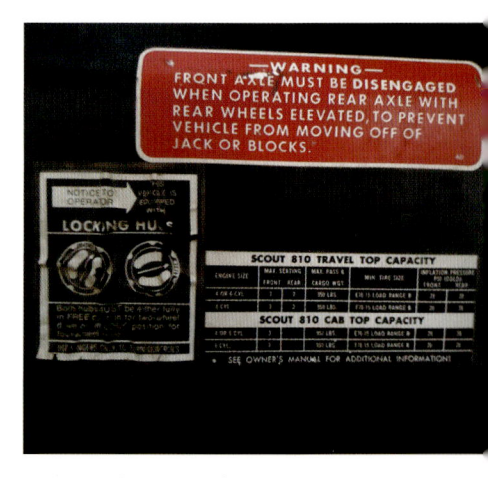

▲ From the start of 1971 production at least into September of 1971, the glovebox door GVW plate marked the model as a Scout 810. From that point, these tags were marked "Scout II." At the time of this writing, the 810 stickers are not being reproduced. *Michael Onstad*

Appendix 3

PARTS AND SERVICE SOURCES

Scout Light Line Distributors (www.scoutlightline.com) was created in 1991 to take over the Scout and Light Truck parts support business from Navistar. These OEM parts are still available through those Navistar International truck dealers that choose to cater to these customers and through the following SLL-authorized dealers:

INDEPENDENT SCOUT/LIGHT LINE PARTS DEALERS

UNITED STATES

Isa's IH, Kingman, AZ, 661-269-0272

IH Only, Lancaster CA, 661-728-9552

IH Parts America, Grass Valley, CA, 530-268-0864

Federal Valley Motors, Westminster, CO, 303-429-1577

Anything Scout, Ames, IA, 515-233-3020

Scout Connection, Ft. Madison, IA, 319-372-3272

Roedel Bros., Kirkwood, MO, 314-489-2241

Super Scout Specialists, Enon, OH, 937-525-0000

Binder Boneyard, TerraBourne, OR, 503-481-4171

Scoutparts.com, Portland, OR, 503-772-0070

Barnes International, Cochranville, PA, 717-875-2271

Scout Madness, Lubbock, TX, 806-745-7475

Scoutco Products, Hassisonburg, VA, 540-433-5136

CANADA

Southland International Trucks, Lethbridge, AB, 403-328-0808

ScoutpluSS!, Hope, BC, 604-869-3665

EUROPE

Scout Store, Germany http://www.ihc-scout.de/scoutstore/index.htm

ENTHUSIAST RESOURCES

For information regarding clubs or groups, contact one these tech forums or your nearest SLL dealer:

BINDER PLANET
www.binderplanet.com

BINDER TV (VINTAGE FILM AND VIDEO)
www.bindertv.com

CORNBINDER CONNECTION MAGAZINE
www.cornbinderconnection.com/

NATIONAL IH COLLECTORS CLUB
www.nationalihcollectors.com

OLD IHC
www.oldihc.org

MUSEUMS AND ARCHIVES

Auburn Cord Duesenberg Automobile Museum
1600 South Wayne St., Auburn, IN 46706
www.automobilemuseum.org
260-925-1444

IH Scout, Truck and Tractor Museum
6711 Dayton Springfield Rd., Enon, OH 45323
937-525-0000

National Auto and Truck Museum
1000 Gordon M. Buehrig Pl., Auburn, IN 46706
www.natmus.org
260-925-9100

Wisconsin Historical Society
McCormick Collection
816 State St., Madison, WI 53706
www.wisconsinhistory.org
McCormick Collection direct line: 608-264-6410
McCormick Collection email: askmccormick@wisconsinhistory.org

Appendix 4

GUIDE TO GRAPHIC PACKAGES (APPLIQUÉS)

International offered a variety of graphic packages for Scouts. These were decals—referred to as appliqués by IH—that were applied to the truck. Each caption lists the models and years the graphic package shown was offered. The package number is the component code that appears on the Line Setting Ticket (LST). The code "PK" indicates the graphics were part of a package of options without a separate code.

▲ 1969 Aristocrat, code PK: rocker stripe, dark blue

▲ 1970 SR-2, code PK: rocker stripe, belt stripe, hood stripes, white

▲ 1971 Comanche, code PK: rocker, wheel arch stripes, gold, hood stripes, white

▲ 1971 Sno-Star, code PK: rocker stripes, black; belt stripes white/black; door "Sno-Star" graphic white

▲ 1974–75 Traveltop, code 10842: panel appliqué, white

▲ 1974–75 Traveltop, code 10843: panel appliqué, woodgrain

▲ 1974–75 Traveltop, code 10844:
lower body triple stripe, white

▲ 1976–77 Traveltop, code 10842:
split panel appliqué, white

▲ 1976–77 Traveltop, code 10843:
split panel appliqué, Cork (aka woodgrain)

▲ 1976–77 Traveltop, code 10846:
panel Feather design, red/white/black

▲ 1976–77 Traveltop, code 10845PK:
Rallye Version 1, white

▲ 1976 Spirit, Patriot, and SnoStar, code 10876PK:
Spirit appliqué, red/blue, similar to split panel
10842 design

APPENDICES   375

▲ 1977–79 SSII / 1977–79 Baja Cruiser / 1977 Sport, code 10710PK: belt appliqué, black/gold

▲ 1978–79 Traveltop, code 10845PK or 10969PK: Rallye and Off-Road Rallye Version 2, white

▲ 1978–79 Traveltop, code 10865: belt accent stripe, white

▲ 1978–79 Traveltop, code 10992PK: belt accent stripe, gold, Selective Edition (same style as 10865)

▲ 1978–79 Traveltop, code 10842: side panel, white

▲ 1978–79 Traveltop, code 10843: side panel, woodgrain (aka cork)

▲ 1978–79 Traveltop, code 10864: side panel, black

▲ 1980 All models, code 10845PK: Rallye Version 5, white/yellow/orange

▲ 1980 Traveltop, code 10846PK: Rallye Version 4, gold

▲ 1980 Traveltop, code 10833PK: Rallye Version 3, silver

▲ 1980 All models, code 10969PK: Rallye Version 6, black/yellow/orange

▲ 1980 Traveler or Terra, code 10840: split panel appliqué, woodgrain

APPENDICES   377

▲ 1980 Traveltop, code 10864: see-thru flare, orange/white

▲ 1980 Traveltop, code 10865: see-thru flare, orange/black

▲ 1980 All models, code 10843: spear, green/blue

▲ 1980 All models, code 10844: spear, orange/orange

▲ 1980 Traveltop, code 10841: wave, yellow/orange/orange

▲ 1980 Traveltop, code 10842: wave blue/green/yellow

Appendix 5

ABBREVIATIONS AND TERMS

CS - Champagne Series. The 1965 Scout 80 Doll-Up that would be the final one for the 80 line.

CT-399 - The special colors paint-chip book dealers used to order special paint for customers. It usually had about 205 choices above whatever was standard in a particular year.

Doll-Up - An internal IH term for an upmarket Scout package. Most often seen on LSTs but also seen frequently on internal documents. Often abbreviated "dollup."

Gold Star Prototypes - Preselected options packages concocted by International and published in a book for dealers. Dealers could then streamline their ordering process by using a single Gold Star number. In some cases, International used the Gold Star book to send dealers what they though they should have on hand based on regional sales statistics.

LST - Line Setting Ticket. The assembly line "recipe" for a vehicle. It contained the list of what components were to be installed on a particular chassis. An abbreviated copy of the LST included was in the vehicle, attached in various places, but the full version was retained by IH. The dealers also had microfiche that listed the major options and components for every Scout by serial number or VIN.

LHD - left-hand drive

LWB - long wheelbase

MTC - Motor Truck Committee. A planning and decision-making group made up of managers in all the related truck departments. The MTC predated Scout, but insofar as the Scout was concerned, it was replaced by the Scout Product Committee (SPC) in October 1967, which took over the Scout decision-making process.

MT-90 - A parts department paint code listing document, divided by hues.

PPG - Phoenix Proving Grounds. International had a hot-weather testing area near Phoenix, Arizona. It was also used occasionally as a base for dealer and press events.

RCS - Red Carpet Series. The 1965 Doll-Up in the Scout 80 line.

RHD - right-hand drive

SPC - Scout Product Committee. A planning and decision-making group comprising managers from all the relevant Scout departments. It replaced the MTC in the decision-making role just for the Scout product in October 1967 and continued to the end of the Scout in 1980.

SWB - short wheelbase

TSPC - Truck Sales Processing Center. There were several around the country, including one at Fort Wayne, where special installations and modifications took place. This was a catchall overflow area where time-consuming installations or installations on backordered parts could be carried out. The Fort Wayne TSPC was located about a mile southeast of Scout Assembly on Moeller Road.

WHS - Wisconsin Historical Society. The Madison, Wisconsin, location where most of International Harvester's archive is housed within the McCormick Archive. A good deal of the research for this book was done there, along with what is contained in John Glancy's vast collection of IH documents and images.

XLC - Xtra Load Capacity. An acronym chosen to mark International's standard 6,200 pounds GVW for 1975.

Acknowledgments

We have a lot of people to thank, besides each other, for this book. One of the standouts is Howard Pletcher. Howard was carrying the Scout torch long before this book was even considered. His efforts to promote, protect, and clarify the Scout's role in automotive history dates back to before the Scout was in the ranks of the orphan SUVs. As far as this book goes, he took an early interest and was probably the first to step up and offer his services. These have included wise advice on what to cover, fact-checking, and endless, "Hey, Howard?" inquiries.

Along the same lines are George Kirkham and Todd "Doc" Sommer. Both shared their energy, experience, and collections with us above and beyond the level of simple help. In addition, Dick Hatch was a designer involved in the creation of Scout who opened his collection and his mind to us.

A special thanks has to go to the stalwart crew at the Wisconsin Historical Society. Sally Jacobs, chief archivist of the McCormick Archives, pulled out all the stops to help us complete this book and gave us boundless encouragement. Lindsey Hillgartner endured our endless requests for information with great cheer and dispatch. Along the same lines, Sam Julian was a tower of encouragement and a veritable bulldog for digging up lost information. Lee Grady is the former chief archivist of the McCormick Collection, and even though he had moved on to other work at WHS, he was still willing to go beyond his responsibilities and lend his decades of experience to the cause.

John wishes to acknowledge the role his late father played in his love of all things IH. Bob Glancy began working for IH in the early 1940s and finished his career as a district truck sales manager in 1982. He was a Travelall man and the Glancys always had a company Travelall in the driveway. After the end of the Travelall, Bob drove company Scout Travelers. In the 1970s, he was the branch manager of the Springfield Truck Sales Center, located just across from the IH body plant. This gave him an advantage over other dealers because if a customer came in that day with certain need and it wasn't in stock, Bob would walk across the street, find an LST that fit, and "borrow" that vehicle to sell when it was built later that day or the next. This story, and others like it, show a kind of dedication to customer service that has always been an inspiration to John.

Special mention should also be made of John's mother, Donna J. Glancy, who always supported John's business aspirations, and in founding the Annual IH Scout and Light Truck Nationals Show and Swap Meet in 1990. Mrs. Glancy was the "T-Shirt Lady" and sold the event shirts to the participants every year. She passed away just days before the 1995 show, and the staff, which included nearly all the Glancy family, was devastated but pressed on. The late Scout and International Motor Truck Association recognized Mrs. Glancy at the event with a special award given annually in her name to a person or persons showing outstanding dedication in helping out with the show.

Finally, our gratitude goes out to the men and women who worked at International Harvester at all levels and who made the Scout possible. You gave us a great subject for a book!

THE SCOUT ALL-STARS

Bruce Armstrong, IH/Scout collector
Jeff Bade, Binder Planet, IH/Scout collector
Sherman Balch, Scout Race Team member

Brendon Barnes, Barnes International, IH/Scout collector
Jon Bill, ACD archivist and educator
Betsy Blume, granddaughter of Ted Ornas
Michael Bolton, Scout collector and historian
Jerry Boone, Scout Race Team member
Rick Boone, Scout racer
Rickie Bordelon, LLOL, IH/Scout collector
Steve Bostic, brother of Jim Bostic
Wes Brent, Scout collector
Chris Brooks, Scout collector and Scout Camper restorer
Jeff Calfa, chief intellectual property attorney, Navistar, Inc.
Patrick Caraher, Scout Collector
Tom Clark, Navistar keeper of the flame
Tony Clark, SCW, IH/Scout restorer
Dale Clifford, SIMTA, IH/Scout collector
John Comer, Gryphin Racing, IH/Scout collector
Phil Coonrod, Coonrod's IH, Scout assembly line
Duane Creek, Ft. Wayne IH Engineering, IH/Scout collector
Fred Crismon, International truck book author
Earl Culley, IH Truck salesman, IH/Scout collector
Don Davis, Glassic Annex
John Donnelly, Binder Planet/Diesel Specialist
Mark Drake, Scout Madness, IHC collector
Lynn & Heather Faeth, Scout Connection, IH/Scout restorers/collectors
Joel Faircloth, Glassic Motors
Guy Fay, author, IH historian and collector
Bob Glancy Jr., sSs, Parts salesman
Keith Glancy, IH/Scout collector
Rick Glancy, sSs, Parts salesman
Mike and Brendan Grieble, Scout Collectors
Don Grogg, NATMUS skipper
Terry Hankins, former IH engineer at the Indianapolis Engine Plant

James Hirchberg, photographer
Frank Howarth, Scout Race Team member
Herb Huddle, IH collector
Roy Ireland, Scout collector and historian
Jeff Ismail, IHPA, IH/Scout collector
Mike and Debbie Ismail, IH Only Racing
Jeff Jamison, IH/Scout collector
Lance Jones, Jimmy Jones historian and son
Larry Kelly, K-Bar Sheetmetal, IH Truck Plant
Carl & Mary Kindberg, IH/Scout collectors
Rex Kirkingberg, Glassic collector
Eason Lilly, Scout collector and historian
Rob Lykins, sSs, parts salesman
Mike Moore, ScoutCo, IH/Scout restorer/collector
Munson Family, Keepers of the Shawnee Scout
Jerry Muzillo, NATMUS chief engineer
Dick Nesbitt, IHC, CVI designer
Steve Nieman, Scout collector
Michael Onstad, Scout Collector
Rod Phillips, Gidd'um up Scout, SLL, IH/scout collector
Jim Poiry, Former Scout Plant Manager
Jim Potter, sSs, D-Line expert, IH/Scout collector
John (JP) Prosser, SLL, sSs, general manager
Jim and Wanda Ray, IH Engine Plant/IH/Scout collector
Rick & Paulette Riley, IH/Scout collectors
Gene Robinson, IH/Scout collector
Kent Robinson, IH/Scout collector
Jason Roedel, Roedel Brothers, IH restorer
Michael Schmudlach, IH historian and collector
Lynn Smith, formerly of IH Sheet Metal Development Department
T.J. Stevens, former IH employee
Tom Thayer, IH collector
Bill Thebert, IH/Scout collector
Aaron Warkenton, ACD curator
Bob West, IH/Scout collector
Wayne Wilkomm, Glassic collector

Specials thanks goes to the staff of Scout/Light Line Distributors, Inc. and Super Scout Specialists, Inc. for the support, assistance, and patience while John was focused on the writing and research of this book (www.scoutlightline.com, www.superscoutparts.com).

Special thanks, too, to the Annual IH Scout and Light Trucks Nationals Show & Swap meet (www.midnitestar.org) for a very small grant of funds for research and travel expenses.

The images included in this book came from disparate sources. They are marked with their source at the end of the caption. Those without such a notation come from the authors' private collections, mostly John Glancy's vast and awe-inspiring archive of original Scout and International material. Some also came from Jim Allen, who photographed extant originals and restored vehicles, and had a few stockpiled vintage images as well.

Those with a "WHS" come from the magnificent McCormick Collection at the Wisconsin Historical Society, which originated at Navistar International. Navistar placed most of their old files with WHS and donates funds annually for the maintenance of this collection. Take note of the image number after the WHS prefix, which is your key to finding that image in the archives. For a fee, all of which goes to support the non-profit collection, any of the images so-marked are available to you from WHS. In our many research hours at WHS, we had the honor to unearth some of these for the first time in decades. These newly-discovered images were then cataloged, numbered, and made available for others to view on their online catalog and use for other projects.

Other items came from the collections of well-known Scout experts, such as Howard Pletcher, George Kirkham, Todd "Doc" Sommer, Dick Hatch, and Dick Nesbitt, to name just a few. Our appreciation to all those who gave us the use of images from their private collections is boundless and we ask that you take note of their names in the acknowledgments.

Index

Accessories/Conversions
Ayr Way bed topper, 157
brush guard, factory, 35, 81, 257, 262, 265, 267
Dreamer camper, 65
dump truck Scout II, 129
fire truck conversion of Scout 80, 46
fire truck conversion Scout II, 128
garbage truck Scout II, 129
IH snowplow kit, Scout 80, 42
Meyer snowplow kits for Scout 800, 62, 236–239, 241
Mini-Shelter, 135
Ramsey winch, 70, 78, 115
Utemco conversions, 39
Warn winch, 78, 201, 242–243
Whitco Safari Tops, 76

Companies
Case IH, 11
Creative Industries, 15–16, 19, 110
Custom Vehicles International (CVI), 170–171
Gilmore International, 124
Good Times Inc., 273, 278–279, 282–283
Hurst, 302–304, 306
Midas International, 257
Navistar, 9, 11
Scout Light Line dealers, 33, 255, 3011, 299, 362–363

Data and Reference Study
abbreviations and terms, 379
line setting tickets, 363–369
paint and paint codes, **364–365**
parts groups, 369
serial numbers and VIN locations, 64, 371–372
Scout 80 data section, 48–52
Scout 800 data section, 84-89
Scout 800A and B data section, 90–95
Scout II, Traveler and Terra data section, 174–187

Engines
4-152, 21, 23–25, 54–59, 65,
4-152T (turbo), 44, 56, 59, 64–65, 67, 71, 74
4-152E, 4-196E, V-304E, 105–106
4-196, 54, 59, 64, , 67, 76, 96–97, 103,105–107, 123, 188–191
6-232, 76, 101,105, 124, 126, 191, 228–229, 237
6-258, 126, 129, 137, 180–181, 191, 341
6-33 Nissan diesel, 143, 189, 197, 271
6-33T Nissan turbo diesel, 99, 301, 316
A-55 Austin prototype engine, 20–22, 53–54
V-266 introduction, 54, 59, 68–69,
V-304, 54, 57, 76–77
V-304A introduction, 102–104
V-345, 54, 106, 117, 124, 129, 132, 136, 191–195
V-392, 54, 103, 107, 117, 188, 190–191, 194, 350
X-133, X-152, X-176, X-196 prototype engines, 54–56
AMC 343 V8 considered for Scout, 70
contoured vs notched vs flattop pistons, 105–107
Jeep 225 V6 considered for Scout, 64
PT-6, see 6-232

General Topics
Appliqués, 374–378
Campermobiles, 65
Canada built Scouts, 83, 84
Ford Bronco inspired by Scout, 38, 97
introduction of automatic transmission, 67, 76–77
Motor Truck Product Committee (MTC), 12, 15, 39, 54, 56-7
round side reflectors vs rectangular lights on 800A, 75, 77, 79
Scout Product Committee (SPC), 77, 117, 219, 231

Non-Scout International Models
AM-80 Metro-Mite van, 19–20, 53, 56
DF-808 6x6 truck, 34
100 Pickup, 147

People
Bakkom, Dick, 248, 302, 336–339, 352, 359
Balch, Sherman, 339–340, 352–357, 361
Barris, George, 74
Blume, Betsy, Ted Ornas granddaughter, 26, 29
Bolton, Mike, 173, 200, 202–203
Boone, Jerry, 339, 354, 357–362
Boone, Rick, 366
Bostic, Jim, 161, 171, 279,297, 302, 308
Buzard, Ralph M., 16–18
Colton, Jerry, 362
Ehlers, Larry, 248–249, 338–339
Felber, Willi, 272–273
Garst, Steve, 173, 203
Glancy, John, 167, 263, 267. 286, 334, 351, 370
Hankins, Terry, 104, 189, 349, 351
Hatch, Dick, 26, 165, 248
Hayashi, Kikuyo, 72, 205
Howarth, Frank, 251, 339–344, 348–351
Ismail, Mike, 33, 362
Jones, Jimmy, 251, 337–348
McCahill, Tom, 31–32
McGrew, Chuck, 12–13, 16
Monteverdi, Peter, 270–271
Nesbitt, Dick, 171, 278–280, 296, 298
Nicklin, Larry, 143–144, 158, 319–320, 333, 346
Nodwell, Bruce, 24–25
Ornas, Theodore "Ted," 11–18, 26–29, 203, 275, 321–322, 325, 327, 338–339
Pletcher, Howard, 26, 29, 118, 247–249
Poiry, Jim, 241
Reese, W. D., 12–13
Russell, Ed, 170–171, 278–279, 296
Sasser, Dick, 340, 343, 346
Windecker, Leo, 27, 29, 152–153, 325–327, 334

Prototypes Not Reaching Production
1981 model year, 316–324
Baja Binder, 341–343, 348
"Brown Door," see Trailblazer
Coachman, Royal Coachman, 222, 225
Comanche, 1961 proposal, 24
"Greenhouse Club" (Club Cab) Terra, 120, 143–144
HVS (High Volume Scout), 335
Long-wheelbase Scout 80, 120–122
Sahara package, 134
Scout 85, 321
Scout 90 LWB, 62–63
Scout 800A fiberglass top, 71
Shawnee, 224, 302–309
SSV, 311, 322–335
Sunriser, 296
Trailblazer, 205, 224, 226
Trailstar SSII, 293

Racing
Baja 1000, 337–343, 347–348, 352–357, 359–360
NORRA, 337–338, 341, 351
racing, 337–362
SCORE, 337, 343, 345, 349, 352–353

Scout Models
Scout 80 Topics
1961, 31–35
1962, 35–40
1963, 40–44
1964, 44
1965, 44–47
4.88:1 axle ratio introduction of, 38
50,000th Scout, 38
100,000th Scout, 44, 211, 215
axle update from Dana 27 to Dana 27A, 39
Cab-Top, early production focus on, 21, 32
Dana 44 (RA-9/24) axle debut, 38-39
data section, 48–52
developing the concept of Scout 80, 25
early prototypes, 20–24
early sales, 32–33
early sketches of, 10–18
end of production, 47
exhaust updates for 1963, 44
first chassis layout and prototypes, 18–21, 23

Scout Models (cont.)
Scout 80 Topics (cont.)
    full-size clay model, 19, 24
    first production line models, 3–32
    first road tests of, 31
    Hi Sport Top, 36
    military Scout 80, 35
    naming of the model, 23–25
    Nodwell Scout, 25
    Panel-Top, 30-31, 39
    right hand drive (RHD) introduction of, 39–40
    roll-up windows, 40, 42, 63
    rust issues, 41–42
    tailgate logo changes, 40
    Travel-Top design, 17, 22, 38
    Valley Forge Military Academy, use of, 37
    walk-through ("walk-thru") design, 36, 38–41

Scout 800 Topics
    1966, 66–67
    1967, 68–70
    1968, 70
    800Q suffix, 70
    Custom Doll-Ups, 70–71, 74, 221
    data section, 84–89
    development of, 60–65
    Dreamer camper, 65
    fold-down windshield option, 74
    introduction of Dana 20 transfer case, 63, 65
    introduction of T-45 (Warner T-18C) 4-speed option, 65
    military Scout 800, 81
    Postal Service Scout 800, 67, 69, 76
    Scout 90 interim Scout, 63
    Sportop, 65, 67, 71, 218–220

Scout 800A and 800B
    200,000th Scout, 76
    automatic transmission introduction, 76–77
    Custom option, 73
    data section, 90–95
    round side reflectors to rectangular marker lights transition, 75, 79
    Safari Top, 76
    safety upgrades, 68

Scout II Topics
    1971, 123–126
    1972, 126–128
    1973, 128–130
    1974, 130, 132–134
    1975, 134–137
    1976, 137, 140, 142–144, 146–147, 149
    1977, 149–150, 152–154, 156
    1978, 156–157, 162
    1979, 158, 160–161
    1980, 161, 163–165
    3.07:1 axle ratio replaces 3.31:1, 136–137
    810 designation, 117–118
    armored Scout II, 152-155
    automatic transmission for 4-196 engine, 147
    brown Terra and Traveler tops, 137
    Cabtop (A.K.A. Cab-Top, pickup), 124, 126, 128-129, 132, 135, 137
    cruise control, 152
    data section, 174–187
    debut of Scout II in Arizona, 118, 121
    Deluxe and Custom Trim, 125
    development of Scout II, 109–118
    development of Terra and Traveler, 118–123
    diesel engine option debut, 140, 194, 197
    dual exhaust, 129
    Economy Scout II, 144, 154
    ending production of, 166–169
    finned front brake drums 1973, 130
    Hy-Vo (chain drive) transfer case, 120–130
    last vehicle off the production line, 171–173, 200–203
    military, 125
    Panel Top, 129–130
    Pillow top seats, 163, 203
    Polycast wheel, 163–164, 170, 173
    RHD Traveler and Terra, 149
    rust issues, 165, 196
    Scout Product Division, 161
    Scout Terra and Traveler development of, 118–123, 137, 140, 142–148
    Solar Yellow paint issue, 144
    Spicer 30 open knuckle axle, 115
    T-19A (T-428) transmission intro, 136
    T-19 (T-427) transmission intro, 136
    Terra Suntanner, 156, 158, 160, 161
    towing, 133, 137, 146–147
    turbo diesel introduction, 163
    two-tone paint, 122, 132–133, 144, 157
    XLC models, 135
    X-Scout, 78–79, 112–113, 116–118
    X-Scout export pickup 122 inch wheelbase, 118, 120

Special Models
    434, 844, 298–299
    Aristocrat, 76–77, 102, 105, 205, 222–227
    Baja Cruiser (SSII), 253–255, 263, 265, 267
    Brush Buster (SSII), 251–252, 254
    Camper, 206–210
    Canada Dry, see Midas Canada Dry, variant of the Midas SSII
    Champagne Series (CS), 70–71, 211–212, 215–217
    Comanche, 231, 233, 235
    CVI Specials, 278–281
        5.6 Litre, 287
        Billy Kidd Ski Scout, 294
        Classic, Classic Sounds, 286
        GMS, 294–290–291
        Hot Stuff, 284
        Midnitestar, 278–280
        Police Special, 295
        Raven and Shadow, 292
        Sportstar, 288
        Terrastar, 282
        Trailstar, 289
        Travelstar, Travelstar XL, 283, 285
    Felber Oasis, 272–273
    Glassic, 276–277
    Midas International and the "Midas Touch" Scouts, 257–258
        Midas Baja Scout SSII, 263–267
        Midas Canada Dry, 265-267
        Midas Cruiser/Family Cruiser, 258–262
        Midas Street Machine, 258
        Midas Off-Road Vehicle Package, 257
    Monteverdi Safari and Sahara, 270–273
    Outlaw, 274
    Patriot, 241, 246–247
    Python and Python II, 74,–75
    Rancher Special (SSII), 252, 254,
    Red Carpet Series (RCS), 211–215
    Sno-Star 800B, 236–241
    Sno-Star Scout II, 247
    Special Edition RS, 299–301
    Spirit of '76, 241, 244–246
    Sport (SSII), 250, 252–254
    Sportop, 218–220
    SR-2, 79, 228–230
    SSII, 78, 154, 156, 158, 162, 164, 222, 224, 248–256, 297, 338, 340, 343–344, 347, 352, 358–359
    U.S. Ski Team, 241–244
    Van American, 275